The synergetic approach towards analysing and controlling the collective phenomena in multi-agent systems

Von der Fakultät Informatik,
Elektrotechnik und Informationstechnik der Universität Stuttgart
zur Erlangung der Würde eines Doktors der Naturwissenschaften (Dr.rer.nat.)
genehmigte Abhandlung

Vorgelegt von

Olga Kernbach

aus Kislovodsk

Hauptberichter:	Prof. Dr. habil. P. Levi
Mitberichter:	Prof. Dr. Dr. h.c. mult. H. Haken
Tag der mündlichen Prüfung:	22 Dezember 2011

Institut für Parallele und Verteilte Systeme
der Universität Stuttgart

2011

D93 (Diss. Universität Stuttgart)

Bibliografische Information der Deutschen Nationalbibliothek

Die Deutsche Nationalbibliothek verzeichnet diese Publikation in der Deutschen Nationalbibliografie; detaillierte bibliografische Daten sind im Internet über http://dnb.d-nb.de abrufbar.

ISBN 978-3-8325-3260-4

Logos Verlag Berlin GmbH
Comeniushof, Gubener Str. 47,
10243 Berlin
Tel.: +49 (0)30 42 85 10 90
Fax: +49 (0)30 42 85 10 92
INTERNET: http://www.logos-verlag.de

Acknowledgment

The work and experiments presented in this thesis is the result of my research in the Institute of Parallel and Distributed Systems (IPVS) at the University of Stuttgart. This work would not have been possible without the support and kindness of many people to whom I would like to express my undying gratitude.

First of all I would like to thank my department chair and thesis advisor Prof. Paul Levi for his openness, encouragement and giving me the chance to work on such challenging projects as "Transformable Business Structures for Multiple-Variant Series Production" and "Intelligent Small World Autonomous Robots for Micro-manipulation (I-Swarm)". I also thank for his guidance, good advice and helpful comments to this work.

I am grateful especially to Prof. Herman Haken for inspiration and active support. The essential ideas for this work emerged during our useful discussions at the Institute of Theoretical Physics I.

Extra special thanks go to my colleague and husband Serge Kernbach for an endless list of reasons: for his understanding, personal support, tolerance, for his close reading of this work, valuable comments and pleasantry remarks.

I also thank our secretary Ute Gräter for her tireless help under nearly any circumstances.

A big thanks goes to my parents Walentina and Alexander for their love, the warm encouragement and patience. I appreciate their essential support and many other contributions more than I can express in any acknowledgments here.

Next, I would like to thank all people at the Image Understanding Department at the IPVS for the friendly working atmosphere.

Last, but most assuredly not least, I have to thank all the students for an incentive and a significant help around the Open-source micro-robotic project http://www.swarmrobot.org as well as many external research partners for their interest in it.

Finally, of course, I would like to thank the German Academic Exchange Service (Deutsche Akademische Austauschdienst) and the Germany National Research Foundation (Deutsche Forschungsgemeinschaft) for financial support.

Stuttgart, November 2010 *Olga Kernbach*

It was a long winding road to science ...

To see a world in a grain of sand
And a heaven in a wild flower,
Hold infinity in the palm of your hand
And eternity in an hour.
William Blake
Auguries of Innocence.

There are more things in heaven and earth,
Horatio, than are dreamt of in your philosophy.
William Shakespeare
The Tragedy of Hamlet, Prince of Denmark
(Act I, Scene5).

I want to know how God created this world.
I am not interested in this or that phenomenon,
In the spectrum of this or that element.
I want to know His thoughts; the rest are details.
Albert Einstein.

Abstract

This thesis deals with theoretical aspects of creating purposeful self-organization in collective autonomous systems. These autonomous systems are represented by corresponding analytical models in the form of dynamical systems. The core of the work is closely related to a qualitative dynamics of these models and to the reductive synergetic approach.

Changes of qualitative dynamics can be associated with, for instance, collective decisions in distributed autonomous systems. Through controlling qualitative dynamics by using different control approaches, we are able to create dedicated decision-making mechanisms for real autonomous systems.

An essential aspect of the thesis is devoted to the design of interactions between autonomous systems. It is demonstrated that spatial interactions, during collective decision-making, can be represented as coupled map lattices with fixed or random couplings. By using the normal form reduction, or the center manifold approach, lattices can be reduced to "order parameters".

This work considers three different forms of order parameters, which correspond to decision-making processes such as voting, bargaining, and selection. These decision processes are implemented in two real collective systems: first, agent-based process planning in an industrial environment, and second, in a swarm of micro-robots during collective energy foraging.

The work demonstrates the advantages of applying analytical mechanisms to micro-systems: simplicity and reliability of decision mechanisms, minimal requirements on hardware resources and energy consumption, and the predictability of collective dynamics.

Zusammenfassung

Seit vielen Jahren befasst sich die Robotik-Forschung mit der Entwicklung von autonomen mobilen Robotern, die den Menschen die Arbeit erleichtern oder gar abnehmen. Es existiert schon eine Vielzahl von Robotersystemen, die in den Bereichen eingesetzt werden, wo für die Menschen eine potenzielle Gefahr besteht oder Dienstleistungen erbracht werden. Der Nachteil von diesen Systemen ist, dass sie groß, teuer, kompliziert und sehr anwendungsspeziell sind.

Die zahlreichen technischen Systeme, die aus mehr als 100 kleinen autonomen Robotern bestehen, nennt man Roboterschwärme. Diese Roboter agieren miteinander in der Mikro-Welt und unterscheiden sich von anderen kollektiven Gruppen von Robotern hinsichtlich der Koordinations-, Synchronisations-, Betätigungs-, Wahrnehmungs-, Kommunikationsmechanismen.

Als interdisziplinärer Forschungsbereich hat sich die Schwarm-Robotik in der letzten Zeit aus der Notwendigkeit heraus entwickelt, die kollektiven Schwarmsysteme aus der Natur zu untersuchen, um die gewonnenen Mechanismen und die Strategien der Problemlösung auf künstliche Systeme übertragen zu lassen. Es wurde dadurch motiviert, dass der Schwarm als Ganzes erstaunliche Fähigkeiten hat, die durch lokale Wechselwirkungen und synchronisierte Verhalten von einzelnen Individuen entstehen. Der Schwarm hat eine kollektive Intelligenz, die die einzelnen autonomen Agenten (z.B. Bienen oder Ameisen) nicht haben. Die Forscher wurden auf der Suche nach einer Antwort auf die Frage "Auf welche Weise erscheinen diese kollektiven Phänomene?" inspiriert.

Die vorliegende Arbeit befasst sich mit der Untersuchung von theoretischen und praktischen Aspekten, die zu den Selbstorganisationsprozessen in kollektiven autonomen Systemen, hauptsächlich in Multi-Agenten und Multi-Roboter Systemen, führen. Wie soll sich jeder einzelne Agent verhalten, damit eine perfekt funktionierende Einheit zusammenhält und die Störungen nicht beeinflussen können? Aus welchen Regeln entsteht das System, das sich selbst organisiert, sich steuert und dabei kollektive Entscheidungen trifft? Welche Synchronisierungsmechanismen und Interaktionen ermöglichen dem Schwarm, auf die neuen Situationen zu reagieren? Diese autonomen Systeme werden durch entsprechende algorithmische und analytische Modelle in Form von dynamischen Systemen dargestellt.

Das Hauptziel der Arbeit besteht in der Untersuchung von qualitativer Dynamik von diesen Modellen mit Hilfe eines synergetischen Ansatzes. Die Änderungen der qualitativen Dynamik werden auf der Systemebene als Prozess der kollektiven Entscheidungsfindung betrachtet. Die Steuerung von qualitativer Dynamik anhand von verschiedenen Steuerungsmechanismen er-

möglicht dabei die Entwicklung bestimmter Mechanismen für die Entscheidungsfindung in realen autonomen Systemen.

Ein weiterer wesentlicher Aspekt dieser Arbeit widmet sich der Fragestellung bezüglich der Modellierung der Wechselwirkungen zwischen den autonomen Systemen. Es wird gezeigt, dass die räumlichen Interaktionen während der kollektiven Entscheidungsfindung in Form von Coupled Map Lattices (CML) mit festen oder zufälligen Kopplungen dargestellt werden. Die Anwendung von verschiedenen Reduktionsmechanismen, z.B., den Normalformen oder der zentralen Mannigfaltigkeit, ermöglicht die Vereinfachung der Lattices bis zur Ordnungsparametergleichung.

In dieser Arbeit werden verschiedene Formen von Ordnungsparametern behandelt, die dem Entscheidungsfindungsprozess wie Abstimmung, Handeln und Auswahl entsprechen. Diese Mechanismen für die Entscheidungsprozesse werden in zwei kollektiven Systemen implementiert: auf die industrielle agentenbasierte Prozessplanung und in einem Schwarm von Mikrorobotern für die kollektive Energieversorgung.

Die Arbeit demonstriert die Vorteile der Anwendung von analytischen Mechanismen auf kollektive Systeme: Einfachheit und Zuverlässigkeit von Entscheidungsmechanismen, minimale Anforderungen an Hardware- und Energieverbrauch, und die Voraussagbarkeit der kollektiven Dynamik.

Der wesentliche wissenschaftliche Beitrag der vorliegenden Dissertation liegt:

- *in der Untersuchung und Findung des interdisziplinären Ansatzes, welcher die Reduktion der Systemkomplexität ermöglicht;*
- *in der Entwicklung eines methodologischen Modellierungskonzepts, welches die verschiedenen hierarchischen Modellierungsebenen verknüpft, die Änderungen des qualitativen Verhaltens des Systems auf jeder Ebene mittels Änderungen von Ordnungsparameter ermittelt, und nachher diese Änderungen bis in das ursprüngliche System hinein propagiert;*
- *in der Analyse von zeitdiskretem dynamischem Systemverhalten und Steuerung mithilfe von Kopplungsmodifikationen.*

Kapitel 1 gibt einen Überblick des Robotik-Bereichs und beschreibt das Entstehen der kollektiven Robotertechnik:

- *Einführung des Begriffs der kollektiven technischen Systeme nach der Analogie mit den kollektiven Systemen in der Biologie, die sich nach dem Prinzip der Selbstorganisation verhalten;*
- *Charakterisierung zweier Vorgehensweisen zur Analyse der selbstorganisierenden Systeme (makroskopische und mikroskopische) und Vergleich miteinander;*
- *Erläuterung der Probleme, die bei der Analyse der kollektiven Phänomene entstehen;*
- *Erklärung des vorgeschlagenen Ansatzes der Komplexitätsreduktion (bzw. systematische Reduktion von Multi-Agenten Systemen bis zum Ordnungsparameter) für die Analyse der kollektiven Phänomene;*
- *Darstellung der Struktur der Dissertationsarbeit.*

Kapitel 2 vermittelt die zum Verständnis des vorgeschlagenen Konzepts notwendigen Theoriekenntnisse aus zwei verschieden Forschungsbereichen: Informatik (KI, VKI bzw. Schwarmintelligenz) und Nichtlineare Dynamik (Chaostheorie, Synergetik, CML):

- *Erläuterung der wichtigsten Grundbegriffe der agentenbasierten Modellierung: Elementar-Agent, Autonomiezyklus, Multi-Agenten-System, Kooperationsstrategien, Verhandlungs- und Entscheidungsmechanismen, verteilte Problemlösung, und die charakteristischen Eigenschaften von Systemen mit der Fähigkeit zur Selbstorganisation;*
- *Beleuchtung des phänomenologischen und methodologischen Konzeptes der Synergetik und seiner zugrunde liegenden mathematischen Methoden: kontinuierliche und diskrete dynamische Systeme, Bifurkation, Modenamplituden, Versklavungsprinzip, zentrale Mannigfaltigkeit, Ordnungsparameter, Normalformen;*
- *Beschreibung, wie CML für die Modellierung benutzt werden, um die Reduktion der Systemkomplexität zu ermöglichen. Dabei sind die theoretischen Grundlagen der CML, Klassifikation von gekoppelten Systemen und eine Übersicht der räumlich-zeitlichen Effekte gegeben, die durch die bestimmten Kopplungen entstehen;*
- *Beschäftigung mit den mathematischen Mechanismen, die zur Bifurkation- und Chaoskontrolle geeignet sind.*

Kapitel 3 befasst sich mit den mathematischen Methoden zur lokalen Analyse vom analytischen Agent, der durch ein diskretes dynamisches System dargestellt ist.

- *Fokussierung auf die mathematische Reduktionstechnik wie zentrales Mannigfaltigkeit Verfahren, Normalformen und adiabatische Elimination, die auf dem Beispiel von Hénon map demonstriert und miteinander die Genauigkeit und die Berechnungskomplexität vergleicht;*
- *Ableitung der Ordnungsparametergleichung für den einzelnen Agenten, der an kollektiven Prozessen teilnimmt;*
- *Begründung solch wichtiger Faktoren wie die ursprünglichen dynamischen Systeme und die Kopplungen zwischen ihnen, die den Ordnungsparameter stark beeinflussen.*

Kapitel 4 ist den Lösungsverfahren von Ordnungsparametergleichung für aufeinander wirkenden Agenten und den Modifikationsmechanismus von Ordnungsparametern gewidmet und wendet sich der Analyse von homogenen und nichthomogenen Kopplungen zu, die die Struktur von Ordnungsparameter modifizieren.

Kapitel 5 behandelt die Anwendung von Kontroll-Mechanismen für die Ordnungsparametergleichung. Gewöhnlich ist die Dimension von der Ordnungsparametergleichung kleiner als die Dimension der ursprünglichen Systemgleichungen. Deshalb ist es angemessen, die Kontrollmechanismen in erster Linie in die Ordnungsparametergleichung einzuführen, um dann die erhaltenen Ergebnisse für die Kontrolle der ursprünglichen Systeme zu verwenden. Die Kontroll-Mechanismen wurden für den einzelnen Agenten sowie für die

Gruppe von Agenten mit dem Ziel angewendet, das gewünschte Verhalten zu gewinnen. Die folgenden Aktionen werden benutzt:

- *Verschiebung von lokalen Bifurkationen, um die Stabilisierung der nichtperiodischen und periodischen Bewegung (sogar mit verschiedener Periode) zu ermöglichen;*
- *Kontrolle von Chaos mithilfe von zeitverzögerter Kopplung;*
- *Umwandeln verschiedener Typen von Bifurkationen.*

Kapitel 6 und 7 enthalten Beispiele der kollektiven Entscheidungsfindung im Schwarm von Robotern und in der Gruppe von Produktionsagenten basierend auf den Ordnungsparametergleichungen.

In Kapitel 6 wurden die Verhandlungsstrategien dargestellt, die dem gesamten System ermöglichen, gegenseitige Abmachungen in einer endlichen Zeit zu treffen. Diese Strategien wurden in einen Szenario integriert, in dem die Schwarm-Roboter die kollektive Homöostase beibehalten sollen. Sie müssen die individuellen Entscheidungen über Prioritäten und Reihenfolgen synchronisieren, um Engpässe rund um die Ladestation zu vermeiden.

Vom Standpunkt der synergetischen Methodologie ändert die Entscheidungsfindung das Verhalten von Agenten Koalition qualitativ. Diese qualitative Änderung wird anhand von Ordnungsparametern beschrieben. Durch die Modifizierung von lokalen Wechselwirkungen in autonomen kollektiven Systemen ändert sich der Ordnungsparameter, der entsprechend zur Veränderung der Verhandlungsstrategie führt.

In Kapitel 7 wird ein zweites Szenario betrachtet, in dem die Produktionsagenten die Belegungspläne erstellen. Die erzeugten Pläne werden als Muster in spezifischen Systemen von Eigenvektoren angesehen und die Agenten treffen die kollektive Entscheidung bezüglich Pläne, die die geforderten Optimierungskriterien erfüllen.

Kapitel 8 fasst die Ergebnisse dieser Arbeit zusammen. Der Anhang enthält Einzelheiten und Rechnungsdetails.

Contents

Chapter 1

Introduction

1.1 Motivation

The important thing in science is not so much to obtain new facts as to discover new ways of thinking about them.
W. Bragg

1.1.1 From mechanical to autonomous mechatronic systems

Since antiquity, humans have been interested in constructing machines and mechanisms to solve different practical tasks. The progress of human society and growing material and cultural needs promoted bartering and the expansion of trade between different tribes, cities, and later, countries. The challenges and problems faced by humans required the development of many technical devices that facilitate the production of goods and services. The accumulation of technical innovations, new production methods and growing scientific knowledge led to the era of industrial revolution which began in the second half of the eighteenth century. This revolution can be characterized by the replacement of hand work by machines, the introduction of conveyer processes, the creation of new engines, new communication systems, and so on. Thus technological and economic progress facilitated changes in human environments.

Researchers were always interested in the creation of autonomous machines - robots and vehicles - that are capable of performing complex tasks and working independently of humans. This idea, expressed even in the time of Babbage, was only fully implemented in the twentieth century. Mechanical engineering, robotics and the rapid progress of computer technologies contributed essentially to the appearance of new autonomous devices. Generally, these systems are very important in hazardous or dangerous environments, for increasing the performance and quality of production, as well as in assisting elderly or disabled people. Couriering and transportation, health care, ship maintenance, and autonomous environmental cleaning (for example, ocean cleaning by underwater robots) are just a few applications that are important in human environments and service industries [1], [2]. Autonomous systems already perform planetary surface missions [3] and their applications in space will grow in future.

There is an essential benefit in using autonomous robotic systems. Operating in a dynamically changing environment, robots continuously interact with each other and can build heterogeneous groups. Some of these groups may be specialized for solving specific tasks, others may be general-purpose. The common collective structure of these groups can be viewed as societies of interacting basic systems without central coordination [4], [5]. Cooperation within and between these groups allows better performance in solving complex problems. Even if an individual component of such systems fails, the whole system will still work. Generally, such systems are referred to as technical collective systems .

The high degree of miniaturization and increasing requirements for efficiency and compactness of hardware and software contributed to the appearance of a new branch of robotics: swarm robotics. According to [6], swarm robotics is the study of how a large number of simple physically-embodied agents can be designed to allow a desired collective behavior to emerge from interactions between the agents and their environment. A group of simple robots is able to solve tasks beyond the capabilities of a single robot [7]. Micro- [8], [9] and nano-robotic systems [10] open new applications, for example, microbiology, medicine (for example, microsurgery), or industrial micro-manufacturing.

1.1.2 Natural and artificial collective systems

By *collective systems* we understand a collection of independent computation components that appears as a single coherent system [11]. The components are connected by a communication medium and interact with each other to achieve a common goal. We can observe many interesting phenomena in collective systems. Simple collective systems can be found in physics; the laser is a classic example. Laser light, arising from the collective behavior of atoms and photons [12], [13], has considerable practical importance and is used in many branches of modern technology. Many independent atoms emit light waves, which are normally uncorrelated. However, in certain defined conditions, the atoms begin to act in a well-ordered way, thereby emitting laser light and may be said to be self-organized.

Interesting collective phenomena can be found in biology [14], [15] and microbiology, such as colonies of bees, ants or termites, groups of bacteria or even the human immune system (see [16]). These biological systems can spontaneously organize themselves - or rather self-organize - to build complex spatial patterns. The swarm-like behavior of living creatures has become a metaphor for self-organization in recent years (see [17], [18]). The spontaneous emergence of collective behavior (see Fig. 1.1) is seen in flocks of birds, shoals of fish, swarms of bees, colonies of ants, and herds of sheep [19].

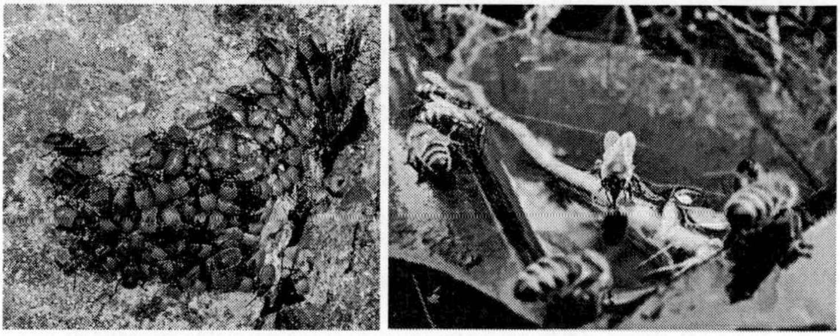

Figure 1.1: Examples of collective behavior in swarm of beetles and bees. Image of bees is taken from [20].

When avoiding danger, or changing course, birds and fish react in an elegantly synchronized manner. Sometimes, the swarm or shoal behaves as if it were a single enormous organization or single giant animal. There is no "head fish," "bird leader," or "ant boss" that coordinates the others and tells them how to behave [21]. It seems that the controlling structures in these collective phenomena are distributed over the whole system.

The observation of natural collective phenomena demonstrates the high reliability and robustness of the underlying mechanisms as well as their ability to dynamically adapt to changes in the environment. Therefore, an investigation of these emergent mechanisms (such as aggregation and segregation, foraging, collective gathering, cooperative transportation, flocking and navigation in formation, source localization, task allocation, mutual exclusion, and so on) is fundamental for applying them to technical collective systems. Primarily, this depends on developing an analytical understanding of these processes in terms of their structure and the dynamics of their interactions. This attempt represents one of the main objectives of this work.

1.2 Challenges and problem definition

1.2.1 Collective behavior

As mentioned the previous section, the behavior of collective systems has many interesting properties. These can be explained by the nature of collective interactions, which are substantially nonlinear. Due to these nonlinear interactions, the degree of freedom of the whole collective system is much higher than the degree of freedom of the non-interacting elements. Therefore, the whole system exhibits new collective properties. The ordered spatial-temporal structures, macroscopic behavior or even the emergence of new functions are examples of such collective properties. Since there is no central element coordinating the system, this process, leading to the appearance of new collective phenomena, is commonly called *self-organization*.

Collective phenomena have considerable practical importance in many branches of science [22] and engineering [23], [24]. Investigating self-organizing processes, as well as understanding the mechanisms of self-organization, is a prerequisite for further applications. However, the difficulty of performing such investigations lies in the nonlinear interactions among systems. Due to these interactions, the overall complexity is much higher than can be treated analytically. Without such an analysis, effective prediction, design, and accordingly control is enormously difficult and sometimes impossible.

Technical collective systems, whose elements behave in an ordered way, must possess distributed mechanisms guaranteeing synchronization among their different parts [25]. In the case of autonomous robots, this synchronization is a form of collective decision-making about the currently executed activity. The collective behavior arising can be viewed as a decision pattern, where decisions of all elements are first uncorrelated, then, in a result of self-organization, become synchronized [26].

For instance, every micro-robot in a swarm can autonomously recharge without considering the other micro-robots. This strategy can be the best for an individual robot, however, it can be very poor for the whole swarm, because it can lead to bottlenecks around the docking station. A better strategy involves negotiations about priorities in accessing the docking station. This can increase the swarm's performance. Similar collective strategies are used for example, in soccer-playing robots [27].

As demonstrated by experiment, such coordinated teams have more chances of success than many non-coordinated robots. The problem of coordinated behavior in autonomous systems lies in negotiation. On the one hand, the negotiation among the robots cannot be directly programmed, because the group will be then restricted to the pre-programmed cases. On the other hand, when robots have unlimited alternatives for negotiation, they may not achieve a common decision because of endlessly long negotiations. Therefore, it is necessary to design a self-organizing process, which can lead to the desired decision-making without limiting the collective degree of freedom. This is one of the points discussed in this work.

Not only the design, but even the analysis of collective phenomena, emerging through self-organization, presents problems. Neurons, considered separately, can demonstrate simple chaotic behavior. However, neurons combined into a neural network are synchronized through their cooperative activity. Investigations of this phenomenon contribute directly to our understanding of the origin of neuronal regularization. However, as pointed out by many researchers, and primarily because of their high complexity, neural networks can be investigated only on the basis of simplified models. Considering collective phenomena in physical systems, we can come to the same conclusion. To return to the example of a laser, a direct microscopic investigation of self-organization on the atomic level is hardly possible, and an application of specific approaches is needed. Such a non-reducible complexity speaks to the common nature of collective phenomena.

1.2.2 Levels of self-organizing systems

The macroscopic theory of self-organization is known from theoretical physics. During recent decades, the study of how ordered structures emerge from disorder has become a major topic in various fields of research such as biology, sociology, neurology, information technology, and so on. In this work, we follow the definition given by Bushev [28]: "Self-organization is a process by which global external influences stimulate the start of internal processes within the system mechanisms, which bring forth the origin of specific structures."

Long-term investigations of nonlinear many-body systems demonstrate that the whole system cannot be described in terms of linear causality. These investigations also reveal specific approaches for describing self-organization. These can be roughly divided into two groups: restricting the level for example, macroscopic, mesoscopic, microscopic, and so on) and restricting the dynamics of systems (for example, local, global, chaotic, attracting, and others). These different theoretical approaches can successfully be applied to the investigation of collective phenomena.

The first macroscopic approach was developed in statistical physics, originating from the research into gas dynamics. Systems consisting of gas molecules cannot be described by a microscopic model with a small number of degrees of freedom. The proposal instead was to introduce the term "noise", in order to take into account residual degrees of freedom [29]. Even when a system has many degrees of freedom, it can be described by characteristic values (the so-called "macroscopic order parameters") with a corresponding noise term. As well as successful applications in lasers [30], this approach has been adopted in many other domains, for example traffic dynamics (see [31]). A traffic system is a classic collective system, in which a car represents a locally interacting system. The traffic system does not possess a central control element, and collective phenomena appear in the form of (for example) traffic jams, which are undesirable and to be avoided [32], [33]. Macroscopic models (in the sense mentioned above) allow the understanding and optimizing of instabilities in traffic flows, estimating amount of consumed fuel, making short-term predictions, and so on. Unfortunately, this macroscopic approach generally does not provide any information about the behavior of individual cars: this may be viewed as a restriction.

In contrast to macroscopic approaches, the development of powerful computers initiated a completely new, microscopic, approach, known as a *computer simulation*. The major focus of computational research is the study of interacting elements (generally denoted as agents). The behavior of these elements involves strategies adopted by individual agents in response to interactions with other agents or the environment. Depending on context, the agents in these networks may represent a population of algorithms, cells in the human immune system, genes involved in regulatory networks, individual species in ecological systems, individual consumers, or any of a wide variety of other examples. This approach represents a powerful modeling technique for incorporating realistic data (for example, differences in individual characteristics, details of spatial distributions, and so on) that are difficult to address when using traditional theoretical approaches (see [34]).

Swarms are one of many self-organizing systems that are now studied through computer simulations (see [35], [36], [37]). Whereas it was very difficult to mathematically model systems with many degrees of freedom, the advent of inexpensive and powerful computers made it possible to construct and explore model systems of various degrees of complexity. This method underlies the new domain of "complex adaptive systems", which was pioneered in the 1980's by a number of researchers associated with the Santa Fe Institute in New Mexico. The biologist Stuart Kauffman studied the development of organisms and ecosystems. Through computer simulation, he has tried to understand how networks of mutually activating or inhibiting genes can give rise to the differentiation of organs and tissues during embryological development. Kauffman proposed that the self-organization exhibited by such networks is an essential factor in evolution, complementary to Darwinian selection. Through simulations, he showed that sufficiently complex networks of chemical reactions will necessarily self-organize into autocatalytic cycles, the precursors of life [38], [39].

Another example are markets, where different producers compete and exchange money and goods with consumers (see [40], [41]). Although markets are highly chaotic, non-linear systems, they usually reach an approximate equilibrium in which the many changing and conflicting demands of consumers are satisfied.

Generalizing the microscopic approach, we can point to its efficiency. Computer simulation has considerable future prospects in science and engineering. However, two problems typically arise: first, the so-called problem of "parameters," which arises when the model has n-control parameters; their successive variation can lead to *NP-hard* problems. Second, the problem of understanding the results obtained by simulation. Therefore, some analytical investigations need to be performed before simulation, in order to avoid the "parameters" problem and to understand at least the qualitative behavior of the models under consideration.

As mentioned above, a third approach also exists, whose contribution into the investigation of collective phenomena is enormously important. In statistical or simulative approaches, the dynamics of a distributed system are considered in the whole; only the level of consideration is restricted. Thus, macroscopic models deal with systems described at the macroscopic level, whereas simulation treats systems microscopically, that is, at the level of initial systems. The third approach proposes instead that, rather than restricting the consideration level, the dynamics should be restricted to some elements of interest (see [42]). This restricted consideration allows the reduction of the number of degrees of freedom to those that can be treated analytically. Such an approach allows researchers to study local interactions between systems and to understand the conditions where collective behavior qualitatively changes. The dynamics at a remove from local instabilities can still be treated numerically and statistically. The investigation of collective phenomena undertaken is, in fact, based on this approach, therefore it is discussed in further detail.

1.3 Suggested approach and goals of the thesis

The most important question in considering collective phenomena is how do they arise? Which collective phenomena does the self-organization process lead to? How may we avoid undesired collective phenomena? These and similar questions require the understanding the origin of self-organization in collective systems. From the viewpoint of dynamics, the emergence of collective phenomena demonstrates changes in qualitative behavior. For instance, coupled chaotic oscillators are synchronized. Due to cooperative activity, their behavior is changed from chaotic to (for example) periodic. The appropriate theory to explain such qualitative changes of dynamics is the *mathematical theory of dynamical systems*. This branch of mathematics studies dynamical systems and develops concepts that may improve our understanding of modeled real-world-systems. The primary objective of synergetics lies in investigating the qualitative dynamics of collective systems and understanding the common characteristics of self-organization [43].

The interdisciplinary domain dealing with cooperative phenomena was named synergetics by H. Haken. It began in the investigation of lasers and was later generalized to other kinds of collective phenomena [44]. Thus, it was emphasized that all collective phenomena possess a common nature. As shown both theoretically and in experiment, only a few quantities control collective phenomena. Following the tradition of physics, they are denoted the order parameters. The order parameters are not explicitly present in collective systems, but have to be derived using analytical methods (see [45]). Using this concept, the dynamics of interest is restricted only to such regions where collective changes occur and order parameters may be derived.

Order parameters allow the understanding of the arising of collective phenomena. They describe changes of environment, or the internal structures of systems, which are required for the emergence of collective behavior, that is, they are useful for analysis. Physical examples, in which collective phenomena were first investigated, are determined by molecular and atomic interactions. They generally do not allow any radical modification of the collective phenomena. Many biological and chemical systems also do not allow any essential modification of the phenomena.

In contrast to natural systems, technical systems are artificially designed and so their basic systems and interactions can easily be modified. The creation of the desired self-organization in these systems, besides being of scientific interest, has considerable practical importance.

Remark.

Order parameters, as well as their applications in analysis, can also be used to control and design cooperative phenomena in artificial collective systems. Fusion of analytical mechanisms with algorithmic models represents a focus of this work.

The structure of the suggested approach is shown in Fig. 1.2. Since the focus of

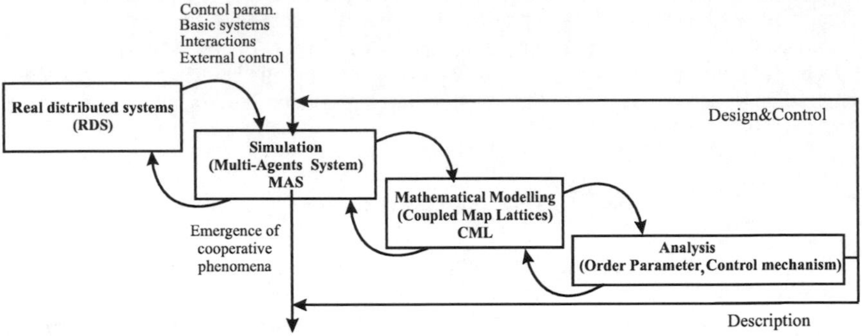

Figure 1.2: Step-wise simplification approach and obtaining the order parameter equation (OP) for prediction, control and designing the collective phenomena.

the work presented here is primarily on technical collective systems, they must first be transformed into models with a number of degrees of freedom such that they can be analytically investigated. This successive complexity reduction is performed in three steps:

- first, it is assumed that technical collective systems have the structure of multi-agent systems (MAS). They can be simulated in a software environment, and real-world details (for example, hardware constraints, energy supply, etc.) can be ignored. As a result, we realize an idealized multi-agent system, free of implementation details;
- second is algorithmic simplification, in which MAS is mapped into the coupled map lattices (CML);
- third, using analytical reduction procedures, the required order parameter may be derived from the mathematical models obtained.

Advantage of this step-wise simplification consists in preservation of system's structure on all levels of simplification (in sense of basic systems and interaction). It is a prerequisite for successful analysis and control. The obtained mathematical models possess a discrete nature. Local analysis, primarily computational techniques, of time-discrete systems is investigated not enough.

An advantage of this step-wise simplification is the preservation of the systems structure at all levels of simplification (in the sense of basic systems and interactions). This is a prerequisite for successful analysis and control. The mathematical models obtained are discrete in nature. Local analysis, primarily computational techniques, of time-discrete systems is insufficiently investigated.

Remark.

The essential point of this work is to improve and to extend the computational methods for deriving the order parameter for time-discrete systems.

Information about collective behavior, provided by the derived order parameters, can be used for creation the desired collective activities. In this work, two such approaches are considered. First, an external control governing the cooperative activity. Second, the structural design of collective systems such that they possess the desired collective properties. This approach is illustrated by several examples of collective decision- making originating from manufacturing planning and from collaborative actuation in micro-robotic swarms.

The derivation of the order parameters, using analytical methods, imposes some restrictions on the suggested approach:

- First, the dimension of CML and correspondingly the dimension of the collective systems. This is a general restriction, but as has been seen, homogeneously coupled systems allow additional dimension reduction, the so-called "dimension scaling." Multi-dimensional complex systems can be treated using this technique, but the restriction on the use of homogeneous couplings remains imposed.

- Second, as previously mentioned, we also assume that real systems possess the structure of multi-agent systems. Many technical collective systems satisfy this assumption [46], but generally this structure is arbitrary.

- Finally, we restrict ourselves only to such elements of dynamics for which the order parameter may be obtained. In this way, we can deal with collective behavior, but the quantitative properties of the behavior, which are observed away from bifurcations or in chaotic domains, cannot be treated. Investigation of these quantitative properties is not the focus of this work.

Remark.

The main result of this thesis can be understood as a development of a few low-level interaction patterns (for example, collective decision-making). These interaction patterns drive the group to a few desired types of cooperative activities. However, the development of these interactions does not represent the main focus of this thesis. We are primarily interested in understanding the origin and basic mechanisms of self-organizing processes among agents. The understanding of common mechanisms represents a higher-priority goal, because it allows the development of arbitrary low-level interaction patterns, as shown in Chap. 6 through the procedure of collective decision-making.

Summarizing the aspects mentioned above, the main objective of this work lies in investigating the basic theoretical mechanisms underlying self-organization in artificial collective systems. It is intended to reveal the synergy between three fields of research: artificial systems (robots inspired by natural systems), multi-agent systems (behavioral simulation for autonomous systems), and non-linear dynamics (mathematical tools for studying and analyzing artificial systems). The specific contribution of this work lies in the mathematical methodology for investigating nonlinear dynamical systems that are discrete in time.

1.4 Structure of the work

The work presented here is divided into eight chapters and structured thus: **chapter** 2 is devoted to a brief overview of the concepts and theories used. The goal is first to survey recently-published methods and concepts in the vast literature on the subject, and second, to introduce the definitions and ideas used later. Since the work primarily covers analytical approaches as applied to multi-agent systems, these are the main focus of this overview. In addition, the foundational concepts of coupled map lattices, as well as control of bifurcation and chaos, are briefly considered.

Chapter 3 deals with the mathematical methods of local analysis applied to discrete systems. This chapter is necessary because these methods were primarily developed for time-continuous systems. Some discrete calculation techniques, required later, are not advanced enough. The focus of consideration is on the center manifold approach, the normal norm reduction and some other techniques, such as adiabatic elimination. These techniques are compared from the viewpoint of accuracy and computational complexity. As an example, we consider first a single agent, whose mathematical model is given by a two-dimensional Hénon map. For this model, the order parameter equation is derived. It is demonstrated that even a single agent, considered as a participant of collective phenomena, can possess the order parameter.

The order parameters, which describe the cooperative activity of collective systems, are influenced by two main factors: the initial systems and the couplings between them. Since the initial systems are usually fixed by various restrictions, only couplings can exert influence, as discussed in **chapter** 4. The chief point, as shown by calculation, is that the order parameters can be completely modified by couplings. Therefore, the collective phenomena can be controlled and, moreover, designed, by the appropriate choice of couplings. Additionally, the problem of dimension reduction is discussed, in which the reducible and non-reducible cases are shown. The derived order parameters contain information about the macroscopic behavior of a collective system. This information can be used in an external control feedback. The time-delayed control mechanism applied to discrete systems is discussed in **chapter** 5. Developed bifurcation control can be used for instabilities with unstable eigenvalues $+1, -1$; the case of complex-conjugate unstable eigenvalues is also considered. Special attention is paid to the control of periodical motion with different periods.

The approach is illustrated by several examples. The first is given by the single agent considered in chapters 3 and 5. This example is intended to show only some computational aspects that underlie the derivation of order parameter equations. The more complex two-agent system originates from the study of production systems, whose computer simulation is performed by $Matlab^{©}$. This elementary collective system demonstrates the influence of interactions on a cooperative activity. The mathematical model, denoted as an OLL map, is shown in chapters 4 and 5.

Finally, two complete examples of collective decision-making based on synergetic concepts are given in **chapter** 6 for a real robot swarm and in **chapter** 7 for the assignment problem in industrial manufacturing. The main point is first, to show that the design of cooperative activity can be performed by the modification of couplings (interactions), and second, to demonstrate successive complexity reduction from the level of real systems to multi-agent systems and then to coupled map lattices. Following that, the modification provided by order parameters, are propagated backwards from CML to real systems.

Chapter 6 considers collective energy homeostasis in a swarm of "Jasmine" micro-robots. During collective recharging, swarm robots must take decisions about priorities and orders of recharging. The individual decisions of all robots must be synchronized, to avoid bottlenecks around the docking station and increase energetic efficiency. A procedure for collective decision-making was developed, based on the proposed synergetic approach and tested in several simulated and real experiments. It is shown that a collective strategy allows an almost doubling of the energetic efficiency of swarm robots.

Finally, the methods developed within this thesis are applied to the problem of industrial planning in **chapter** 7. The planning approach considered should map different manufacturing steps to available machines, taking into account multiple constraints. The main part of this problem is solved in an algorithmic way by cooperating agents and has been discussed in [48] and [49]. That multi-agent system produces multiple plans, which can be viewed as planning patterns in specific systems of eigenvectors. Agents should perform collective decision-making towards (a) a plan with optimal cost and (b) the avoiding of all forbidden plans (for example, a plan that oversteps a time limit). Chapter 7 addresses this problem, based on analytical systems of dynamical equations.

This thesis is concluded in **chapter** 8, which summarizes all achieved results and draws several conclusions towards appearance of group-related social laws and collective intelligence.

Chapter 2

Outline

2.1 Autonomous agents and multi-agent systems

Coming together is a beginning.
Keeping together is progress.
Working together is success.
Henry Ford

This section gives an overview of autonomous agents and multi-agent systems. The concept of "agent" involves such notions as intelligence, autonomy, degree of "free agency", and others. This overview, based on recently-published works, introduces these notions as working definitions that can be used further. In particular, we discuss the agents paradigm, which is both a programming technique and simultaneously, a useful modeling approach. Essential notions about agent autonomy, rationality and intelligence are introduced. Special attention is paid to multi-agent systems and the specific problems arising in them.

2.1.1 The paradigm of an autonomous agent

Recently, the concept of autonomous agents and agent-based modeling has advanced in many branches of science (see [50], [51], [52]) and engineering (see [53], [54], [55], [56]). The application domains in which agent-based solutions are applied or investigated include workflow management, telecommunications network management, air-traffic control, business process reengineering, data mining, information retrieval management, electronic commerce, education, personal digital assistants, e-mail filtering, digital libraries, command and control, smart databases, scheduling and diary management, and others. Furthermore, agent-based modeling and simulation enables and aids understanding of complex systems such as manufacturing.

On the one hand, this concept extends the ideas of object-oriented programming and distributed artificial intelligence (see [57], [58]), and on the other hand represents new trends in the understanding of the phenomenon of complex systems (see [59]).

Unlike traditional computer programs, agents exhibit different degrees of such capabilities as autonomy, reactiveness, proactiveness and social abilities. An agent can either be static or mobile. A mobile agent is able to migrate with its associated data, state, and logic to interact with local resources and other agents, to perform given tasks. Several agents can collectively form a multi-agent system with decentralized data, and varying degrees of global system control.

Similar objects in the real world possess the same degree of autonomy. Therefore, the agents concept can support understanding in the real world by computer simulation and analysis, and offers convenient and efficient tools for control. As mentioned in Chap. 1, powerful simulations using the agent concept have recently been performed, for example studying the behavior of collective insects [60], [61], and the dynamics of markets [62]. The results obtained are worth studying, to understand the origin of regulation in real systems that do not have centralized coordination.

The agent paradigm refers primarily to agent-based simulations, often called multi-agent systems, which can be viewed as a programming technique. However, this technique has a deep equivalence to real phenomena; it may be described as "the world within the world." This equivalence is further used in making possible an algorithmic complexity reduction of collective phenomena and verifying the suggested approach by simulation.

However, if the agent concept were only a programming technique, it would never evoke the interest in agent-based approaches that is observed in different branches of engineering [63] and manufacturing [64].

2.1.2 General notions of an agent

To introduce the most important notions about an agent, we must answer the main question of interest: what is the key difference between an agent and other algorithmic structures? It might be expected that the main distinction concerns the principles of agent architecture and activity. Unfortunately, there is no agreement about such principles that would be accepted by all researchers; there is only a general consensus that autonomy is a central notion of an agent. Below are some definitions concerning agent-based systems used in this work.

- **Autonomy**. The term "autonomy" brings to mind the sense of something "not under the immediate control of a human" and relates to behavior. In a strong sense, the agent's behavior needs to be determined in the context of its environment (whether virtual or physical). We define an object in an environment *as autonomous if its behavior is determined by its own experience, and it exhibits behavior without being operated upon by another object in this environment* [5], [58]. In general, agents can be driven from outside and so depend on others. However, they are not completely controllable, and always try to achieve their own goal. In other words, an autonomous agent exhibits control over its own behavior. When a mobile robot or an "autonomous land vehicle" is not driven by a human driver, we refer it to "autonomous", even though the vehicle is constructed and maintained by humans.

The notion of autonomy is closely related to another, no less significant notion: rationality. Rationality complements autonomy, in that an agent has its own goals and tries to increase its own profit.

- **Rationality**. *A rational agent performs whatever action may be expected to maximize its performance, on the basis of the evidence provided by the percept sequence and the built-in knowledge the agent has* [58]. In this way, an agent is first, able to estimate own performance, and second, an agent can perform some action to achieve a goal that, due to autonomy, was determined by the agent itself. Using rationality, the difference between a *C++* object and an agent can be formulated as follows: if the object does something for free, the agent will do the same if it thereby obtains a profit [65]. Moreover, we refer to rationality when *an agent accepts rules existing in the environment or used by other agents*; it is a natural restriction of an agent's behavior. This is one of the necessary conditions for the consolidation of agents into a society.

Since an agent is acting in an environment, its abilities of perceiving and acting are a matter of course, on which all researchers agree.

- **Perceiving and acting**. Perceiving and acting are natural abilities, whereby an agent is "incorporated" into its environment and is able to meet its design objectives [66]. Perceiving and acting refer more to the question of equipment than to the question of being an agent. The processing of perceived data and the planning of actually executed actions, especially the ability to do the right thing, are certainly properties of an intelligent agent.

Processing and planning abilities, as well as the ability to infer, are the ones that make an agent intelligent. However, their definition, like intelligence, is not easy, and is a topic on which researchers' opinions vary. Following [66], we define an intelligent agent thus:

- **Intelligence**. *Intelligence is the ability to solve concrete and abstract problems, as well as to adapt oneself adequately to changes in one's structure or in the environment, and accordingly to relatively new situation. An intelligent agent is one that is capable of flexible autonomous activity in order to meet its design objectives.*

  Flexibility in this context means: **reactivity** - intelligent agents are able to perceive and act; **pro-activeness** - intelligent agents are able to exhibit goal-directed behavior by taking the initiative; and **social ability** - intelligent agents are capable of interacting with other agents (including human agents).

The social ability of an agent is very important; it is the means by which agents are able to associate into a society.

- **Social ability of an agent**. the social ability of an agent is not a trivial concept and goes beyond a simple information exchange between agents. An agent can achieve its own goals only by cooperation with other agents, therefore agents must negotiate, cooperate or compete within a society via some kind of agent communication language. *This type of social ability is much more complex, and much less well understood, than the simple ability to exchange binary information* [65].

The ability to form societies enormously enhances the potential of autonomous agents. Societies are capable of defining roles within the group, where the role itself consists of set of duties. When an agent joins a society, it takes on a role (or several roles) and is committed to fulfill the associated duties. The roles which agents follow in the group may be cooperative and imply social dependencies in achieving a common goal. The problem of how agent-level (individual) activity and group-level (societal) rules interact and depend on each other is known in sociology as the micro-macro problem [67].

Sociology studies natural societies, their habits, and the interaction between individuals that lead to well-defined organizational structures. These societies are, by their existence, successful. That is why researchers try to find inspiration in and extract new ideas from nature. Learning from simple creatures offers several advantages. Most of the time micro-robots work in a team, a so-called "swarm" of robots, which offers extended flexibility and robustness. A swarm can outperform single robots where tasks are complex, due to natural properties such as emergent behavior and collective decision-making. The absence of a centralized control allows successful task determination even when failures in some members occur [68].

2.1.3 Structure of an agent

Before introducing the common structure of an agent,we should describe the known types of systems that can be treated in a sense of agents. Russell and Norwig [58] define an agent in the following way: “an agent is any entity that can be viewed as perceiving its environment through sensors and acting upon its environment through effectors”. Generalizing this structure of an agent, we derive the main modules required for the *architecture* of each agent.

The best-known type of agent-like structure is presented by control systems. Control mechanisms, such as in the simplest case the stabilization of feedback, provide control over behavior, pointing to the autonomy of such systems [65]. Of course, more complex environment control systems possess rich decision structures, allowing the choice of control strategies dependening on external conditions. Essentially, every control system has sensors, modules for data processing and control strategy (even in a simple form) and, finally, actors. Examples are shown by the structure in Fig. 3.2, shown in Chap. 3.

Most software demons can be also viewed as agents. These small programs, which run in the background, for example monitor incoming mail, the correct use of licenses or a variety of other tasks. They are more generally known as software agents [65] or agent programs [58]. In this description, a software agent is an autonomous piece of software that exists for a clearly-defined purpose, is situated in and is aware of its environment.

These agents work in the program environment that they perceive by means of a defined protocol or interface, that is, they have some kinds of sensors and actors. Their ability for information processing, planning and communication with other agents is much richer than in control systems. A typical example of a software agent is shown in the framework of the simulation software package *Breve*, shown in Chap. 6.

The final example of agent architecture considered here is autonomous robotic systems [69]. Robots have to perceive and manipulate different objects, for example, to perceive a cup of tea on the table and the capability to grasp it with a manipulator. Information processing allows them to distinguish the cup and the table from other objects, and the robot must successfully plan how to operate the manipulator in order to take the cup from the table. Such systems are usually referred to as “hardware agents”. There are many other agents, such as agent language, agent expert systems, or spy agents, but they are outside the framework of this representation.

Generalizing these examples, four modules are required in the structure of an autonomous agent: sensing, information processing, planning and acting. These four steps are executed cyclically, therefore this common structure is often called the *autonomy cycle* [69] of an agent and is shown in Fig. 2.1.

- **States**. The environment and an agent can be characterized by a set of states. At any given instant, the environment and an agent are assumed to be in one of these states. If two different environment states are mapped to the same percept, these different states can be indistinguishable to an agent. Its sensors may be unable to distinguish them, or the world model may be restricted. In this case, the concept of a self-learning agent, able to update its own world model, must be introduced.

- **World model**. A world model is expected to contain information about the environment, the agent's construction, its goals, and so on. The agent uses these information in both a declarative as well as procedural sense, that is, whether as data or programs. Without loss of generality, the whole autonomy cycle can be considered as the procedural knowledge possessed by an agent.

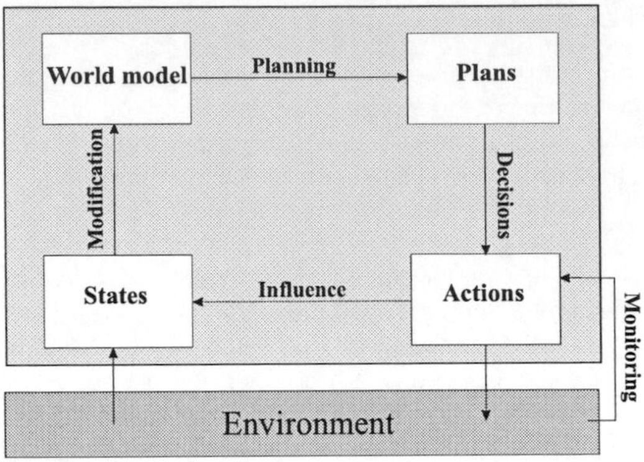

Figure 2.1: Generalized structure of an autonomous agent. Cyclical execution of four elements (with environment loop) called autonomy cycle.

- **Plans**. An agent, perceiving its environment, processes information. The result of this information processing is a statement of whether the agent is content with its current state. If its current state must be altered, the agent makes an internal decision in favor of one of various predefined plans. If a suitable plan does not exist, a planning phase must be executed. The result of planning must be a simple action or a sequence of actions. Often, in simple agents, information processing and planning are implicitly determined in the form of function(s).

- **Actions**. A derived sequence of simple actions, intended to modify the environment, is finally executed by off-the-shelf actors. The actions module can be viewed as a simple agent having its own autonomy cycle. It monitors the activities performed, and if required, can undertake the correction of a currently performed action.

An agent is purely reactive if it simply responds to its environment without reference to its history. Otherwise, an agent can accumulate a history of environment states. This is very similar to finite-state automata with and without memory. Moreover, each module, as shown in Fig. 2.1, can be represented by separate agents that are hierarchically associated in a global agent architecture.

2.1.4 Multi-agents systems

Multi-agent systems, most commonly viewpoint, can be characterized as a system of interacting agents, in which all keep their own autonomy. Successful interaction requires the ability to cooperate, coordinate, and negotiate with each other. The agents can synchronize their behavior, share knowledge and so on. For an agent group to achieve a common goal requires the coherent behavior of all the agents. One of the central problems for distributed systems is that such coherence has to be exhibited without explicit global control. Since the agents remain autonomous within a group, there are only two possible coordination mechanisms: cooperation and competition (see Table 2.1).

Goal	Resources	Skills	Types of situation	Relationship
Compatible	Sufficient	Sufficient	Independence	Indifference
Compatible	Sufficient	Insufficient	Simple collaboration	Cooperation
Compatible	Insufficient	Sufficient	Obstruction	Cooperation
Compatible	Insufficient	Insufficient	Coordinated collaboration	Cooperation
Incompatible	Sufficient	Sufficient	Pure individual competition	Antagonism
Incompatible	Sufficient	Insufficient	Pure collective competition	Antagonism
Incompatible	Insufficient	Sufficient	Individual conflicts over resources	Antagonism
Incompatible	Insufficient	Insufficient	Collective conflicts over resources	Antagonism

Table 2.1: Categories of interactions among agents within a group (from [25]).

Definition 1 (Cooperation and competition).

Cooperation is coordination among non-antagonistic agents, while negotiation is a co-ordination among competitive or simply self-interested agents [70] (see Fig. 2.2).

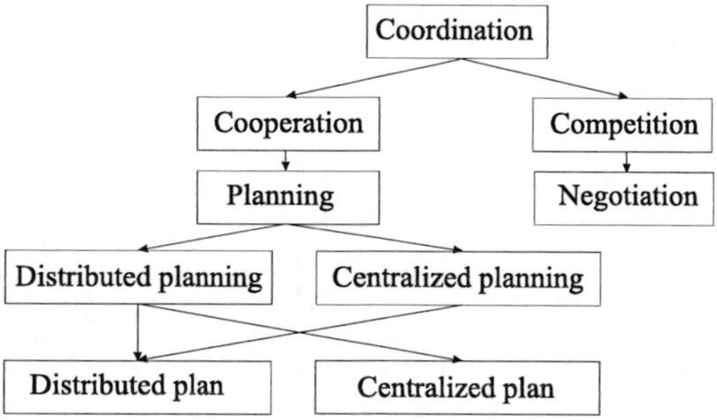

Figure 2.2: A taxonomy of some ways of coordination among agents within a group (from [70]).

To build a multi-agent system in which agents "do what they should do" is particularly difficult. Two main approaches exist to determine how agents should cooperate or compete in order to fulfill their design objectives [5]:

- **bottom-up:** define single agent (individual) capabilities that result, through appropriate interactions, in the desired group behavior;
- **top-down:** extract specific group level rules that appropriately constrain the individual's action set.

This is the problem of distributed and centralized approaches in system design. Centralized solutions are traditionally very efficient. Considered historically, this is the main problem-solving method used by humans. However, distributed problem-solving is sometimes easier to understand and to develop, especially when the initial problem is also distributed. Moreover, there are problems that cannot be treated by centralized approaches, for example, when information belongs to independent organizations that want to keep this information private and secure. The distributed approach is more efficient if the problem being solved [70]:

- is distributed, for example geographically;
- has many components;
- has huge content, both in the number of concepts and in the amount of data about each concept;
- has a broad scope, that is, covers the major portion of a significant domain;
- has to be distributed from the viewpoint of, for example, security;
- has to guarantee a high degree of reliability.

Characterizing the multi-agent environment, we note that it:

- provides an infrastructure specifying communication and interaction protocols;
- is typically open and has no centralized coordinator;
- contains agents that are autonomous and distributed and may be self-interested and/(or)cooperative.

Agents operating in a multi-agent environment may be expected to be able to cooperate in problem solving, to share expertise, and to work in parallel on a common problem. Therefore the first problem is to provide a method for *problem-solving in distributed systems*. To coordinate their activities, agents have to communicate (interact) within the group as well as with the environment. These *interaction procedures* have to guarantee, as a minimum, that the group achieves a common agreement in a finite time. This is one of the problems considered in this work. Finally, questions such as "what exactly should agents know in order to solve the problem successfully?" and "how can one agent access and use the knowledge of another agent?" compound the problem of *distributed knowledge* considered in this overview.

2.1.4.1 Distributed problem-solving

The distribution of resources such as knowledge, capability, information, and so on among agents means an individual agent is unable to accomplish its tasks alone but can achieve more success through working with other agents. In this case, the capability of a group to solve a problem is much greater than the ability of a single agent. Solving distributed problems assumes that agents not only work together but also that they know how to work together. Sometimes the distributed problem being solved is to design a plan. In many respects, each problem includes a planning phase, for example, agents are planning how to work together, how to split the main problem into subproblems, and so on.

The first and perhaps the most popular problem arises when an agent has many tasks to do. In this case, it enlists the help of other agents with few or no tasks. Such a distributed problem-solving strategy is called "task sharing" in the literature (see [72]) and consists of the following main steps:

- **Task decomposition**. The agent looks for tasks being potentially passed together or splits large tasks into subtasks.
- **Task allocation**. This assigns the derived or created subtasks to appropriate agents.
- **Task accomplishment**. Each agent accomplishes its assigned task.
- **Result synthesis**. When all the subproblems are accomplished the results obtained are composed into the overall solution. It is assumed that agents know how to do this.

The case in which there is only one agent and many tasks is supplemented by the case in which there are many agents and only one task. The results of the problem-solving procedures performed by different agents may also be expected to be different. Through sharing these results, the whole group can improve its performance by executing a combination of the following ways:

- **Confidence**. Agents can use independently-derived results for the same tasks to collaborate with each other, yielding a collective result, and having a higher degree of confidence in being correct.
- **Completeness**. The results of separately performed problem-solving procedures cover a more complete portion of the overall task, for example the monitoring of geographically distributed objects.
- **Precision**. The path of problem-solving, especially in the perceiving and specification phases, accomplished by one agent can be useful for other agents; sharing allows increasing common precision.
- **Timeliness**. Solving the subtasks in parallel can make the overall solution faster. Moreover, if some agents cannot be ready before a deadline, others can use a more timely solution.

As mentioned above, distributed planning can be thought of simply as a specialization of distributed problem-solving. In the vast literature on the subject, planning and the plan itself are distinguished; both can be centralized as well as distributed. The first (centralized planning for centralized plan) is not especially interesting; this is the issue of AI's traditional approaches. However, we can consider the remaining three.

- **Centralized planning for distributed plans**. Plans that are executed in a distributed manner can, none the less, be formulated in a centralized manner. This case arises when one agent takes the role of plan maker, while the others are doing something else and therefore are busy during result sharing. Sometimes, if the communication channels needed for the execution of distributed plan are slow, undependable, or very expensive, it might be better to form a more efficient centralized plan. These and many other options, must be taken into account by a plan-maker-agent.
- **Distributed planning for centralized plans**. Formulating a complex plan, executed by one agent equipped with adequate actors, might require collaboration among a variety of cooperative agents that are planning-specialists in different areas. Both task-sharing and result-sharing strategies are involved into this approach.

For some types of problem, the interaction among the planning agents might be thought of as an exchange of partially-specified plans. In other cases, such an interaction, a plan has a very complex form, in which many subtasks and intermediate results must be shared.

- **Distributed planning for distributed plans**. The most challenging type of distributed planning is when both the planning process and its results are performed within an agent society. The minimum requirement is that the agents should not conflict with each other when executing the plan, and preferably should help each other to achieve their plans. This planning approach primarily requires negotiations among agents.

2.1.4.2 Negotiation among agents and decision-making

The case of cooperative agents leads to the problem of plan-making discussed above. Therefore the focus of this brief consideration concerns competitive agents. According to [73],

> **Definition 2** (Negotiation).
>
> Negotiation is a form of interaction, where a group of agents, with conflicting interests and a desire to cooperate, try to come to a mutually acceptable agreement on the division of scarce resources.

In other words, when agents are faced with a selection of actions or opinions and have to make one choice, the decision-making process begins. The process of decision-making takes place on two levels: the first concerns only a single agent and the second one concerns collective decision-making. Therefore, a decision emerges from the numerous interactions among individuals and between the individuals and their environment.

As described in [71], decision-making mechanisms are important features of an intelligent agent, as they make it possible to display different behaviors as a function of environmental situations and in relation to its beliefs and desires. For example, animals collect information about the quality of a food source while foraging. Depending on this information, they can decide whether to stay in the same area or to search for a more profitable one.

However, agents may have different goals and each tries to maximize its outcome without concern for the global group outcome. A self-interested agent will choose the best strategy for itself, which cannot be explicitly imposed from outside. Therefore, the interaction protocol needs to be designed using a non-cooperative strategic perspective: the main question is what social outcomes will guarantee that each agent's desired local strategy is best for that agent - and thus all the agents will use it [26]. In our representation, the agents will make a binding deal regarding coordination in order to achieve a collective goal. This case refers to so-called "cooperative" games in contrast to non-cooperative ones, in which agents cannot make a binding deal [74].

We further distinguish other types of interactions, such as voting, auctions, bargaining, contacting, and coalition formation. These mechanisms can be evaluated according to different criteria. The overview below closely follows Sandholm [26]. The evaluation criteria are:

- **Social Welfare**. The global good for an agent group is measured as the sum of all agents' payoffs in a given solution.

- **Pareto Efficiency**. A solution x is Pareto optimal if there is no other solution y such that at least one agent is better off in y than in x and no agent is worse off in y than x.

- **Individual Rationality**. The agent's payoff in the negotiated solution is no less than the payoff that the agent would get by not participating in the negotiation.

- **Nash equilibrium**. Sometimes an agent has the best payoff using a specific strategy, no matter what strategy other agents use. However an agent's best strategy often depends on the strategies chosen by other agents. In the Nash equilibrium, each agent chooses a strategy that is a best response to the strategies of other agents.

- **Computational Efficiency**. The interaction mechanism has to be designed so that computations are required as little as possible.

These evaluation criteria are used by the following interaction mechanisms:

- **Voting**. All agents add their statements to a interaction mechanism, and the result derived by this mechanism is a solution for all agents. This outcome is enforced, so that all the agents must execute the solution prescribed by this mechanism.

- **Auctions**. The result of auctions is usually a deal between two agents playing the roles of auctioneer and potential bidder. Both want to maximize their profit, while the auctioneer tries to sell an item at the maximum possible price and bidder wants to acquire an item at the lowest possible price.

- **Bargaining**. During bargaining, the agents must make a mutually beneficial agreement, but have a conflict of interests about the kind of agreement that has to be achieved. The mutually desired beneficial agreement is, in fact, a type of equilibrium, for which, for example, the Nash equilibrium can be used.

- **Contracting**. Implementations of general equilibrium market mechanisms require the presence of a single central mediator, which might become a potential point of failure for the whole distributed system. Therefore it makes sense that the agents have direct control of who receives their sensory information, instead of posting the information to a mediator controlling its dissemination.

- **Coalition Formation**. In many domains, the self-interested agents can save cost by coordinating their activity with other parties. This can be done through coalition-formation by means of specific negotiations performed with each party. Agents are expected to have a common opinion about what the cost is that they are intending to save. Moreover the procedure of coalition formation has to offer coordination that each agent will accept for itself.

2.1.4.3 Distributed knowledge

Reasoning about knowledge plays a fundamental role in distributed systems. Any coordinate activity, performed by such a system, requires every agent knows what any other agent is currently doing. Each agent should possess at least the knowledge about common goal, availability of actors, methods of problem solving and so on. However the distribute nature of agent's coalition does not allow building any centralized database being accessible for each agent. The common-system-knowledge, in sense of permanent, fundamental as well as temporal information, has to be divided among separated agents.

The way of knowledge distribution that would make efficient reasoning, decision making, coordination etc. within a coalition, still remains a problem causing many discussions [75].

Definition 3 (Common knowledge).

Let G be a group of agents, whereas φ be a fact that an agent i can know (denoted as $K_i\varphi$). There are two restrictions, firstly, agent's knowledge at given time must depend only on its local history, secondly, only true things are known. Then:

- $D_G\varphi$: **The group G has distributed knowledge of** φ, meaning that someone who knew everything that each member of G knows would know φ. For example, if one member of G knows ϕ and another knows that $\phi \supset \varphi$, the group G may be said to have distributed knowledge of φ;
- $S_G\varphi$: **someone in G knows** φ or more formally $S_G\varphi = \bigvee_{i \in G} K_i\varphi$;
- $E_G\varphi$: **everyone in G knows** φ or more formally $E_G\varphi = \bigwedge_{i \in G} K_i\varphi$;
- $E_G^k\varphi$: φ **is E^k knowledge in** G if everyone in G knows that everyone in G knows that ... that everyone in G knows that φ is true, where the phrase "everyone in G knows that" appears in this sentence k times.
- $C_G\varphi$: φ **is common knowledge in** G if φ is E_G^k-knowledge for all $k \geq 1$ or more formally $C_G\varphi = E_G^1\varphi \bigwedge E_G^2\varphi \bigwedge ... \bigwedge E_G^m\varphi \bigwedge ...$

Clearly, the mentioned knowledge forms a hierarchy

$$C\varphi \supset ... \supset E^{k+1} \supset ... \supset E\varphi \supset S\varphi \supset D\varphi \supset \varphi \tag{2.1}$$

Common knowledge is a prerequisite for agreement, convention, and coordinate action. Possessing common knowledge allows more efficient communication among agents. Since these goals are frequently sought in distributed environment, the problem of how to attain common knowledge becomes crucial. To get a flavor of the issue involved in attaining common knowledge, consider the coordination attack problem, originally introduced by Gray [76].

Two divisions of an army are camped on two hilltops overlooking a common valley. In the valley awaits the enemy (see Fig. 2.3). It is clear that if both divisions attack the enemy simultaneously, they will win the battle; whereas if one division attack, it will be defeated. The divisions do not initially have plans for launching an attack on the enemy, and the commanding general of the first division wishes to coordinate a simultaneous attack (at some time the next day). Neither general will decide to attack unless he is sure that the other will attack with him. The generals can only communicate by means of a messenger. Normally, it takes the messenger one hour to get from one encampment to the other. However, it is possible that he will get lost in the dark or, worse yet, be captured by enemy. Fortunately, on this particular night, everything goes smoothly. How long will it take them to coordinate an attack? Suppose General A sends a message to General

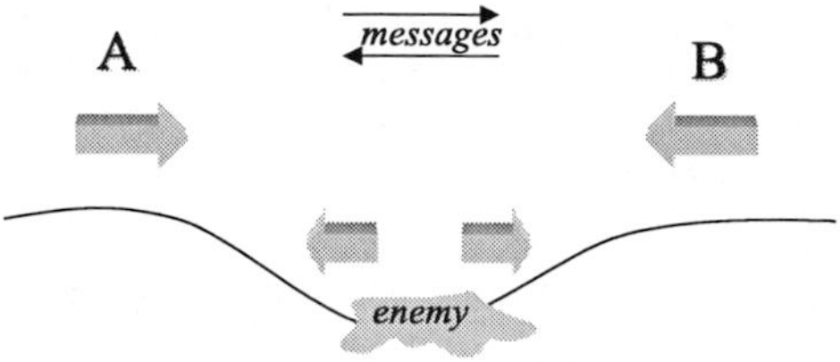

Figure 2.3: Coordination Attack Problem.

B saying "Let's attack at dawn", and the messenger delivers it an hour later. General A does not immediately known whether the message succeeded in delivering the message. ... Consequently, B sends the messenger back to A with an acknowledgement. ... Since B knows that A will not attack without knowing that B received the original message, he knows that A will not attack unless the acknowledgement is successfully delivered. Thus B will not attack unless he knows that the acknowledgement has been successfully delivered. ... A must sent the messenger back with acknowledgement to the acknowledgement. ... There is no fixed number of acknowledgements, acknowledgements to acknowledgements, etc., suffices for the generals to attack [75].

Considering this problem, we arrive at the conclusion that a common knowledge of this system is that the generals are attacking, anyway, at any correct protocols for the coordinated attack. However, there are no protocols that guarantee generals will ever attack, in other words, system never attains the common knowledge. In this way paradox has arisen that, in precise sense, reaching agreement and coordinating actions, a distributed system requires attaining common knowledge of certain facts. This kind of knowledge *cannot be attained in practical distributed systems* [75]. Lately many partial solutions of attaining common knowledge has been proposed, e.g., to introduce concurrent common knowledge [77], probabilistic common knowledge [78], polynomial-time common knowledge [79] or simply to restrict simultaneousness of actions.

Summarizing this overview of autonomous agents and their societies, we would like to add the following remarks. First, not all material about agents is considered here. The number of publications within the domain of autonomous intelligent systems seems to be growing exponentially in recent years and there is no hope of finding common definitions of agents. Therefore here, the work of Weiss and his co-authors [5], devoted to autonomous agents and multi-agent systems, is the primary source. Second, we would like to emphasize the main ability of an agent - its autonomy - that in fact causes the phenomenon of collective behavior, whose analysis and control represent the focus of this thesis and will be treated further.

2.2 Swarms of agent-based micro-robot systems

Swarm robotics focusses on designing collective intelligent systems comprising large numbers of robots. A typical swarm intelligence system has the following properties [80]:

- it is composed of many individuals;
- the individuals are relatively homogeneous (that is, they are either all identical or they belong to just a few typologies);

- the interactions among individuals are based on simple behavioral rules that exploit only local information that the individuals exchange either directly through antennation, trophallaxis (food or liquid exchange), mandibular contact, visual contact, chemical contact, and so on, or via the environment (stigmergy);
- the overall behavior of the system results from the interactions of individuals with each other and with their environment, that is, the group behavior self-organizes.

The fascination of swarms and its theoretical foundations originate in the studying and understanding of group behavior in animals and insect societies. Natural collective systems demonstrate that many individuals cooperate when it is profitable for each of them. Examples include cooperative hunting in predatory animals, group-based foraging, or nest building in social insects. In these, and many other examples, animals get together when it provides better opportunities for foraging, for defense, or generally for surviving in their environment. Individual participants can be weak, with limited sensory/actuator capabilities (for example, only local sensing and interactions), however collectively they can build a strong group with extended capabilities. It is expected that in a similar way, a group of relatively simple and cheap robots may solve a large task that is beyond the capabilities of a single robot [81]. The challenge of research is how to transform the swarm intelligence metaphor to engineering methodologies adapted to the current technology and how to design and control collective systems consisting of up to thousands of units?

The design of micro-robotic systems confronts the researcher with many unsolved problems. Some are typical of a distributed system, for example information transfer, cooperation and coordination between team members, or collective energy management. All these open questions require answers, but collective energy management is one of the most critical. What is the point of developing a system in which agents cooperate perfectly with each other to achieve some useful activities, when after a while the robots stop work because they run out of power? Their actions, which probably needed great efforts to be exactly coordinated, will be interrupted and only direct human intervention can bring the robots back to work. Recharging might take a lot of time and during the recharge, the swarm will not be able to executing any useful activity. The challenge of energy management changes greatly in situations where energy sources are rare within the "living" area of the swarm and agents have to compete to use them. In their attempts to reach the energy source, they struggle and hinder each other. As a result, the energetic bottleneck in the area is increased, reducing the surviving abilities of the swarm.

A foraging strategy that regulates the energetic needs of the swarm and negates competitiveness among the agents would help the swarm to maintain its energy homeostasis. To demonstrate this strategy, we consider a collective energy foraging scenario for "Jasmine" micro-robots (see Sec. 6.2).

2.3 Dynamical systems and synergetics

The mathematical sciences particularly exhibit order, symmetry, and limitation; and these are the greatest forms of the beautiful.
Aristotle

In this section, the synergetic concept and its underlying mathematical methods are discussed. The main goal of this representation is to give a brief review of methods that can be applied to an investigation of the local behavior of nonlinear dynamical systems. All these methods are conceptually combined by the concept of "synergetics " [44], to which we restrict ourselves. The primary problem of synergetic analysis can be viewed as obtaining the order parameter equations that describe the behavior of whole system in the neighborhood of instability. Deriving these equations, we use approaches such as the center manifold theory, the adiabatic elimination procedure, the normal form reduction and other methods. Since this thesis is an interdisciplinary work, this review begins with a consideration of dynamical systems and synergetics in a nonmathematical fashion, useful for experts from other scientific domains. After that, we consider rigorous analytical methods.

2.3.1 Survey of dynamical systems

The phenomenon of *complex systems* is often related back to Aristotle (384-322 v.Chr.) [85], who described the *holistic approach* to systems. Such an approach looks at complex systems as a whole individual object, instead of a collection of separate parts. From this viewpoint, the system represents a fusion of different interacting elements and has to be investigated as a common interacting structure.

The *interactions* between the components of a system that are, for example, influenced from outside, are constantly changing. Such a modification of interactions changes *the state of a system.* Every system existing in nature undergoes a modification of its own states over time, that is, possesses *dynamics*. Therefore, considering the dynamical processes in such systems, we refer to these systems as *dynamical systems*.

Understanding the dynamics of complex systems is one of the most challenging problems in many branches of modern science, such as biology, physics, engineering, economics, computer science, social sciences, and so on. However real dynamical systems, because of their enormous complexity, cannot be treated as the whole and must be broken down. This problem was discussed by, for example, the computer pioneer, M. Minsky, who has argued that the core of this problem lies in the limited capability of the human brain [86]. Finding a compromise between considering a system in the whole and the necessity of breaking it down is a systematic problem in modern science.

In solving the practical problems, we are in fact forced to *analyze* the dynamical system, breaking it down into various components, aspects, and so on. In other words, the simplified *models* are built and then considered with already known methods. Accordingly, investigating the dynamics of such models is an attempt to understand and predict the behavior of real systems, and to develop suitable control methods. In this case, the notion of dynamical systems more often refers to the *mathematical models* of real systems, than to the systems themselves.

Originally, from the later nineteenth century, the term *"dynamical system"* meant only mechanical systems whose motion is described by differential equations, derived in classical mechanics [1]. Basic results using such dynamical systems were obtained by Lyapunov [88] and Poincaré [89]. Their work has been continued by Dulac [90] and Birkhoff [91], among others. Later, it became clear that this notion is useful for the analysis of various evolutionary processes, studied in different branches of science and described by ordinary or partial differential equations (ODEs, PDEs), or by iterated maps. The modern period in dynamical systems theory started with the work of Kolmogorov [92], Smale [93]-[95], Anosov [96] and Arnold [97].

Continuing the overview of dynamical systems, we come to classification. Since dynamical systems, as mathematical models, represent real systems, it makes sense to build their classification on the underlying processes of real systems. The use of time and energy, the reversibility of the dynamics, and so on (see Table 2.2) form the main criteria of this classification. There are further special cases of real systems that can be classified in this system but they require special types of mathematical models. Some of these systems are summarized in Table 2.3. Moreover there are special cases of

Properties of real systems	Characteristics of mat. models	
Effluent of energy [1]	dissipative	conservative
Effluent of time	continuous	discrete
Reversibility	reversible	irreversible
Nonlinearity	linear	nonlinear
Degree of freedom	finite-dim.	infinite-dim.
Stochasticity	deterministic	stochastic
Influence of time on dynamics	autonomous	no autonomous

[1] This class includes also dispersive systems

Table 2.2: Some main characterizations of dynamical systems.

real systems, that can be also featured by the mentioned classification, but they require the special types of mathematical models. Some of such systems are summarized in Table 2.3.

Special cases	Mathematical models
Clas. mech. systems	Systems with Hamiltonian H
Selections systems	Selection equations
Reaction-diffusion systems	Reaction-diffusion equations
Self-reproduction	Cellular automatons
Distributed systems	Coupled map lattices, Artificial neural networks

Table 2.3: Some special cases of mathematical models possessing a specific dynamics. Left column shows the real systems, right column represents the equivalent mathematical models.

The classes of dynamical systems that will be discussed in this work are:

- **Dissipative systems.** Open, nonlinear, irreversible systems, existing far from thermodynamic equilibrium and demonstrating a spontaneous emergence of stable spatio-temporal structures. These kind of real systems and also mathematical models called dissipative [98], [99] and are often used in nonlinear dynamics.

[1] following [87].

- **Discrete in time.** Time-continuous systems are the most prevalent type of mathematical models described by ODEs or PDEs. This is connected with real systems, where time is continuous in nature. Time-discrete systems were first obtained by Poincaré section [89] of time-continuous systems and consequently were not often used in nonlinear dynamics. However, the importance of such systems has grown because of the appearance of new real processes, mainly in technical areas and computer science, that also use discrete time.

- **Distributed systems.** The systems we are interested in are distributed, which means they consist of autonomous subsystems (components) behaving independently of each other and without centralized coordination; synchronizing the behavior of the subsystem participants is achieved by means of interactions. Examples of distributed systems can be found in biology, social systems, robotics, and so on. As a result of the self-organization processes arising in such systems, the whole distributed system exhibits the macroscopic pattern, in which the components behave in an ordered way. Investigation of these so-called "collective phenomena" is the main focus of the synergetics considered later.

2.3.2 The phenomenological concept of synergetics

The main conceptual idea of synergetics was "*die Lehre vom Zusammenwirken* first formulated by Haken in 1969 and based on the investigation of the laser ([100]-[102]). When the atoms that the laser consists of are excited from outside, they emit light waves. At low power, the waves are entirety uncorrelated, as in a normal lamp, whereas by increasing the power, the individual atoms begin acting in a completely correlated way.

Discipline	**System**	**Subsystems**	**Ordered Structure**	**External Parameter**
Physics	Laser	Atoms	Laser light	Pump rate
Physics	Ferro-magnet	Elementary magnets	Magnetization	Temperature
Chemistry	Liquid e.g. BZR[1]	Molecules	Spatio-temporal pattern	Supply of raw material
Hydro-dynamics	System of Bénard	Molecules	Roll pattern	Heat supply
Biology	Population	Creatures	Natural coexistence e.g Swarm	Availability of raw material
Biology	Cells	Molecules	Distribution of concentration	Temperature, light, nutrient supply
Medicine	Brain	Nerve cell	Function	External stimulus
Sociology	Society	Individual	Communication	Language
Computer science	Multi-Agent System	Agent	Interactions pattern	Sensory data

[1] Belousov-Zhabotinsky-Reaction

Table 2.4: Phenomenological consideration of some dynamical systems (partly from [103]).

This collective ordered behavior of atoms at the microscopic level causes the emergence of coherent laser light at the macroscopic level. A similar type of behavior is observed in many other systems (see Table 2.4), therefore we can speak of a phenomenology of dynamical systems.

From the *phenomenological* viewpoint, synergetics deals with systems composed of many components and focuses its attention on the spontaneous, that is, self-organized emergence of new qualities which may be structures, processes or functions. The characteristic values (for example the energy of separate atoms, motion of fluid molecules, and so on) of these components represent the *microscopic level*. The interactions between these subsystems, as well as the behavior of separate elements, being influenced from outside, change. This change modifies the global behavior of the whole system observed at the *macroscopic level*. Thus, a qualitatively new behavior of a system at the macroscopic level is provided by the *collective ordered* motion of subsystems at the microscopic level (see Table 2.5).

Level of consideration	Qualitative characterization
Micro-level	internal structure of the system, where only the characteristic values of separate elements are taken into account
Meso-level	one considers the characteristic values of the whole system instead of separate elements
Macro-level	the system is viewed as one joined object that interacts with similar co-systems in the network

Table 2.5: Different levels of the system consideration.

As pointed out by Haken [104], the choice of level considered, as well as the characteristic values at these levels remains nontrivial. The example of laser light demonstrates the emergence of a new quality arising at the macroscopic level. Sometimes, this emergent quality back-influences elements at the microscopic level. Such an example is given by a distributed system making collective decisions. Interactions among subsystems causes the emergence of determined social laws in the group. On the basis of accepted social laws, the whole system takes a decision that already belongs to the macroscopic level. The collective decision correspondingly modifies the behavior of separate subsystems, taking part in the procedure of decision-making. However, most systems generally do not demonstrate new qualities at the macroscopic level. The spatio-temporal pattern exhibited by these systems undergoes only some modifications that can be observed on the macroscopic level.

As proposed by synergetics, macroscopic behavior and the corresponding new structures arising can be described by the derived macroscopic quantity, denoted as the order parameter. Derivation of the order parameter from microscopic equations is known as the *bottom-up* approach or the direct problem. Since this quantity, at the macroscopic level, presents the most common properties of the overall system, modifying this quantity implies the deepest change of the system being considered. Propagation of this modification from the macroscopic to the microscopic level is denoted as the *top-down* approach or the inverse problem. The difficulty of the inverse approach is that the information eliminated in the bottom-up procedure has to be reconstructed in the top-down method.

This work presents a combined approach. First, the complexity reduction from the real systems to their mathematical models is performed then, we derive the order parameter from these mathematical models. This methodology corresponds to the bottom-up strategy. By introducing control elements into the order parameter or modifying its structure, we are going back to initial systems. In order to solve the problem of inverse procedure, two techniques have been proposed. In the first, the desired control methods are introduced into the order parameter and, simultaneously, into the initial system. Then, comparing the results obtained, control in the initial system is tuned, until they coincide or, at least, are similar to each other. In the second variant, the eliminated information is fixed, assuming that all modifications in initial systems caused by the changed order parameter will be created by modifying the couplings between subsystems. The changes of couplings required can then be determined, in the way described in Chap. 6.

2.3.3 Methodology of synergetics

The above discussed concept can be expressed in mathematical terms underlying the synergetics. In mathematical terminology, each system is described by its characteristic value $q(t)$ that generally depends on the time. The set of all conceivable states of system is called the *state space*. Collecting m components of the system, we get the vector $\underline{\mathbf{q}}(t) = (q_1(t), q_2(t), ..., q_m(t))$ that represents a *states vector* of the whole system with $q_i \in \mathbb{R}$ for $i = 1, 2, \ldots, m$, where m usually denoted as the number of degrees of freedom. In view of it, the notion of state space is strongly connected to the notion of phase space, mentioned hereinafter. Then the dynamics of this system is described by the change of the state vector, defined in a corresponding state space $\mathcal{D} \subset \mathbb{R}^n$ for some positive $n \in \mathbb{N}$, in the course of time, i.e. by the following *equations of motion*

$$\dot{\underline{\mathbf{q}}}(t) = \underline{\mathbf{N}}(\underline{\mathbf{q}}(t), \{\alpha\}), \tag{2.2}$$

where $\underline{\mathbf{N}}$ is a nonlinear vector function and $\{\alpha\}$ is a set of control parameters. The *control parameters* are the methodologically essential point that should be separately mentioned. Qua them we understand such parameters that affect the system from outside. The examples can be given by a temperature, influx of energy and so on. Changing the control parameters up to critical values, the system undergoes some quantitative modifications of the behavior. Reaching these critical values, the system turns over into a new qualitative state. This transition is often denoted as a *bifurcation*.

Describing the qualitative behavior of system, we are mainly interested in two significant questions: firstly, the determination of the bifurcation point and, secondly, what type of behavior is expected after this critical point. Answer on both questions can be given by the bifurcation analysis, performing in the small area (vicinity) of bifurcation point. Therefore we can speak about the *local character* of synergetics methodology. Generalizing aforesaid, we point out that the *analysis of dynamical behavior in the vicinity of bifurcation can be viewed as the main goal of synergetic analysis* in sense of the Bottom-Up approach.

Considering the topological structure of trajectories of the system (2.2) in a neighborhood of an instability, it needs to remark that different "internal components" (as e.g. different terms of the Taylor decomposition) exert different influences on the structure of these trajectories. It can be distinctly seen, decomposing the right part of the system (2.2) into the Taylor series. Since all $\underline{\mathbf{q}}(t)$ are small, the linear terms of this series make the most intensive contribution into the topological structure of trajectories in a neighborhood of bifurcation. Therefore, if we are able to "foliate" the system (2.2) into such components, its dynamics could be reduced only to dynamics of the relevant components.

In order to extract these relevant components, we first introduce new state variables $\underline{\xi}$

$$\underline{\mathbf{q}}(t) = \underline{\mathbf{q}}_{st} + \sum_i \xi^i(t)\underline{\mathbf{v}}_i, \tag{2.3}$$

where $\underline{\mathbf{q}}_{st}$, $\underline{\mathbf{v}}_i$ are coefficients. Performing the coordinate transformation that decouples the linear part of Eq. (2.2) from each other, we get the following equations

$$\dot{\underline{\xi}}^u = \underline{\underline{\mathbf{\Lambda}}}_u \underline{\xi}^u + \underline{\tilde{\mathbf{N}}}_u(\underline{\xi}^u, \underline{\xi}^s, \{\alpha\}), \tag{2.4a}$$

$$\dot{\underline{\xi}}^s = \underline{\underline{\mathbf{\Lambda}}}_s \underline{\xi}^s + \underline{\tilde{\mathbf{N}}}_s(\underline{\xi}^u, \underline{\xi}^s, \{\alpha\}). \tag{2.4b}$$

Here $\underline{\xi}^u$, $\underline{\xi}^s$ are so-called mode amplitudes (that just represent the "internal components"), and $\underline{\underline{\mathbf{\Lambda}}}_u$, $\underline{\underline{\mathbf{\Lambda}}}_s$ are the matrices of eigenvalues satisfying

$$\lambda_u \geq 0, \lambda_s < 0, \tag{2.5}$$

where indices s, u mean respectively *stable* and *unstable*.

The concept of *mode amplitudes* is the next essential point of synergetic methodology. Generally, paradigm of "modes" in physics relates to the superposition of modes which can represent the initial system, like the Fourier modes [105]. They separately are an abstract mathematical quantity and accordingly are not encountering in real systems. However, the mode amplitudes play a key role in the theory of nonlinear dynamical systems. The reason lies on the decoupled linear terms of the corresponding mode equations (2.4), where these terms present the first terms of the Taylor decomposition of initial Eq. (2.2). Therefore, the most intensive influence on the local dynamics of the mode amplitude is given by the eigenvalues. It means, that these components possess "own" rate of changes and in a dependence of the eigenvalues they can be divided into the slow (with $\lambda \approx 0$) and fast components (with $\lambda < 0$). This kind of dynamics is often denoted as *intrinsic* (own) *dynamics* of amplitudes. Thus, extending the expression (2.3), the state vector $\underline{\mathbf{q}}(t)$ can be decomposed in the vicinity of an instability as follows [43]

$$\underline{\mathbf{q}}(t) = \underline{\mathbf{q}}_{st} + \sum_u \xi^u(t)\underline{\mathbf{v}}_u + \sum_s \xi^s(t)\underline{\mathbf{v}}_s, \tag{2.6}$$

where $\underline{\mathbf{v}}_i$ are in general spatial dependent coefficients and $\underline{\xi}$ are the mode amplitudes.

At the instability point, just a few collective modes becomes unstable and determine the whole dynamics of a system. These so-called equilibrium configurations [101] or mode amplitudes may either compete, so that only one survives, or coexist, stabilizing hereby each other. It reveals, that in a neighborhood of bifurcation a time of the internal relaxation of slow amplitudes is apparently very long. Because of the coupling in the nonlinear terms (so-called *external dynamics* of amplitudes), the remaining fast components are forced to follow the slowest components (*the time-scale hierarchy*), so to say, they are "enslaved" by the slowest components

$$\underline{\xi}^s = \underline{\mathbf{h}}(\underline{\xi}^u). \tag{2.7}$$

This functional dependence is generally called as the "*slaving principle*" whose strong formulation is given by such mathematical methods as e.g. the center manifold reduction [106]. Decoupling these slowest components from the rest of a system (e.g. substituting (2.7) into (2.4a)), we get the characteristic value of qualitative behavior, denoted in terms of synergetics *the order parameter*, given by the following equation

$$\dot{\underline{\boldsymbol{\xi}}}^u = \underline{\underline{\boldsymbol{\Lambda}}}_u \underline{\boldsymbol{\xi}}^u + \tilde{\underline{\mathbf{N}}}^u(\underline{\boldsymbol{\xi}}^u, \underline{\mathbf{h}}(\underline{\boldsymbol{\xi}}^u), \{\alpha\}). \tag{2.8}$$

This low-dimensional *order parameter equation* no longer depends on the stable modes and describes the qualitative change of dynamical behavior of initial system (2.2) in the vicinity of bifurcation.

The order parameter is the most significant concept of the synergetics because it methodologically demonstrates the process of *self-organization* in the distributed systems. The unstable modes, possessing unstable linear part, cause an instability at the critical values of control parameters. However, the successive dynamics of the initial system (2.2), after the bifurcation, is defined by the nonlinear terms of the order parameter equation (2.8). These terms are influenced by stable modes by means of the slaving principle (2.7). Accordingly, new dynamical states of the distributed system are determined by a cooperation between the stable and unstable modes. *This process of self-organization between the internal components of system leads to the emergence of ordered macroscopic behavior after bifurcation.* Therefore, the synergetics represents an adequate mathematical description of the collective behavior of many subsystems, moreover it shows the interdisciplinary character and role of *self-organization process* in dynamical systems.

The order parameter equation (2.8) can be derived by different mathematical methods. In the following sections we begin a mathematical introduction into such techniques.

2.3.4 Local analysis methods of dynamical systems

After the nonmathematical introduction into the main points of dynamical systems and their synergetic treatment, we start to consider a mathematical matter of the mentioned concepts. The begin of this representation is given by the dynamical systems interpreted geometrically. Such a consideration, accepted originally in physical disciplines, allows understanding and representing a complex behavior of dynamical systems in a graphical way. Our discussion follows closely Wiggins [42], Kuznetsov [87] and Arnold [97]. The focus of successive sections lies on the methods of local analysis, expressing the synergetic concept. The first of them concerns the linear stability analysis and derivation of mode equations, the further ones are the reduction techniques. Finally, we discuss the problems of equivalence and finite determinacy of reduced systems.

The given work deals primarily with time discrete systems. However, many notions used in the theory of nonlinear dynamical systems, such as e.g. nonlinear fields, manifolds and so on, are developed for time continuous systems. The nonlinear maps, obtained as Poincare sections, obviously use the same or similar notions. If there are known time discrete formulations of these notions, we demonstrate them explicitly, otherwise we consider only the time continuous case.

2.3.4.1 Geometrical presentation of dynamical systems

Dynamical systems are mentioned here, in sense of mathematical models describing the evolution of existing real physical, biological, and other systems and processes [107]. The first step towards analyzing dynamical systems consists in determining a space, where such an evolution occurs, and its metrics. This space contains all possible states of the considered dynamical system and is usually denoted as the *states space*. Often, the state space is called a *phase space*, following a tradition from classical mechanics. The notion of phase space narrows down the analysis of evolutional processes to investigation of geometrical curves and manifolds defined by vector fields.

At each time moment, the dynamical system occupies a defined spatial position in the states space given by a point denoted as the *phase point* or the representing point of this dynamical system. Motion of the phase point in the states space is governed by some evolution law reflecting the dynamics of original system. In case the future state of system can be predicted by knowing its present state and the governing law, such a dynamical system describes the deterministic process and refers to the *deterministic dynamical systems*. Provided that the governing law does not change in time, the behavior of such a system can be considered as completely defined by its initial state. Thus the specification of a initial state $q_0 = q(t_0)$ at time t_0 uniquely determines the state $q(t)$ at any time t. By assumption $t_0 = 0$, it is convenient to write $q_0 = q(0)$.

It is important to remark, that a equation, describing a relationship between one state of point in time and a previous state of point in time, where the size of time step is finite (the time must be an integer value respectively), is called a time discrete dynamical system or *a discrete map*. If time step is infinitesimal small (in other words the time can be any real value), resulting continuous dynamical system is expressed as *a differential equation*.

Considering a compact subset $\mathcal{M}$ of phase space, consisted of an infinity number of points, it is reasonable to introduce the evolution operator φ_t, that switches each state $q_i \in \mathcal{M}$ in the new state $\varphi_t(q) \in \mathcal{M}$, where $t \in \Theta$ and $\Theta \in \mathbb{R}$ or $\Theta \in \mathbb{N}$

$$\varphi_t : \mathcal{M} \to \mathcal{M}, \; \varphi_0(q) = q, \; \forall q \in \mathcal{M}, \; \forall t \in \Theta. \tag{2.9}$$

where $\varphi_0 = id$, denoting the identity map for all $q \in \mathcal{M}$. Property 2.9 expresses that q is the prescribed state at the time zero. A family of mappings $\{\varphi_t\}_{t\in\Theta}$ in $\mathcal{M}$ is denoted as *one-parameter group of transformations*. The $(\mathcal{M}, \{\varphi_t\}_{t\in\Theta})$ is denoted as *phase flow*, whose elements are phase points. The following basic defining properties hold for the one-parameter group of transformations

$$\varphi_t = \varphi_t \circ \varphi_0, \quad \varphi_0 = \varphi_{-t} \circ \varphi_t, \quad \varphi_{t+s} = \varphi_t \circ \varphi_s. \tag{2.10}$$

Last property of (2.10) means, that the result of the evolution of original system in the course of $t + s$ units of time, starting at a point $q_0 \in \mathcal{M}$, is the same as if the system were first allowed to change from the state q_0 over only t units of time and then evolved over the next s units of time from the resulting $\varphi_s(q_t)$. This property implies that the law governing the behavior of system does not change in the time, the system is "autonomous" [87] (see Fig. 2.4(a)). In the same way (2.10) implies that

$$\varphi_2 = \varphi_1 \circ \varphi_1 \text{ and for } t \in \mathbb{N} \;\Rightarrow\; \varphi_t = \underbrace{\varphi_1 \circ \varphi_1 \circ \cdots \circ \varphi_1}_{t-times}. \tag{2.11}$$

By means of these formulas is defined time discrete dynamical system corresponds to the iteration of a map. If $q \in \mathcal{M}$ is a some phase point, then the mapping

$$q(t) = \varphi_t(q), \; q : \Theta \to \mathcal{M}, \tag{2.12}$$

for all t is called a *motion* of the point under an influence of the phase flow (see Fig. 2.4(b)). A derivation

$$\frac{d}{dt} q(t) \overset{def}{=} \dot{q}(t) = N(q(t)), \tag{2.13}$$

denotes a *phase rate* of this flow in point $q \in \mathcal{M}$, where $N(q(t))$ is a *vector field* on $\mathcal{M}$ and t is the time variable. Let $\mathcal{M} \subset \mathbb{R}^n$ be the state space, and $N : \mathcal{M} \to \mathbb{R}^n$ is the vector field on $\mathcal{M}$. Then the equation

$$\dot{q}(t) = N(q(t), \{\alpha\}), \tag{2.14}$$

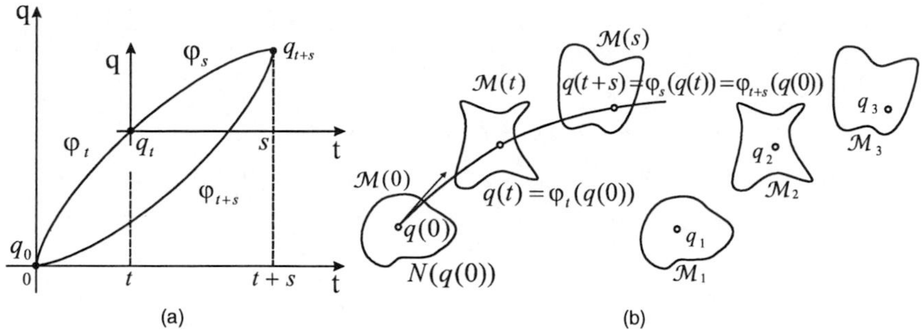

Figure 2.4: (a) Evolution operator of autonomous system (group property); **(b)** Evolution of a volume with the flow (left) and with an iterated map (right).

is referred to as *differential equation*, that is defined by the vector field N on $\mathcal{M}$, where $\{\alpha\}$ is a set of control parameters. In the case of a discrete time the dynamics is a map with the discrete time n

$$q_{n+1} = N(q_n, \{\alpha\}). \tag{2.15}$$

The path in phase space, traced out by a solution of (2.13) or of (2.15) with determined initial value $q(0)$ or correspondingly q_0 is called an *orbit* or *trajectory* of the dynamical system. If the state variables take real values in the space $\mathcal{M}$, the orbit at $\Theta \in \mathbb{R}$ is a a continuous curve parametrized by the time variable, while the orbit at $\Theta \in \mathbb{N}$ is an ordered sequence of points in this space enumerated by increasing integers (see Fig. 2.5). Correspondingly the first case refers to the time-continuous system and the second one to the time-discrete system. The *phase portrait* of a dynamical system is composed of

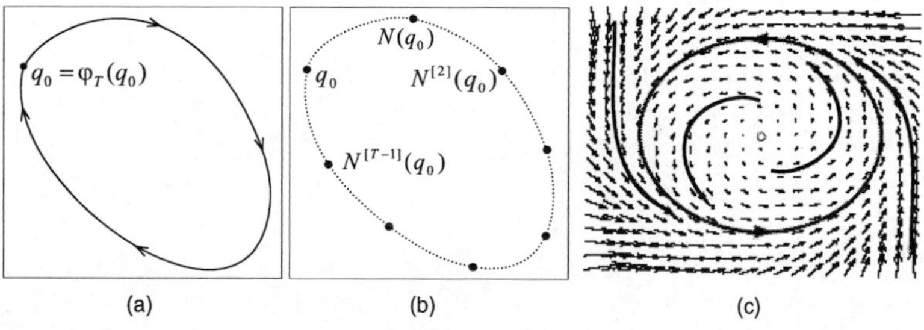

Figure 2.5: Example of periodic orbits; **(a)**orbit of time-continuous system; **(b)**orbit of time-discrete system; **(c)**phase portrait of periodic orbit (limit cycle).

these orbits. The simplest orbit is called *equilibrium point* that is also known as *fixed point*, *steady state* or *stationary point* and correspond to stationary solution [42]. For nonlinear systems there it is typically no hope of finding the trajectories analytically. Even when explicit formulas are available, they are often too complicated to provide much insight. Instead, we will try to determine the qualitative behavior of the possible trajectories (solutions) respectively. We classify here the possible trajectories as:

- *The fixed point.* Fixed point of the vector field satisfies $N(q_{st}) = 0$ and of the map satisfies $N(q_{st}) = q_{st}$, and corresponds to equilibria or steady states of the system.

- *The closed orbits.* Closed orbit is a nonequilibrium orbit and correspond to periodic solutions, i.e., solutions for which $q(t+T) = q(t)$ for all t and for some given minimal period $T > 0$.

Some of the most salient features of any phase portrait are:

- The arrangement of trajectories near the fixed points and closed orbits.
- The stability of the fixed points and closed orbits.

2.3.4.2 Stationary states

The point $q^{fix} \in \mathcal{M}$ is called equilibrium or fixed point if $\varphi_t(q^{fix}) = q^{fix}$. Considering the case $t \in \mathbb{R}$, the point q^{fix} with $N(q^{fix}) = 0$ is the singular point of the vector field N. The vector field in the vicinity of this point discontinuously changes own direction. Every singular point q^{fix} of vector field N in Eq. (2.14) generates a stationary solution $q(t, \{\alpha\}) \equiv q^{fix}(\{\alpha\}) \equiv q_{st}(\{\alpha\})$. In the following we will denote the stationary solution as the stationary state q_{st}. Hereby the stationary state is such a state of system, that this system behaves independently of time in, i.e.

$$N(q_{st}, \{\alpha\}) = 0 \tag{2.16}$$

The condition (2.16) for $t \in \mathbb{N}$, meaning the time discrete system, has the following form

$$N(q_{st}, \{\alpha\}) = q_{st} \tag{2.17}$$

In common, we can distinguish in the solutions set of (2.14), among others, a non-periodical, and periodical solutions. In this work the following definition of periodical solutions is used:

Definition 4 (Periodic orbit).

A periodic orbit (cycle) is a nonequilibrium orbit, such that any point q_0 satisfies $\varphi_{t+T}(q_0) = \varphi_t(q_0)$ with the same $T > 0$ and all $t \in \mathbb{R}$. The minimal T having this property is called the period.

Definition 5 (Periodicity of vector fields).

A solution of (2.14) through the point q_0 is said to be periodic of period T if there exist $T > 0$ such that $N(t, q_0) = N(t+T, q_0)$ for all $t \in \mathbb{R}$.

Definition 6 (Periodicity of maps).

The orbit of $q_0 \in \mathbb{R}^n$ is said to be periodic of period $[k] > 0$ if $N^{[k]}(q_0) = q_0$. Here periodic orbits are finite sets of points of the form $\{q_0, N(q_0), N^{[2]}(q_0), \ldots, N^{[T-1]}(q_0)\}$ with $N^{[T]}(q_0) = q_0$ for a integer period T.

Note that fixed points and periodic orbits are *non-wandering* and the set of all non-wandering points of a map or flow is called the non-wandering set of that particular map or flow.

From a geometric point of view the properties of system trajectories around an equilibrium point is related to stability of systems. A non-wandering set may be *stable or unstable.* In the following some definitions of notions concerning stability of non-wandering set are given not in an exact sense but in a way, that they can be used and referred to in the remaining work.

Definition 7 (Lyapunov stability).

Non-wandering set is called stable if every orbit starting in a neighborhood of the non-wandering set remains in a neighborhood.

Definition 8 (Asymptotic stability).

Non-wandering set is called asymptotically stable if it is stable and every orbit in a neighborhood approaches the non-wandering set asymptotically.

Asymptotically stable non-wandering sets are also called *attractors.* The *basin of attraction* is the set of all initial states approaching the attractor in the long time limit. In order to to analyze the stability of a non-wandering set the linear stability analysis can be employed.

2.3.4.3 Linear stability analysis

By means of linear stability analysis of flows and maps we get the basic information about local dynamics. The obtained stationary solution q_{st} of the differential equation (2.14) represents point where the system behaves independently of time. The system, starting exactly from this point, will forever remain there. However, the stationary states do not provide any information about the system motion in case the system starts in their neighborhood. In order to get insight into this essential topic, it needs to investigate the attraction of nearby solutions to these points i.e. to investigate the stability of stationary solution. The stability of q_{st} means the solutions starting "close" to q_{st} at given time remain close to q_{st} for all later times. The q_{st} is asymptotical stable if nearby solutions actually converge to q_{st} as $t \to \infty$.

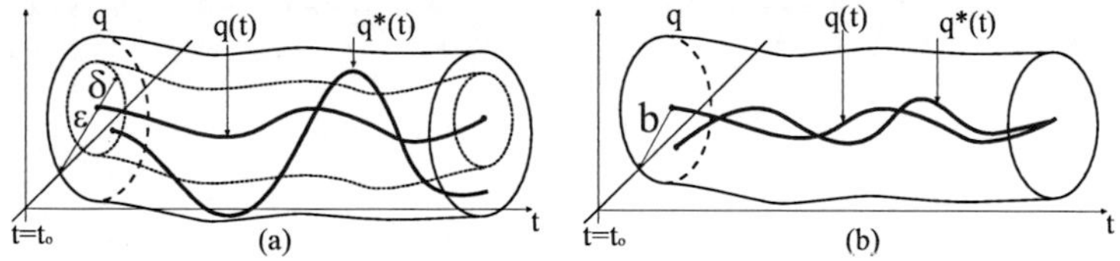

Figure 2.6: Equilibrium solutions: stability; **(a)**Liapunov stability; **(b)**asymptotic stability. These Figures are taken from [107] and modified.

Formalizing above-said, we have the following definition of local stability [42].

Definition 9 (Liapunov stability).

Solution $q(t) = q_{st}$ is said to be stable (or Liapunov stable) if, given $\epsilon > 0$, there exists a $\sigma = \sigma(\varepsilon) > 0$ such that, for any other solution, $q^*(t)$, of (2.14) satisfying $|q(t_0) - q^*(t_0)| < \sigma$, then $|q(t) - q^*(t)| < \varepsilon$ for $t > t_0,\ t_0 \in \mathbb{R}$.

Definition 10 (Asymptotic stability).

Solution $q(t)$ is said to be asymptotic stable if it is Liapunov stable and if there exists a constant $b > 0$, such that, if $|q(t_0) - q^*(t_0)| < b$, then $\lim_{t\to\infty} |q(t) - q^*(t)| = 0$.

Considering the stability of fixed point, we assume the system motion is slightly perturbed from the stationary states q_{st} on the value Δq, i.e.

$$q(t) = q_{st} + \Delta q(t), \tag{2.18}$$

and hereinafter we investigate the behavior of this small deviation

$$\Delta\dot{q}(t) = N(q_{st} + \Delta q(t), \{\alpha\}), \tag{2.19}$$

where q_{st} changes smoothly with $\{\alpha\}$, i.e. $q_{st} = q_{st}(\{\alpha\})$. Because $\Delta q(t)$ is small we may expand the right-hand side of (2.19) into a power series up to second order with respect to $\Delta q(t)$. Then Eq. (2.19) gets

$$\Delta\dot{q}(t) = N(q_{st}, \{\alpha\}) + \lambda(\{\alpha\})\Delta q(t) + O(|\Delta q(t)|^2) \tag{2.20}$$

where coefficient $\lambda(\{\alpha\}) = \frac{dN}{dq}|_{q=q_{st}}$. In the multi-dimensional case $\underline{\underline{\mathbf{L}}}(\{\alpha\})$ is the matrix of partial derivatives of vector field $\underline{\mathbf{N}}$

$$\underline{\underline{\mathbf{L}}} = \left.\frac{\partial N_i}{\partial q_k}\right|_{\underline{\mathbf{q}}=\underline{\mathbf{q}}_{st}}, \tag{2.21}$$

or so-called Jacobian, evaluated on the stationary state $\underline{\mathbf{q}}_{st}$, where $i, k = 1, ..., m$ and m is the dimension of system.

The essential idea is that the high-order terms $O(|\Delta q(t)|^2)$ of (2.20) in the neighborhood of Δq_{st} do not affect on the topological structures of trajectories. Therefore in the first approximation we can neglect these terms. Since $N(q_{st}, \{\alpha\}) = 0$ the equation (2.20) gets the following form

$$\Delta\dot{q}(t) = \lambda(\{\alpha\})\Delta q(t). \tag{2.22}$$

Then the solution of (2.22) is given by

$$\Delta q(t) = e^{\lambda(\{\alpha\})t} q_0 \text{ for } \forall\, t \in \mathbb{R}. \tag{2.23}$$

Analytically calculating the roots of a polynomial of order higher than two is very tedious and is impossible if its order is greater than five. Thanks to the Routh-Hurwitz theorem the question on stability for time-continuous system can be answered without explicitly calculating the eigenvalues [108]. This theorem can be used to determine whether all eigenvalues have negative real parts or not and is useful even in the case of a two-dimensional phase space where the characteristic polynomial is quadratic.

Definition 11.

Both eigenvalues have negative real part if and only if the determinant det $N(q_{st})$ is positive and the trace of det $N(q_{st})$ is negative. A real polynomial is said to be stable if all its roots lie in the left half-plane.

Apparently if

$$Re(\lambda(\{\alpha\})) < 0 \tag{2.24}$$

the motion of $\Delta q(t)$ will converge to zero, therefore any deviation from stationary state q_{st} will decay. Such a fixed point refers to the asymptotical stable one in linear approximation. If

$$Re(\lambda(\{\alpha\})) = 0 \tag{2.25}$$

$\Delta q(t) = q(0)$ and fixed point is stable in linear approximation. In case

$$Re(\lambda(\{\alpha\})) > 0 \tag{2.26}$$

any small deviation from stationary state will grow and such a fixed point is unstable in linear approximation.

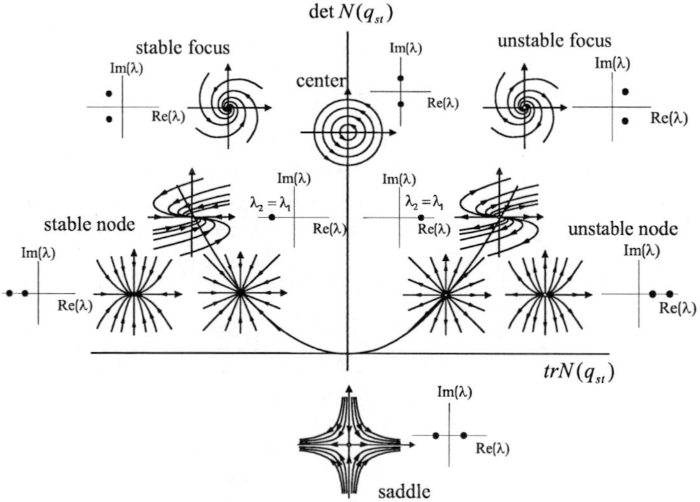

Figure 2.7: Hyperbolic equilibria of two-dimensional phase spaces (modified from [109]).

Definition 12.

Let $q = q_{st}$ be a fixed point of (2.14). The point q_{st} is called a *hyperbolic fixed point*, if Jacobian matrix does not have zero eigenvalues or eigenvalues with zero real part.

When the equilibrium undergoes a bifurcation arise a zero eigenvalue (or eigenvalues with zero real parts) and such equilibria are called *nonhyperbolic*. In this case linear analysis cannot answer the question of stability of a nonlinear system, since small nonlinear (high-order) terms play a crucial role here. Topological classification of the hyperbolic fixed point based on the eigenvalues is shown in Fig. 2.7. For the case of the dynamical systems discrete in time the deviation from the stationary states, discussed above, has for the steps n and $n+1$ the following form

$$q_n = q_{st} + \Delta q_n, \tag{2.27a}$$

$$q_{n+1} = q_{st} + \Delta q_{n+1}. \tag{2.27b}$$

Then analogously to the Eq. (2.20) we get

$$q_{st} + \Delta q_{n+1} = N(q_{st}, \{\alpha\}) + \lambda(\{\alpha\})\Delta q_n + O(|\Delta q_n|^2), \tag{2.28}$$

where $N(q_{st}, \{\alpha\}) = q_{st}$. Correspondingly the linearized system has the following form

$$\Delta q_{n+1} = \lambda(\{\alpha\})\Delta q_n, \tag{2.29}$$

whose solution is given by

$$\Delta q_n = (\lambda(\{\alpha\})^n q_0, \tag{2.30}$$

where q_0 is an initial condition. The fixed point q_{st} is in this case asymptotical stable, stable and unstable if correspondingly

$$|\lambda(\{\alpha\})| < 1, \tag{2.31a}$$

$$|\lambda(\{\alpha\})| = 1, \tag{2.31b}$$

$$|\lambda(\{\alpha\})| > 1. \tag{2.31c}$$

Generalizing aforesaid, we can remark that the linear analysis, firstly, points to the stability of stationary states in a linear approximation. Secondly, the critical values of eigenvalues are advised of instability points in the system behavior. Finally, the values of critical eigenvalues inform about the type of the arising bifurcation and moreover allow constructing the corresponding normal forms (see Sec. 2.3.4.6). In order to infer about the system behavior after the bifurcation, it is necessary to attract for an analysis the nonlinear terms of system (2.14).

2.3.4.4 Mode amplitudes equations

As assumed in the previous sections, the original system intended for investigation is one-dimensional. In this case it was not so difficult to arrive at the conclusion about the behavior of state variable in linear approximation. This information is needed for the further nonlinear analysis, shown in the next sections. However, in case the original system possesses m state variables (i.e. m-dimensional) the information about a behavior of separate state variables in linear approximation cannot be directly obtained. Consequently, the original system has to be transformed in such a form, where the state variables are uncoupled in the linear part and we can make assumptions about the behavior of state variables in linear approximation. This transformation is often denoted as *linear coordinate transformation, the mode amplitudes transformation, transformation to the main diagonals* and so on.

For the simplification of further calculation we use in the following the compact "index notation" closely associated with tensor. Application of the tensor notation for more compact writing, especially in high-order terms, is well-known in scientific literature (see e.g. [110], [111]). Taking into account aforesaid, the equation (2.14) can be rewritten in the form

$$\underline{\dot{\mathbf{q}}} = \Gamma_{(1)} + \Gamma_{(2)}(:\underline{\mathbf{q}}) + \Gamma_{(3)}(:\underline{\mathbf{q}})^2 + \Gamma_{(4)}(:\underline{\mathbf{q}})^3 ... = \sum_{r=0}^{p} \Gamma_{(r+1)}(:\underline{\mathbf{q}})^r, \tag{2.32}$$

where $\Gamma_{(n)}(:\underline{\mathbf{q}})$ is the tensor of rank n, expressed in terms of the corresponding orders of the variables $q^{(i)}$. Equation (2.32) reads in components

$$\begin{aligned} \dot{q}^{(i)} &= (\Gamma_{(1)})_i + \sum_{j_1} (\Gamma_{(2)})_{ij_1} q^{j_1} + \sum_{j_1 j_2} (\Gamma_{(3)})_{ij_1 j_2} q^{j_1} q^{j_2} + ... = \\ &= \sum_{r=0}^{p} \sum_{ij_1 ... j_r} (\Gamma_{(r+1)})_{ij_1 ... j_r} q^{j_1} ... q^{j_r}, \end{aligned} \tag{2.33}$$

where indices above the corresponding variables, e.g. $q^{(i)}$ mean the indexed variables $q^{(1)}, q^{(2)}, \ldots, q^{(i)}$ or the components of a vector $(q^{(1)} q^{(2)} \ldots q^{(i)})$. Indices below the tensors, e.g. $(\Gamma_{(3)})_{ij_1 j_2}$, refer to the corresponding components of these tensors. In the expression (2.33) p is the maximal order of the equation of motion. Inserting $\underline{\mathbf{q}} = \underline{\mathbf{q}}_{st} + \Delta \underline{\mathbf{q}}$ in (2.32), we derive the following equation of motion where $\Delta \underline{\mathbf{q}}$ is the deviation of the state vector from the stationary state $\underline{\mathbf{q}}_{st}$

$$\Delta \underline{\dot{\mathbf{q}}} = \Gamma^L_{(2)}(:\Delta \underline{\mathbf{q}}) + \sum_{r=2}^{p} \Gamma^N_{(r+1)}(:\Delta \underline{\mathbf{q}})^r. \tag{2.34}$$

In Eq. (2.34) Γ^L and Γ^N are the tensors, presenting the linear and nonlinear parts respectively. To proceed further we apply the following linear coordinate transformation, introducing hereby new state variables $\underline{\boldsymbol{\xi}} = (\xi^1, \xi^2, ..., \xi^m)^T \in \mathbb{R}^m$

$$\Delta \underline{\mathbf{q}} = \sum_k \xi^k \underline{\mathbf{v}}_k = \underline{\underline{\mathbf{V}}}\, \underline{\boldsymbol{\xi}} = \Gamma^V_{(2)}(:\underline{\boldsymbol{\xi}}), \tag{2.35}$$

where $\underline{\underline{\mathbf{V}}} = (\underline{\mathbf{v}}_1, ..., \underline{\mathbf{v}}_k)$ is the $m \times m$ matrix of eigenvectors of Jacobian (2.21) and $\Gamma^V_{(2)}$ is the corresponding tensor of eigenvectors. In terms of synergetics the introduced variables $\underline{\boldsymbol{\xi}}$ are usually denoted as mode amplitudes. Substituting expression (2.35) into (2.34), we obtain the following nonlinear equation of motion

$$\Gamma^V_{(2)}(:\underline{\dot{\boldsymbol{\xi}}}) = \Gamma^L_{(2)} \Gamma^V_{(2)}(:\underline{\boldsymbol{\xi}}) + \sum_{r=2}^{p} \Gamma^N_{(r+1)}(\Gamma^V_{(2)}(:\underline{\boldsymbol{\xi}}))^r. \tag{2.36}$$

Now multiplying from the left side with the tensor of inverse eigenvectors $\Gamma^{V^{-1}}_{(2)}$, we get the following mode amplitude equations

$$\underline{\dot{\boldsymbol{\xi}}} = \Gamma^\Lambda_{(2)}(:\underline{\boldsymbol{\xi}}) + \sum_{r=2}^{p} \Gamma^N_{(r+1)}(:\underline{\boldsymbol{\xi}})^r. \tag{2.37}$$

Here $\Gamma^{\Lambda}_{(2)} = \Gamma^{V^{-1}}_{(2)} \Gamma^{L}_{(2)} \Gamma^{V}_{(2)}$ is the tensor [2] of the eigenvalues of the Jacobian (2.21) of system (2.14) in the diagonalized form

$$\underline{\underline{\Lambda}} = \begin{pmatrix} \lambda_u^1 & 0 & \cdots & \cdots & \cdots & 0 \\ 0 & \ddots & \ddots & & & \vdots \\ \vdots & & \ddots & \lambda_u^c & \ddots & \vdots \\ \vdots & & & \ddots & \lambda_s^{c+1} & \ddots & \vdots \\ \vdots & & & & \ddots & \ddots & 0 \\ 0 & \cdots & \cdots & \cdots & 0 & \lambda_s^m \end{pmatrix} \tag{2.38}$$

where $\lambda_u^1, ..., \lambda_u^c, \lambda_s^{c+1}, ..., \lambda_s^m$ are eigenvalues. The tensor $\Gamma^{\tilde{N}}$ presents the nonlinear part of the mode amplitude equations (2.37) and the following identity was used

$$(\Gamma^{V^{-1}}_{(2)} \Gamma^{V}_{(2)})_{ij} = \delta_{ij}, \tag{2.39}$$

where δ_{ij} is the well-known Kronecker-Symbol. Exemplifying (2.37), the tensor $\Gamma^{\tilde{N}}_{(3)}$ reads in components

$$(\Gamma^{\tilde{N}}_{(3)})_{ijk} = \sum_{lmn} (\Gamma^{V^{-1}}_{(2)})_{il} (\Gamma^{N}_{(3)})_{lmn} (\Gamma^{V}_{(2)})_{mj} (\Gamma^{V}_{(2)})_{nk} \,. \tag{2.40}$$

The mode amplitude equations (2.37) have the decoupled linear parts which are given by $\Gamma^{\Lambda}_{(2)}$. From the conditions (2.24)-(2.26) $Re(\lambda(\{\alpha\})) \geq 0$, $Re(\lambda(\{\alpha\})) < 0$ for time continuous case or the conditions (2.31) $|\lambda| \geq 1$ and $|\lambda| < 1$ for time discrete case the eigenvalues are denoted as unstable λ_u and correspondingly as stable λ_s. For simplification, the mode amplitude equations will be denoted as the mode equations and the mode amplitudes $\underline{\boldsymbol{\xi}}$ as the amplitudes. In terms of synergetics, the amplitudes with the stable eigenvalues λ_s in the linear part are denoted as the stable amplitudes $\underline{\boldsymbol{\xi}}^s$ and accordingly with the unstable eigenvalues λ_u as the unstable amplitudes $\underline{\boldsymbol{\xi}}^u$. It is necessary to emphasize, that the terms *"stable"* and *"unstable"* are just refereed to the linear stability analysis only. In many cases, due to the nonlinearities, the amplitudes which were formerly unstable in the linear analysis become stabilized [43].

Since eigenvalues in the diagonalized matrix (2.38) can be separated on the matrix of unstable eigenvalues $\underline{\underline{\Lambda}}_u$ and correspondingly the matrix of stable ones $\underline{\underline{\Lambda}}_s$, we rewrite the mode equation (2.37) in the following form

$$\dot{\underline{\boldsymbol{\xi}}}^u = \underline{\underline{\Lambda}}_u \, \underline{\boldsymbol{\xi}}^u + \underline{\tilde{\mathbf{N}}}_u(\underline{\boldsymbol{\xi}}^u, \underline{\boldsymbol{\xi}}^s, \{\alpha\}), \tag{2.41a}$$

$$\dot{\underline{\boldsymbol{\xi}}}^s = \underline{\underline{\Lambda}}_s \, \underline{\boldsymbol{\xi}}^s + \underline{\tilde{\mathbf{N}}}_s(\underline{\boldsymbol{\xi}}^u, \underline{\boldsymbol{\xi}}^s, \{\alpha\}), \tag{2.41b}$$

where $\underline{\tilde{\mathbf{N}}}_u, \underline{\tilde{\mathbf{N}}}_s$ are nonlinear vector functions of $\underline{\boldsymbol{\xi}}^u, \underline{\boldsymbol{\xi}}^s$.

Both amplitudes $\underline{\boldsymbol{\xi}}^u, \underline{\boldsymbol{\xi}}^s$ have the intrinsic dynamics governed by the eigenvalues and the external dynamics determined by nonlinear terms. Because the unstable eigenvalues are $Re(\lambda_u) \approx 0$ or accordingly $|\lambda_u| \approx 1$ near the bifurcation, the intrinsic dynamics of the unstable amplitudes $\underline{\boldsymbol{\xi}}^u$ varies slowly and therefore $\underline{\boldsymbol{\xi}}^u$ are also called the slow amplitudes and correspondingly $\underline{\boldsymbol{\xi}}^s$ the fast amplitudes. In the following section we show, that the dynamics of slow amplitudes is dominant in the vicinity of instability, this fact can be utilized to eliminate stable modes.

[2] In general the tensor $\Gamma^{\Lambda}_{(2)}$ presents the eigenvalues in Jordan normal form, see [43].

2.3.4.5 Center manifolds with fixed eigenvalues

The overview, considered below, primarily touches upon time continuous systems in case unstable eigenvalues are fixed. Treatment of time discrete systems can be found in Chap. 3, whereas the case of the parameter dependent center manifold we refer to literature (e.g. [42], [45]).

As stated above, the main efforts of nonlinear dynamics are focused in solving the nonlinear equation (2.14), or at least, in understanding its long time behavior. However in the most cases, the exact solution cannot be obtained. A possible way towards the desired solution of this problem is to restrict the consideration on a local behavior in the vicinity of instabilities. In general, two approaches for investigating local dynamics are known: reducing the dimensionality of system and eliminating the nonlinearities. Two rigorous mathematical techniques allowing substantial progress along both lines of reduction approach are the center manifold theory and the method of normal forms [42]. Now we review the first from them. We consider a behavior of system in the subspace of eigenvectors, so-called *invariant manifold* featured first for hyperbolic fixed point.

Let us denote $\underline{\mathbf{v}}_u = (v_u^1, ..., v_u^c)$ as eigenvectors corresponding to eigenvalues of $\underline{\underline{\mathbf{\Lambda}}}_u$ with $Re(\lambda_i) > 0$ and $\underline{\mathbf{v}}_s = (v_s^{c+1}, ..., v_s^m)$ as eigenvectors corresponding to eigenvalues of $\underline{\underline{\mathbf{\Lambda}}}_s$ with $Re(\lambda_i) < 0$ of the linear system (2.22). Then $\mathbb{E}^u$ and $\mathbb{E}^s$ are the unstable and stable subspaces (or so-called manifolds) spanned of eigenvectors $\underline{\mathbf{v}}_u$ and $\underline{\mathbf{v}}_s$ (see Fig. (2.8)(a)).

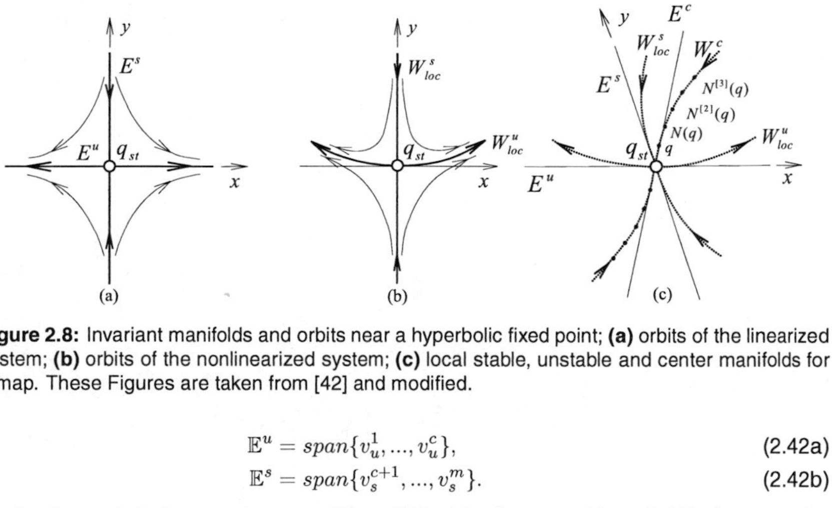

Figure 2.8: Invariant manifolds and orbits near a hyperbolic fixed point; **(a)** orbits of the linearized system; **(b)** orbits of the nonlinearized system; **(c)** local stable, unstable and center manifolds for a map. These Figures are taken from [42] and modified.

$$\mathbb{E}^u = span\{v_u^1, ..., v_u^c\}, \tag{2.42a}$$

$$\mathbb{E}^s = span\{v_s^{c+1}, ..., v_s^m\}. \tag{2.42b}$$

Analogously to linear subspaces $\mathbb{E}^u$ and $\mathbb{E}^s$ of the linear problem (2.22), there are the nonlinear local unstable $\mathbb{W}^u_{loc}$ and stable $\mathbb{W}^s_{loc}$ manifolds of fixed point $\underline{\mathbf{q}}_{st}$ determined as follows (Fig. (2.8)(b))

$$\mathbb{W}^u_{loc} : \{\underline{\mathbf{q}} \in U; \varphi_t(\underline{\mathbf{q}}) \to \underline{\mathbf{q}}_{st} \;\; as \;\; t \to -\infty \;\; and \;\; \varphi_t(\underline{\mathbf{q}}) \in U, \forall t \leq 0\} \tag{2.43a}$$

$$\mathbb{W}^s_{loc} : \{\underline{\mathbf{q}} \in U; \varphi_t(\underline{\mathbf{q}}) \to \underline{\mathbf{q}}_{st} \;\; as \;\; t \to +\infty \;\; and \;\; \varphi_t(\underline{\mathbf{q}}) \in U, \forall t \geq 0\} \tag{2.43b}$$

where $U \in \mathbb{R}^n$ is vicinity of $\underline{\mathbf{q}}_{st}$. Thus, while the invariant manifolds of flows are composed of the unions of solution curves, those of maps are unions of discrete orbit points (see Fig. (2.8)). Unstable and stable manifolds for maps are defined by

$$\mathbb{W}^u_{loc} : \{\underline{\mathbf{q}} \in U; \varphi_n(\underline{\mathbf{q}}) \rightarrow \underline{\mathbf{q}}_{st} \ \ as \ \ n \rightarrow -\infty \ \ and \ \ \varphi_n(\underline{\mathbf{q}}) \in U, \forall n \leq 0\} \tag{2.44a}$$

$$\mathbb{W}^s_{loc} : \{\underline{\mathbf{q}} \in U; \varphi_n(\underline{\mathbf{q}}) \rightarrow \underline{\mathbf{q}}_{st} \ \ as \ \ n \rightarrow +\infty \ \ and \ \ \varphi_n(\underline{\mathbf{q}}) \in U, \forall n \geq 0\} \tag{2.44b}$$

According to Hartman-Grobman theorem (following [45]), there is a homeomorphism assuring the topological equivalence between the linearized system (2.22) and initial nonlinear one (2.14) as shown in Fig. (2.9).

Theorem 1 (Hartman-Grobman theorem for flows).

If the linearized matrix $\underline{\underline{\mathbf{L}}}$ of the nonlinear system (2.14) evaluated on the fixed point $\underline{\mathbf{q}}_{st}$ has no zero or purely imaginary eigenvalues, then there is a homeomorphism $\mathbf{h}$ defined on some neighborhood $U \in \mathbb{R}^n$ of the fixed point $\underline{\mathbf{q}}_{st}$ in $\mathbb{R}^n$ locally taking orbits of the nonlinear flow φ_t of (2.14), to those of the linear flows $e^{\underline{\underline{L}}t}$ of (2.22). The homeomorphism preserves the sense of orbits and can also be chosen to preserve parametrization by time.

When $\underline{\underline{\mathbf{L}}}$ has no eigenvalues with $Re(\lambda_i) = 0, \ i = 1, ..., m$ (hyperbolic fixed point) stability and asymptotic behavior of solutions of (2.14) in the vicinity of fixed point $\underline{\mathbf{q}}_{st}$ is determined by the linearized system (2.22). In this case the phase flow of the nonlinear system (2.14) in a neighborhood of $\underline{\mathbf{q}}_{st}$ is topologically equivalent to the flow of the linearized system (2.22) and both flows can be transformed in each other. But in case any of eigenvalues possesses zero real part (nonhyperbolic fixed point) the stability cannot be determined by linearization and it needs attracting the nonlinear terms. The next theorem tells us that $\mathbb{W}^u_{loc}$ and $\mathbb{W}^s_{loc}$ are in fact tangent to $\mathbb{E}^u$, $\mathbb{E}^s$ at $\underline{\mathbf{q}}_{st}$.

Theorem 2 (Stable manifold theorem for a fixed point).

Suppose that (2.14) has a hyperbolic fixed point $\underline{\mathbf{q}}_{st}$. Then there exist local stable and unstable manifolds $\mathbb{W}^s_{loc}(\underline{\mathbf{q}}_{st})$, $\mathbb{W}^u_{loc}(\underline{\mathbf{q}}_{st})$ of the same dimensions as the eigenspaces $\mathbb{E}^s$, $\mathbb{E}^u$ of the linearized system (2.22) and tangent to $\mathbb{E}^s$, $\mathbb{E}^u$ at $\underline{\mathbf{q}}_{st}$. $\mathbb{W}^s_{loc}(\underline{\mathbf{q}}_{st})$, $\mathbb{W}^u_{loc}(\underline{\mathbf{q}}_{st})$ are as smooth as the function $\underline{\mathbf{N}}$.

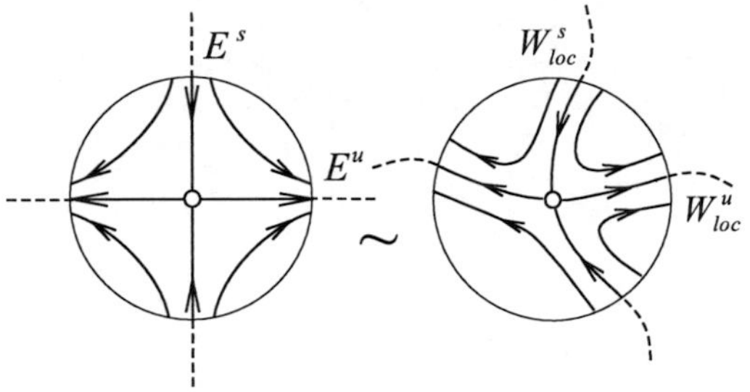

Figure 2.9: Linearization and invariant subspaces. Topological equivalence of the linearized and the nonlinearized system. This Figure is taken from [42] and modified.

As mentioned above, if any one of the eigenvalues has zero real part, then the stability of nonlinear system cannot be determined by linearization. In this case one says that the nonlinear system undergoes the bifurcation, and the further analysis of system (2.14) is performed on the invariant manifold tangent to the center eigenspace as presented in

Fig. (2.8)(c). We conclude that there are three qualitatively different types of hyperbolic points: sinks, which attract all orbits from a neighborhood; sources, which repel all orbits from a neighborhood; and saddles, which attract orbits from one direction, and repel them into another one. These possible solutions are robust to small perturbations.

Theorem 3 (Center manifold theorem for flows).

Let $\underline{\mathbf{N}}$ be a C^r vector field on $\mathbb{R}^n$ vanishing at the origin ($\underline{\mathbf{N}}(0) = 0$) and let $\underline{\underline{\mathbf{L}}} = \frac{\partial \underline{\mathbf{N}}(0)}{\partial \underline{\mathbf{q}}}|_{\underline{\mathbf{q}}=\underline{\mathbf{q}}_{st}}$. Divide the spectrum of $\underline{\underline{\mathbf{L}}}$ into three parts σ_s,σ_c,σ_u with

$$Re(\lambda) = \begin{cases} < 0 & \text{if } \lambda \in \sigma_s, \\ = 0 & \text{if } \lambda \in \sigma_c, \\ > 0 & \text{if } \lambda \in \sigma_u. \end{cases} \tag{2.45}$$

Let the (generalized) eigenspaces of σ_s,σ_c,σ_u be $\mathbb{E}^s$, $\mathbb{E}^c$, $\mathbb{E}^u$, respectively. Then there exist stable and unstable invariant manifolds $\mathbb{W}^s$, $\mathbb{W}^u$, which are of class C^r and tangent to $\mathbb{E}^s$, $\mathbb{E}^u$ at 0 and $C^{(r-1)}$ center manifold $\mathbb{W}^c$ tangent to $\mathbb{E}^c$ at 0. The manifold $\mathbb{W}^s$, $\mathbb{W}^u$, $\mathbb{W}^c$ are all invariant for the flow of $\underline{\mathbf{N}}$. The stable and unstable manifolds are unique, but $\mathbb{W}^c$ need not be.

Let the initial system (2.14) be transformed to the mode equations form (2.41). Since the center manifold is tangent to $\mathbb{E}^c$ we can represent it as

$$\mathbb{W}^c = \{(\underline{\xi}^u, \underline{\xi}^s), |\underline{\xi}^s = \underline{\mathbf{h}}(\underline{\xi}^u)\}; \;\; \underline{\mathbf{h}}(0) = 0, \;\; \frac{\partial \underline{\mathbf{h}}(\underline{\xi}^u)}{\partial \underline{\xi}^u}|_{\underline{\xi}^u=0} = 0 \tag{2.46}$$

Thus the center manifold theorem states in fact that the stable mode is a function of unstable mode that can be interpreted as one of the mathematical formulation of synergetic "slaving principle"

$$\underline{\xi}^s(t) = \underline{\mathbf{h}}(\underline{\xi}^u(t)) \tag{2.47}$$

satisfying the conditions (2.46). The function $\underline{\mathbf{h}}$ can be determined using the stable mode equation (2.41b) in the following form

$$\begin{aligned} &\frac{d\, \underline{\mathbf{h}}(\underline{\xi}^u(t))}{d\, \underline{\xi}^u}[\underline{\underline{\mathbf{\Lambda}}}_u \underline{\xi}^u(t) + \underline{\tilde{\mathbf{N}}}_u(\underline{\xi}^u(t), \underline{\mathbf{h}}(\underline{\xi}^u(t))\{\alpha\})] = \\ &= \underline{\underline{\mathbf{\Lambda}}}_s \underline{\mathbf{h}}(\underline{\xi}^u(t)) + \underline{\tilde{\mathbf{N}}}_s(\underline{\xi}^u(t), \underline{\mathbf{h}}(\underline{\xi}^u(t))\{\alpha\}) \end{aligned} \tag{2.48}$$

Substituting the function $\underline{\mathbf{h}}$ into the unstable mode equation (2.41a) we are able to reduce the dimension of (2.41) and obtain thereby the order parameter equation

$$\dot{\underline{\xi}}^u(t) = \underline{\underline{\mathbf{\Lambda}}}_u \underline{\xi}^u(t) + \underline{\tilde{\mathbf{N}}}_u(\underline{\xi}^u(t), \underline{\mathbf{h}}(\underline{\xi}^u(t)), \{\alpha\}). \tag{2.49}$$

which describes the behavior of the system in the neighborhood of the instability and no longer depends on the stable mode amplitude $\underline{\xi}^s$. The following theorem states the relation between the order parameter equation (2.49) and mode equations (2.41).

Using the procedure of center manifold to the discrete dynamical system (Fig. (2.8)(b)), the interesting results was obtained. In Sec. 3.2 corresponding approach is discussed more detailed. Summarizing the aforesaid, it is necessary to note, that the center manifold theory allows reducing the dimension of original nonlinear system to those equations which possess the unstable eigenvalues in linear approximation.

Theorem 4 (Theorem Henry [112] - Carr [106]).

If the origin $\underline{q} = 0$ of (2.49) is a locally asymptotically stable (resp. unstable), then the origin of (2.41) is also locally asymptotically stable (resp. unstable).

The obtained hereby low-dimensional OP equation demonstrates not only the stability of the nonhyperbolic fixed point, but also points to the behavior of initial system immediately after the bifurcation. Considering the behavior before and after the bifurcation, one can ascertain different types of bifurcations. Moreover, as turned out, each of instabilities has the determined structure of linear and nonlinear parts, which initiate this type of instability. This structure in form of evolution equation is called the normal form (e.g. [42], [45], [113]), and is the focus of two following sections.

2.3.4.6 Procedure of normal forms

The center manifold theorem states that the stable amplitudes are functions of unstable amplitudes in the nonhyperbolic fixed point. In consequence of this theorem, the unstable mode equations can be uncoupled from the stable mode equations and therefore in the vicinity of nonhyperbolic fixed point the system can be reduced only to these unstable mode equations. The second way towards the simplification of the dynamical system is provided by the normal form reduction. The main sense of this procedure is that in some coordinate basis the system may have the "simplest" structure of nonlinear part. Moreover, this structure is determined entirely by the nature of linear part of the vector field. This technique primarily consists of performing the corresponding local nonlinear coordinate transformation (so-called *near identity transformation*) simplifying the nonlinear part, until the specific reduced nonlinearity remains.

Let us start with the mode equations (2.41) written as

$$\dot{\underline{\xi}} = \underline{\underline{\Lambda}}\,\underline{\xi} + \tilde{\underline{\mathbf{N}}}^{(2)}(\underline{\xi}) + \tilde{\underline{\mathbf{N}}}^{(3)}(\underline{\xi}) + ... + O(|\underline{\xi}|^{(3)}). \tag{2.50}$$

Here, $\tilde{\underline{\mathbf{N}}}^{(l)}$ are a nonlinear function of the order l, $\underline{\xi} = \begin{pmatrix} \underline{\xi}^u \\ \underline{\xi}^s \end{pmatrix}$ and $\underline{\underline{\Lambda}}$ is the block-diagonal matrix of eigenvalues of system (2.14). The purpose of normal form is to simplify the nonlinear part of (2.50) by a change of the variables. Now the following near-identity transformation is used

$$\underline{\xi} = \underline{\varphi} + \underline{\mathbf{g}}^{(2)}(\underline{\varphi}), \tag{2.51}$$

where $\underline{\mathbf{g}}^{(2)}$ is second order in $\underline{\varphi}$. Substituting (2.51) into (2.50), we get

$$(id + D\underline{\mathbf{g}}^{(2)}(\underline{\varphi}))\dot{\underline{\varphi}} = \underline{\underline{\Lambda}}(\underline{\varphi} + \underline{\mathbf{g}}^{(2)}(\underline{\varphi})) + \tilde{\underline{\mathbf{N}}}^{(2)}(\underline{\varphi} + \underline{\mathbf{g}}^{(2)}(\underline{\varphi})) + O(|\underline{\varphi}|^{(3)}). \tag{2.52}$$

where "id" denotes the $m \times m$ identity matrix and D is the differential operator. Taking into account the following expressions

$$\tilde{\underline{\mathbf{N}}}^{(2)}(\underline{\varphi} + \underline{\mathbf{g}}^{(2)}(\underline{\varphi})) = \tilde{\underline{\mathbf{N}}}^{(2)}(\underline{\varphi}) + O(|\underline{\varphi}|^{(3)}), \tag{2.53}$$

$$(id + D\underline{\mathbf{g}}^{(2)}(\underline{\varphi}))^{-1} = id - D\underline{\mathbf{g}}^{(2)}(\underline{\varphi}) + O(|\underline{\varphi}|^{(2)}), \tag{2.54}$$

the equation (2.52) gets

$$\underline{\dot{\varphi}} = \underline{\underline{\mathbf{\Lambda}}}\,\underline{\varphi} + \underline{\underline{\mathbf{\Lambda}}}\,\underline{\mathbf{g}}^{(2)}(\underline{\varphi})) + \underline{\tilde{\mathbf{N}}}^{(2)}(\underline{\varphi}) - D\underline{\mathbf{g}}^{(2)}(\underline{\varphi})\underline{\underline{\mathbf{\Lambda}}}\,\underline{\varphi} + O(|\underline{\varphi}|^{(3)}). \tag{2.55}$$

Now we will choose a specific form for near identity transformation (2.51) so as to simplify the second order terms in (2.55) as much as possible. Ideally, all second order terms could be eliminated

$$\underline{\underline{\mathbf{\Lambda}}}\,\underline{\mathbf{g}}^{(2)}(\underline{\varphi})) - D\underline{\mathbf{g}}^{(2)}(\underline{\varphi})\underline{\underline{\mathbf{\Lambda}}}\,\underline{\varphi} = \underline{\tilde{\mathbf{N}}}^{(2)}(\underline{\varphi}). \tag{2.56}$$

All terms, that cannot be eliminated in (2.56) (so-called resonant terms), remain in the equation (2.55), building the second order terms of the normal norm

$$\underline{\dot{\varphi}} = \underline{\underline{\mathbf{\Lambda}}}\,\underline{\varphi} + \underline{\tilde{\mathbf{g}}}^{(2)}(\underline{\varphi})) + O(|\underline{\varphi}|^{(3)}). \tag{2.57}$$

In order to obtain the third order terms for the NF (2.57) the procedure (2.51)- (2.56) has to be repeated for the third order terms of $\underline{\xi}$ and $\underline{\varphi}$; all other high-order terms may be obtained similarly. Now we would like to make some remarks:

- the shown reduction procedure allows us to find all resonant terms i.e. the "simplest" structure of nonlinear part. But this procedure does not allow calculating the coefficients at these nonlinear terms. The improvement of this approach (exact calculation of the coefficients as well as adaptation for time discrete case) is undertaken in the following sections.

- Structure of resonant terms is determined only by the linear part of (2.50), i.e. by the eigenvalues $\underline{\underline{\mathbf{\Lambda}}}$. Note, these eigenvalues are exactly the same in both the mode equation (2.50) and the normal form (2.57).

The procedure of normal form reduction provides the systematic way towards understanding the bifurcation dynamics, however, it does not answer on three important questions:

- Which order of $\underline{\varphi}$ does represent the maximal order, where this procedure has to be finished ?

- Are there any relations between the instability and the structure of resonant terms (between different normal forms) ?

- What does occur with a dynamics of this normal form, if it is perturbed by additional "non-resonant" terms ?

Because this important problematic is again beyond the scope of this work, some calculations techniques, giving the common understanding of the problem, are considered and discussed in the App. A.1, A.2, A.3 in details. For the derivation of the equations and their meaning it is referred to the original works of Golubitsky and Shaeffer [114], [115].

2.4 Coupled map lattices

This section is devoted to a brief overview of coupled map lattices (CML), used in this thesis primarily in two important ways. First, this is a successful modeling concept, enabling a mathematical description and a consequent complexity reduction of distributed systems. Second, CMLs have succeeded in demonstrating different spatio-temporal phenomena and therefore are often employed as sui generis test systems, exemplifying theoretical propositions. The focus of the following discussion is on the theoretical foundations underlying CML, as well as a short classification of coupled systems and an overview of the predominant spatio-temporal effects arising in them.

2.4.1 Theoretical overview

Through investigating the behavior of different real and artificial systems, researchers arrived at the conclusion that the behavior of such systems, in many cases, cannot be represented by models with reduced degrees of freedom [105]. In order to achieve the necessary degrees of freedom in mathematical models, two systematic ways can be used. We can either make a model more complex by introducing additional elements, constraints and so on, or use a model of many simple basic systems coupled by means of uniform couplings. Both approaches are interesting and have their own application fields.

Coupled map lattices (CML) [118] have been suggested as a useful modeling method allowing a substantial progress in the second modeling approach. In other words, the CML concept, which has been studied in a vast literature, proposes that the required degree of complexity can be achieved by combining many elementary systems. Originally, CMLs were introduced for the study of spatio-temporal chaos and pattern formation [119], but since then, CML have it de facto become an analytical representation tool for neural networks [120], some biological processes [121], multi-agents systems [122], and other classes of distributed systems possessing similar structures; many differently coupled basic subsystems.

From the methodological viewpoint, CML represents the technique often characterized as a constructive approach, in contrast to the descriptive approach adopted historically as a methodological basis of investigations. "For example, an equation at a macroscopic level (like the Navier-Stokes equation) is approximately derived from a microscopic level (like the Newtonian equation of many particles), and then numerically simulated. Conventionally, a model equation in physics is believed to have a one-to-one correspondence with the phenomenon concerned" [105]. But actually, in contrast to this methodology, the importance of the constructive approach is continuously growing. "The model, constructed by combining some basic procedures (such as local chaos, diffusion, flow, ...) cannot be derived from a first-principle equation, as in conventional physics, but still has a strong predictive power for the novel phenomenology classes in complex dynamical systems"[105].

The constructive approach can be understood in sense of a mathematical experiment (or computer simulation) giving the information for constructing real models. This important feature can be easily clarified if we assume the CML is a model of some functional structure, like a neural network, in which basic subsystems are fixed. The actual transfer function of such a network is determined by the coupling configurations among the basic subsystems. Therefore, it is natural to investigate the influence of couplings on the transfer function by first using the mathematical model and only then try to create the real network. In this case, the real system in the operable form does not exist at all, until the results of modeling are obtained. Practical examples of such a constructive approach can be drawn from cellular robotics.

Robotics (see for example [123]), having greatly expanded in recent years, shows that properties of biological distributed systems can be also adopted by artificially constructed systems. Autonomous cellular robots [124] are one example of such an approach, where a “CML-like structure” provides greater reliability and autonomy for these robotic systems. However, many open questions arise concerning how to obtain the defined functional properties of a whole system consisting only of several interacting “cell-modules”. Considering analytical models of such robotic systems, we can narrow down this question to a simpler one, of how the basic systems and the couplings between them affect the behavior of the overall system. Whereas the basic systems are mainly determined by hardware specifications or other restrictions, one aspect of the initial problem still remains, namely, the influence of the couplings on the dynamics of CML. The constructive approach allows not only the generation of the required coupling methods, but also the understanding of the underlying processes of self-organization in such distributed systems.

The last point to be made concerns the complexity of coupled maps regarding initial uncoupled maps. As might be expected, by coupling the maps, their degrees of freedom will be increased and, in consequence, the structural and behavioral complexity of the coupled maps is also increased. This is correct in most cases, but sometimes the coupling of m initial maps does not increase their degrees of freedom. This feature can be countered, for example by performing a local analysis of CML. Local properties of CML, built by homogeneous coupling, can be decomposed in local properties of initial maps and the specific influence of the coupling method. Local analysis of the whole high-dimensional system is reduced to a calculation of how the coupling method modifies the local properties of a low-dimensional basic map. Such cases imply that in a non-chaotic state the CML's degrees of freedom remain unincreased.

Finally, comparing CML with other dynamical models, we note they are dynamical systems with discrete time and space, and continuous states. Correspondence to other kinds of dynamical systems is shown in Table 2.6. The two following sections are devoted

Model	**Space**	**Time**	**State**
Cellular automata	D	D	D
Coupled map lattices	D	D	C
Partial differential equations	C	C	C

D-discrete, C-continuous

Table 2.6: Basic structures of three different types of models (from [119]).

to a brief characterization of CML types and the effects often arising in them.

2.4.2 Characterization of CML's types

Per definition, the CML is a dynamical system obtained by coupling m-basic maps. Many dynamical systems are expected to be matched up with this definition. This characterization is undertaken to give the overview of often used basic systems and coupling methods and does not present a full-scale study of this topic.

Let us consider the following nonlinear autonomous system, discrete in time:

$$q_{n+1} = N(q_n, \{\alpha\}), \tag{2.58}$$

that presents the initial or so-called “basic” system. In general case the system (2.58) can be multidimensional. Coupling m such equations by means of nonlinear functions $\underline{\mathbf{F}}$ consisting of multiplicative $\underline{\mathbf{F}}_M$ and additive $\underline{\mathbf{F}}_A$ components, we get the following coupled system

$$\underline{\mathbf{q}}_{n+1} = \underline{\mathbf{N}}(\underline{\mathbf{q}}_n, \{\alpha\})\underline{\mathbf{F}}_M(\underline{\mathbf{q}}_n, \{\beta_M\}) + \underline{\mathbf{F}}_A(\underline{\mathbf{q}}_n, \{\beta_A\}), \tag{2.59}$$

where $\underline{\mathbf{q}}_n = (q_n^1, q_n^2, ...q_n^m)^T \in \mathbb{R}^n$ is a state vector in the m-dimensional space, $\underline{\mathbf{N}}$, $\underline{\mathbf{F}}$ are in general nonlinear functions of $\underline{\mathbf{q}}_n$, and $\{\alpha\}$, $(\{\beta_M\}, \{\beta_A\} \in \{\beta\})$ are the sets of control parameters [3]. This definition assumes that the couplings $\underline{\mathbf{F}}$ do not modify the structure of basic map (2.58). Considering equation (2.59), one can arrive at the conclusion that both the types of the basic system (2.58) and types of the coupling functions $\underline{\mathbf{F}}$ can underlay the undertaken classification. The most typical basic systems and couplings methods are summarized in Table 2.7.

In this table the basic systems are distinguished by a dimension, dissipation and types of underlying times. Reviewing the recently published papers, we remark that the logistic map still remains the most often used basic system. The coupling methods used in CML are very different indeed and can be only approximately classified by spatial formation, functional properties and the time. Based on publications, we emphasize that the most prevailing couplings types are local (so-called nearest-neighbor) and global (so-called all-to-all) couplings.

Since the locally, additively coupled CML are often employed in the given thesis, some linear properties of these couplings types are discussed below.

One-way ring with open boundary

The most simple case arises when each basic system (except for the first system) is additively coupled only with one neighbor system, this is the case of so-called one-way ring map lattice with open boundary (see e.g. [128]). Jacobian of such a coupled system has the following form

$$\begin{pmatrix} \Omega & 0 & 0 & 0 & ... & 0 \\ \varepsilon & \Omega & 0 & 0 & ... & 0 \\ 0 & \varepsilon & \Omega & 0 & ... & 0 \\ 0 & 0 & \varepsilon & \Omega & ... & 0 \\ \vdots & \vdots & \vdots & \vdots & \ddots & \vdots \\ 0 & 0 & 0 & 0 & ... & \Omega \end{pmatrix}, \tag{2.60}$$

where $\Omega = \dfrac{dN(q_n, \{\alpha\})}{dq_n}|_{q_n=q_{st}}$ and ε is small coupling coefficient. Matrix (2.60) is the $m \times m$ triangular matrix, the eigenvalues can be found from the solution of the following equation

$$(\Omega - \lambda)^m = 0. \tag{2.61}$$

Apparently, any elements lying below the main diagonal do not influence the eigenvalues of (2.60). Therefore the couplings can be extended in this direction without any change of local properties of such a CML.

One-way coupled ring map lattice of length m

When the first system is coupled with the last one in the one-way ring of additively coupled maps, we get the one-way coupled ring map lattice of length m (see e.g. [129]). Jacobian

[3] Here and further we distinguish between the control coefficients $\{\beta\}$ introduced by the coupling mechanism and the general control parameters $\{\alpha\}$ of the system (2.58). We denote further $\{\beta\}$ as the coefficients and $\{\alpha\}$ as the (bifurcation) parameters.

		Basic system
dimension	*one-dimensional*	Logistic map $x_{n+1} = ax_n(1 - x_n)$ or in the form $x_{n+1} = 1 - ax_n^2$
		"Toy" model [125] $x_{n+1} = a_\varepsilon(t)(1 - \lvert x_n \rvert) - 1$
		Tent map $x_{n+1} = 1 - a\lvert x_n \rvert$
		Bernuilli map [126] $x_{n+1} = ax_n mod 1$
	two-dimensional	
		Hènon map $\begin{cases} x_{n+1} = 1 + y_n - ax_n^2 \\ y_{n+1} = bx_n \end{cases}$
time	*autonomous*	
	nonautonomous	
	delayed	Hènon map $x_{n+1} = 1 - ax_n^2 + bx_{n-1}$
dissipation	*dissipative*	
	conservative [1]	$\begin{cases} x_{n+1} = x_n - k/2\pi sin(2\pi y_n) \\ y_{n+1} = y_n + x_{n+1} \end{cases}$
		Coupling (for $\underline{\mathbf{F}}_A$ and $\underline{\mathbf{F}}_M$)
spatial regular	*local* [2]	One-way $F^i(q_n^{i\pm1})$ Two-way $F^i(q_n^{i+1}, q_n^{i-1})$
	global	$F^i(q_n^1, q_n^2, ..., q_n^m)$
	spatial nonregular	
	random	
	hierarchical	
lattices	*one-dimensional*	$F^i(\{q_n^j\})$
	two-dimensional [3]	$F^i(\{q_n^{j,k}\})$, $j \neq k$
functional	*linear*	
	nonlinear	
	symmetric	$F^i(\alpha q_n^{i+1}, \beta q_n^{i-1})$, $\alpha = \beta$
	asymmetric	$F^i(\alpha q_n^{i+1}, \beta q_n^{i-1})$, $\alpha \neq \beta$
	derivative	$F^i(q_n^i - q_n^j)$, $i \neq j$
time	*nondelayed*	$\underline{\mathbf{F}}(\underline{\mathbf{q}}_n)$
	delayed	$\underline{\mathbf{F}}(\underline{\mathbf{q}}_{n-\tau})$

[1] the standard map or Taylor-Chirikov map, see e.g. [127]

[2] with open and close boundary

[3] j, k are row and column of a two-dimensional lattice

Table 2.7: Brief classification of coupled map lattices based on the types of basic maps and couplings.

of this CML has the following form

$$\begin{pmatrix} \Omega & 0 & 0 & 0 & \dots & \varepsilon \\ \varepsilon & \Omega & 0 & 0 & \dots & 0 \\ 0 & \varepsilon & \Omega & 0 & \dots & 0 \\ 0 & 0 & \varepsilon & \Omega & \dots & 0 \\ \vdots & \vdots & \vdots & \vdots & \ddots & \vdots \\ 0 & 0 & 0 & 0 & \dots & \Omega \end{pmatrix}. \tag{2.62}$$

Eigenvalues of (2.62) are given by the solution of

$$(\Omega - \lambda)^m - (-\varepsilon)^m = 0. \tag{2.63}$$

For the first eigenvalue we get $\lambda_1 = \Omega + \varepsilon$, the rest of eigenvalues is determined by the roots of the following equation [130]

$$(\Omega - \lambda)^{m-1} + (\Omega - \lambda)^{m-2}\varepsilon + ... + (\Omega - \lambda)\varepsilon^{m-2} + \varepsilon^{m-1} = 0. \tag{2.64}$$

Two-way coupled ring map lattice of length m

Coupling each basic system additively with two nearest neighbors, we get so-called two-way coupled ring map lattice. This kind of the local coupling is often used further because, firstly, the interactions are propagated in lattice essentially faster than in the case of one-way coupled map. Therefore, all transient processes are also finished much faster. Secondly, this coupling is more reliable than one-way ring, because if one of couplings will be broken the whole system still remains operable. The Jacobian of two-way additively coupled ring map lattice has the following form (see e.g. [131])

$$\begin{pmatrix} \Omega & \varepsilon & 0 & 0 & \dots & \varepsilon \\ \varepsilon & \Omega & \varepsilon & 0 & \dots & 0 \\ 0 & \varepsilon & \Omega & \varepsilon & \dots & 0 \\ 0 & 0 & \varepsilon & \Omega & \dots & 0 \\ \vdots & \vdots & \vdots & \vdots & \ddots & \vdots \\ \varepsilon & 0 & 0 & 0 & \dots & \Omega \end{pmatrix}. \tag{2.65}$$

The eigenvalues of symmetric $m \times m$ matrix (2.65) are given by

$$\lambda_i = \Omega + 2\varepsilon \cos\left(\frac{2\pi i}{m}\right). \tag{2.66}$$

Apparently, that eigenvalues (2.66) take values, maximal of that in the absolute magnitude is defined by

$$\lambda_m = \Omega + 2\varepsilon. \tag{2.67}$$

Determination of the maximal of eigenvalues is important, because these eigenvalues are becoming first unstable and cause a local bifurcation in CML. Therefore, the high-dimensional CML can be reduced in the vicinity of such instabilities only to the equations containing these eigenvalues. In case all eigenvalues are equal, the dimension reduction cannot be performed.

2.4.3 Spatiotemporal effects demonstrated by CML

There are many different effects, exhibited by CML, that are interesting not only from purely academic viewpoint, but also because they possess a practical importance, e.g. in control and communication. In this review we restrict ourselves only to such effects that have been most often discussed in the vast literature on the object. These phenomena are primarily demonstrated by test systems published in the original papers.

Arising of synchronization

Dynamical systems are usually called synchronized, if the distance between their states converges to zero for $t \to \infty$. Generalizing this concept further, the systems are called synchronized if a (static) functional relation exists between states of these systems [132]. Recently, the definition of synchronization is expanded to the cases of interactions between nonidentical systems, which lead to locking their phases, whereas their amplitudes remain uncorrelated (e.g. [133], [134]). Much effort has been dedicated to establishing the necessary conditions for the emergence of various synchronizations types [135]. The synchronization phenomena are also observed in many real physical and biological systems, often in the living neurons e.g. as shown in experimental studies of membrane voltage oscillations of isolated biological neurons from the stomatogastric ganglion of the California spiny lobster *Panulirus interruptus* [136]. Important applications of synchronization effects are found among others in the chaos control (e.g. [137]) and secure communication (e.g. [128], [138]).

In order to demonstrate the stabilization of spatiotemporal chaotic pattern and following the original paper of Hu and co-authors [128], we start with the following one-dimensional one-way coupled ring map lattice of length m

$$\begin{aligned} q^i_{n+1} &= (1-\varepsilon) a q^i_n (1-q^i_n) + \varepsilon a q^{i-1}_n (1-q^{i-1}_n), \\ q^{i+m}_n &= q^i_n, \end{aligned} \tag{2.68}$$

where $i = 1, ..., m$, a is a bifurcation parameter and ε is a small coupling parameter. When

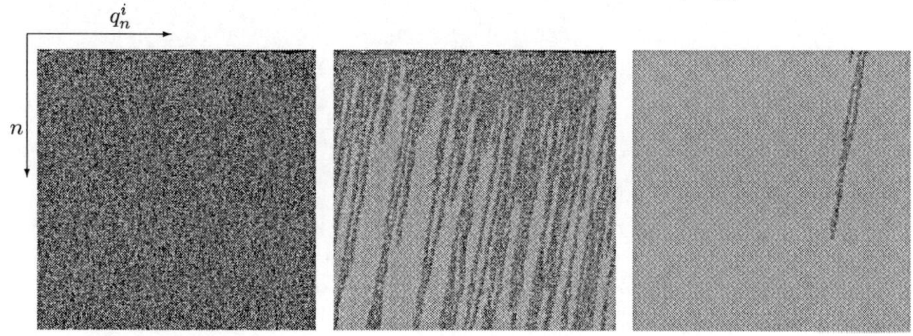

(a) (b) (c)

Figure 2.10: Stabilization of chaotic spatiotemporal pattern, on all figures $q_0^i = 0.1 + 0.0001i$ and $a = 4$ for the system (2.68), where $m = 512$; **(a)** a fully developed chaos at $\varepsilon = 0$, the 0 - 512th time steps are shown; **(b)** arising of synchronization at $\varepsilon = 0.1755$, the 0 - 512th time steps are shown; **(c)** completely synchronized state at $\varepsilon = 0.1754$, the 26500 - 27012th time steps are shown. The given review is based on the work [128].

$a = 4$ the single logistic system is in a fully developed chaotic state that at $\varepsilon = 0$ is shown in Fig. 2.10 (a). In the interval of coupling parameter about $0.16 \leq \varepsilon \leq 0.19$ the logistic systems in (2.68) are synchronized as shown in Fig. 2.10 (b)-(c).

Pattern formation

The pattern-forming phenomena always attracted a serious attention of researchers. The time continuous inhibitor-activator systems are the most known systems exhibiting different patterns as shown e.g. in the beautiful patterns of sea shells described by H. Meinhard in [139]. Recently the discrete, spatially extended, coupled systems have also become in the focus of investigations of the pattern-forming phenomena. Such systems exhibit spatiotemporal chaos in which many stable and unstable modes are embedded. The selection and stabilization of needed modes and correspondingly spatiotemporal patterns require applying the specific control techniques. The following consideration in rough way follows the suggestions of spatial delayed feedback in two-dimensional CML published by Jiang and co-authors in [140].

We consider the following two-dimensional coupled map lattice

$$q_{n+1}^{i,j} = F(\{q_n\}) - k(F(\{q_{n-1}\}) - q_n^{i,j}), \tag{2.69}$$

where k is a control coefficient, $F(\{q_n\})$ is given by

$$F(\{q_n\}) = (1-\varepsilon)f(q_n^{i,j}) + \frac{\varepsilon}{4}[f(q_n^{i-1,j}) + f(q_n^{i+1,j}) + f(q_n^{i,j-1}) + f(q_n^{i,j+1})], \tag{2.70}$$

ε is a small coupling coefficient and $f(q_n^{i,j})$ is usual logistic equation

$$f(q_n^{i,j}) = aq_n^{i,j}(1 - q_n^{i,j}), \tag{2.71}$$

where a is bifurcation parameter. Hereinafter, this parameter will be set on $a = 4$ that corresponds to a fully developed chaos in the logistic map (2.71).

The CML (2.69) possesses the local, nearest-neighbor coupling extended for two-dimensional case. In the study of patterns exhibited by this CML, a few control techniques were suggested, allowing us to stabilize the two-states patterns [140]. But in the given consideration, instead of proposed controls, the simple delayed feedback scheme, first published by Pyragas [141], is used. As it turned out this control method allows also to demonstrate different spatiotemporal patterns.

Two such patterns are demonstrated in Fig. 2.11(a)-(f). In the first one the initial conditions are random in area $[0.1, 0.2]$ that leads to the appearance of the "closed lines" pattern shown in Fig. 2.11(b)-(c). This pattern changes in time. The initial conditions in the second pattern are chosen as $q_0^{i,j} = 0.1 + 0.00001ij$ leading hereby to the "amoeba like" pattern in Fig. 2.11(d)-(f). It in fact does not change in time.

The system (2.69) allows us to emphasize some features of CML that will be further utilized in the presented thesis. From the viewpoint of the CML-structure, the pattern emerged during the time evolution is primarily determined by three factors, namely, the basic system (2.71), the structure of the coupling (2.70) and type of the applied control mechanism in (2.69). Since the basic systems are usually fixed, the most natural way to modify the exhibited pattern consists in the modification of couplings. Moreover, the couplings can be so chosen that the external control become redundant. In this case the control is in fact incorporated into the structure of couplings, hereby allowing to obtain an unconstrained and in the same time a desired behavior.

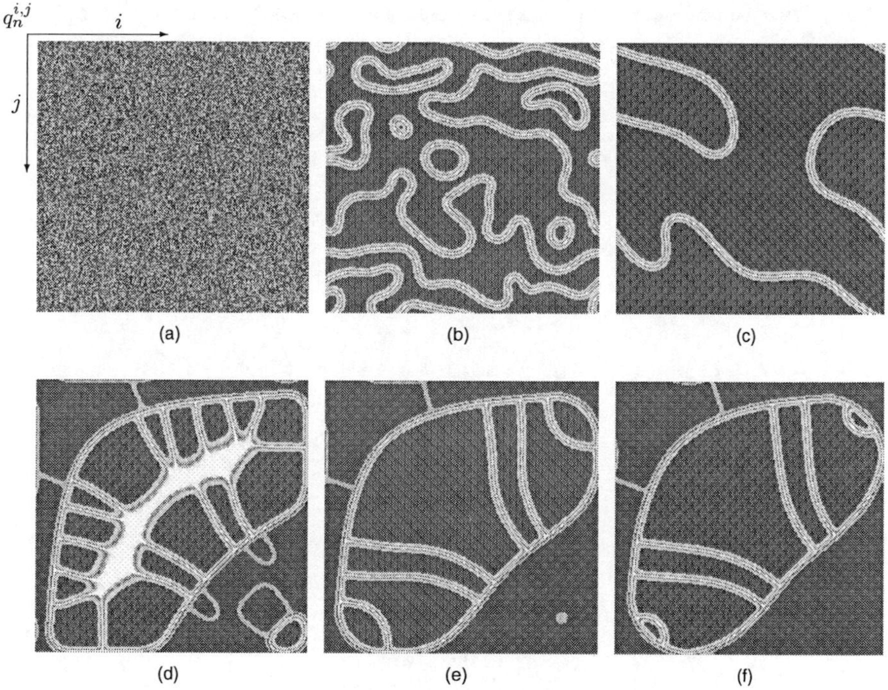

Figure 2.11: Pattern formation demonstrated by the system (2.69) at $a = 4$, $\varepsilon = 0.1$, $k = 0.1$ with 256×256 maps lattice; **(a)** the 1st frame (the 1st time step of 256×256 lattice), random initial conditions in area $[0.1, 0.2]$; **(b)** the 1000th frame with initial conditions as in (a); **(c)** the 10000th frame with initial conditions as in (a); **(d)** the 1000th frame, initial conditions are determined by $q_0^{i,j} = 0.1 + 0.00001ij$; **(e)** the 5000th frame with initial conditions as in (d); **(f)** the 10000th frame with initial conditions as in (d). The given review is based on the work [140].

Riddeld basins of attraction

As shown above, the coupled chaotic oscillators with some values of coupling coefficients attain a state of chaotic synchronization. The important observation is that the basins of attraction for the synchronized states, bifurcations and the attractors themselves undergo qualitative change. In the following consideration we restrict ourselves only on arising riddled basins being a characteristic type of fractal domain of attraction. The given exposition follows closely the work of Yu.L. Maistrenko and others in [142].

The test system emerging the riddled basins consists of two derivatively coupled logistic maps

$$x_{n+1} = ax_n(1 - x_n) + \varepsilon(y_n - x_n), \tag{2.72a}$$
$$y_{n+1} = ay_n(1 - y_n) + \varepsilon(x_n - y_n), \tag{2.72b}$$

where a is a bifurcation parameter, ε is a small coupling coefficient. The first three bifurcation points a_n of homoclinic bifurcations arisen in the logistic map are given by $a_0 = 3.67857351042832$, $a_1 = 3.59257218410697$, $a_2 = 3.57480493875920$. In these points the basins of attraction is numerically investigated. For that the whole area of initial conditions for x_n and y_n is divided into 1000×1000 points. Every of these points leading to the chaotic attractor is plotted as grey point, otherwise the point is left blank. The obtained hereby pictures are shown in Fig. 2.12(a)-(f).

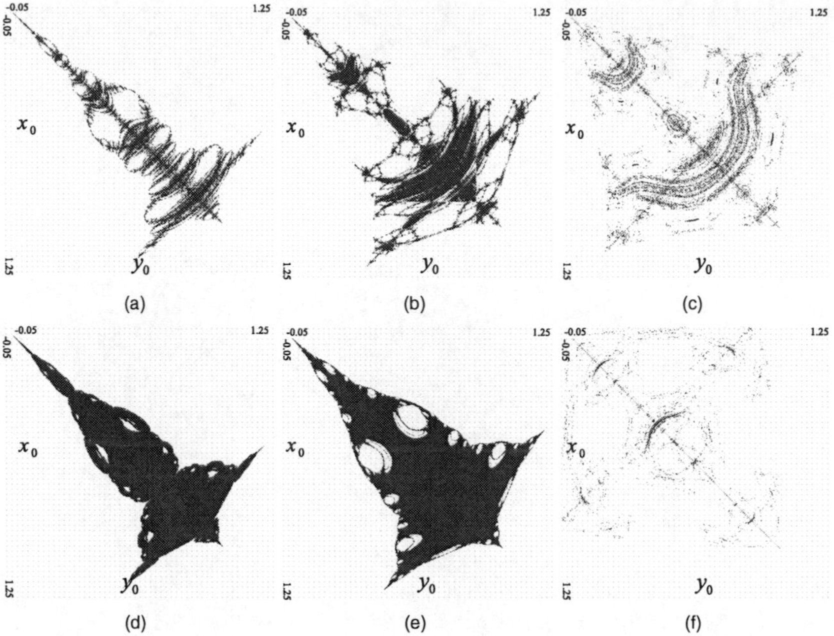

Figure 2.12: Basins of attraction for the system (2.72a) at different values of a and ε; **(a)** global riddling for $a = a_0$, $\varepsilon = -1.4$; **(b)** global riddling for $a = a_0$, $\varepsilon = -1.0$ **(c)** global riddling close to the blowout bifurcation at $a = a_0$, $\varepsilon = -0.5$ **(d)** fractal basin boundary for absolutely stable attractor at $a = a_1$, $\varepsilon = -1.3$; **(e)** locally riddled basin with fractal basin boundary at $a = a_2$, $\varepsilon = -1.14$; **(f)** basin of attraction for $a = a_1$, $\varepsilon = 0.33$. The given review is based on the work [142].

Finishing this brief overview, we have to make the last remark. The CML-branch of nonlinear dynamics is relative young, the active research has been performed only in two last decades. Some features of CML may be yet unexplored, there is no book that would reasonably generalize the made work. The themes, discussed here, do not cover the whole spectrum of CML features. They concern primarily the coupling and the analysis of CML's local dynamics. Some essential points may still remain outside the framework of this overview.

2.5 Control of dynamical systems

The control in different forms refers to the oldest affairs of the human race. Generally, to control an object means to influence its behavior so as to achieve a desired goal. With the development of technical civilization the control has been focused first on the simple mechanisms driving by water or air energy. Later, the main problems of control were concentrated on mechanical devices using the steam energy. Arising of the electrical devices demanded such a control that can be applied to the electrical schemes. Lately, the successive control principles are often applied to the economical and social areas. This development of the control theory has facilitated the development of mathematical foundations, underlying this theory, that made the control principles universal, i.e. independent of the nature of controlled system. The given consideration briefly reviews some mathematical principles as well as control techniques, referred to the control of bifurcation and chaos, that are utilized further in the thesis.

2.5.1 Problem of optimal control

Generally speaking, there are two main aspects of the problem, that the control theory is dealing with. One of these assumes the model of the investigating object is available and the goal is to optimize its behavior at specified conditions. Another aspect is based on the constraints imposed by uncertainty on the model or on the environment, that the object operates in. In order to correct deviations from the desired behavior, different control schemes are used.

Optimal control deals with the problem of finding a control law for a given system such that a certain optimality criterion is achieved. To illustrate the problem of optimal control, we assume that the dynamics of the uncontrolled system is given by

$$\dot{\underline{\mathbf{q}}} = \underline{\mathbf{N}}(\underline{\mathbf{q}}, t) + \underline{\mathbf{u}}(t) + \underline{\mathbf{E}}(t), \tag{2.73}$$

where $\underline{\mathbf{N}}$ in a nonlinear vector function of the state variables $\underline{\mathbf{q}} \in \mathbb{R}^n$, $\underline{\mathbf{E}}(t)$ represents an external influence being generally a stochastic value. The additive control term $\underline{\mathbf{u}}(t)$ satisfies the determined restrictions

$$\underline{\mathbf{u}}(t) \in \mathbb{U} = \{u_p : u_p(t) \in [u_{p\ min}, u_{p\ max}] \ \forall\, t \in [t_0, t_k],\ \ p = 1, ..., m\}. \tag{2.74}$$

In other words, it is required to find the optimal control law $\underline{\mathbf{u}}^{opt}$ on the set (2.74)

$$\underline{\mathbf{u}}^{opt} = \underline{\mathbf{u}}(\underline{\mathbf{q}}(t),\ \underline{\mathbf{E}}(t)), \tag{2.75}$$

transforming the system (2.73) from initial state $\underline{\mathbf{q}}(t_0)$ with $t_0 = 0$ in the determined state $\underline{\mathbf{q}}(t_k)$ where $\underline{\mathbf{q}}^{opt}(0) = \underline{\mathbf{q}}_0$ and $\underline{\mathbf{q}}^{opt}(t_k) = \underline{\mathbf{q}}_{t_k}$. The equation (2.75) is in fact the feedback, that performs the control, being optimal by the chosen criterion. The choice of this criterion depends on a goal of control, the features of the controlled system and so on. The performance, accuracy, length of transient process, efficiency and so on are often utilized as criteria in schemes of the optimal control [143].

In the following overview this generic notion about optimal control is restricted to the cases of bifurcation and chaos control. These control techniques have succeeded in modifying local properties of dynamical systems. We expect that these methods can be also applied to the problem of collective behavior. In the following representation we demonstrate only three most quoted mechanisms of bifurcation and chaos control.

2.5.2 Static and dynamic feedback control

Static and dynamic feedback control is historically the oldest control techniques originated from classical control theory. The idea of this approach in the most simple form can be presented in the following way. Let

$$\dot{\underline{\mathbf{q}}} = \underline{\mathbf{N}}(\underline{\mathbf{q}}, t, \{\alpha\}) + \underline{\mathbf{u}}(\underline{\mathbf{q}}, \{\beta\}), \tag{2.76}$$

be a system with e.g. static feedback $\underline{\mathbf{u}}(\underline{\mathbf{q}})$ where $\{\alpha\}$, $\{\beta\}$ are sets of control coefficients. Being intended to modify the local properties of this system by means of the feedback, we point to the normal form (here one-dimensional) describing the corresponding local instability

$$\dot{\xi} = \lambda(\{\alpha\}, \{\beta\})\xi + \gamma_2(\{\alpha\}, \{\beta\})\xi^2 + \gamma_3(\{\alpha\}, \{\beta\})\xi^3 + ... + O(\xi^r), \tag{2.77}$$

where eigenvalue λ as well as coefficients γ depend on both $\{\alpha\}$ and $\{\beta\}$, and correspondingly on the form of nonlinear feedback functions $\underline{\mathbf{u}}$. The Eq. (2.77) can be derived by means of standard techniques, like the normal form reduction and can represent the normal form and its unfolding as well. Each of λ, $\{\gamma\}$ determines some element of local dynamics that initial system (2.76) exhibits. Solving the following system

$$\lambda(\{\alpha\}, \{\beta\}) = \Omega_1, \tag{2.78a}$$
$$\gamma_1(\{\alpha\}, \{\beta\}) = \Omega_2, \tag{2.78b}$$
$$\gamma_2(\{\alpha\}, \{\beta\}) = \Omega_3, \tag{2.78c}$$
$$...$$

where $\{\Omega\}$ are mathematically formulated conditions for desired dynamics and taking into account the boundary conditions for $\{\Omega\}$, λ, $\{\gamma\}$ we get the explicit control law in the form of feedback function $\underline{\mathbf{u}}$.

Control of chaos, using this approach, can be performed by shifting the corresponding local bifurcations and "stretching" the whole dynamics, or by obtaining instead of chaos some previous non-chaotic states. The problem of the mentioned approach is that the function $\underline{\mathbf{u}}$ is in fact introduced into all terms of normal form (2.77), therefore modifying only one of them, others are modified too.

To illustrate this method for a discrete system, let us consider a one-dimensional discrete map

$$q_{n+1} = N(q_n), \tag{2.79}$$

where $N(q_n)$ is a nonlinear function. The equation (2.79) can be written in the form of logistic map by the following constant feedback control suggested by Parthasarathy and Sinha in [144]

$$x_{n+1} = a x_n(1 - x_n) + k, \tag{2.80}$$

where k represents the strength of the feedback which can take both negative and positive values. The bifurcation diagrams of this system at different numeric values of k are shown in Fig. 2.13(a)-(f).Varying k, we change all elements of the dynamics, even create such ones that are not existed in the initial dynamics. Lately, many improvements as well as modifications of the static and dynamic feedback have been suggested (e.g. [145]).

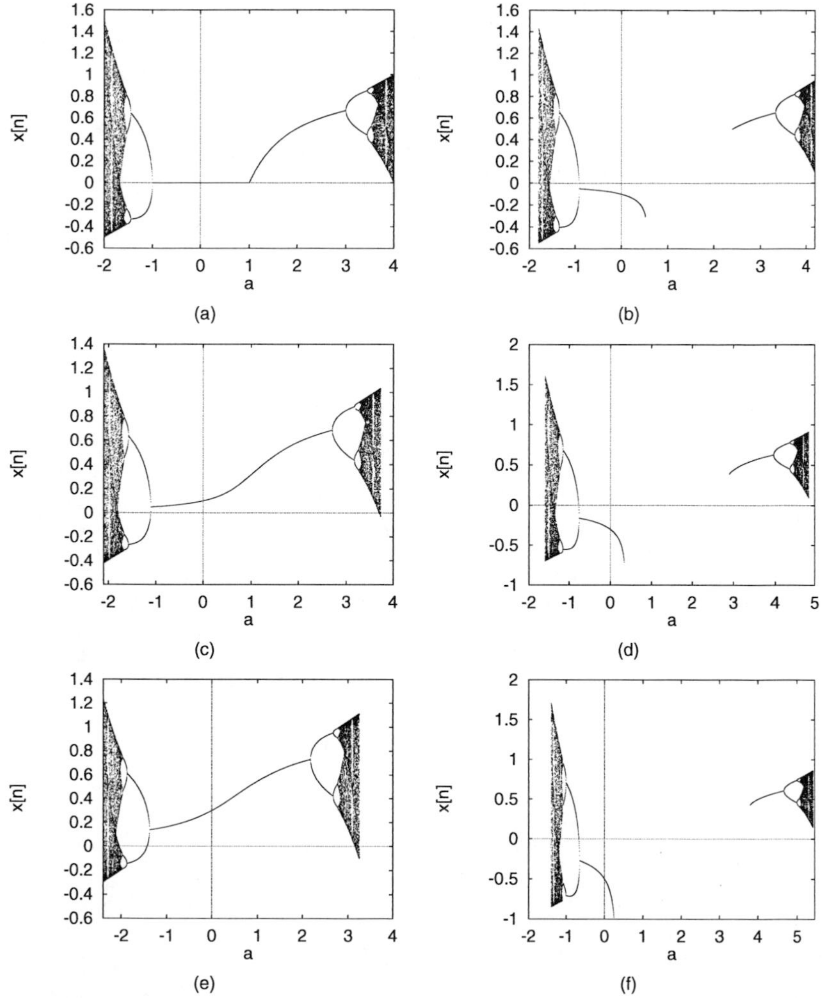

Figure 2.13: (a) Bifurcation diagram of system (2.80) at $k = 0$; **(b)** $k = -0.1$; **(c)** $k = 0.1$; **(d)** $k = -0.3$; **(e)** $k = 0.3$; **(f)** $k = -0.5$; $x_0 = 0.4$ for all figures. Varying coefficient k all elements of initial dynamics shown in Fig. 2.13(a) undergo perturbations.

2.5.3 OGY scheme

One of the most successful ideas about a control of chaos was suggested by Ott, Gregory and Yorke in [146]. This publication caused dozens of various improvements, modifications, practical implementation schemes and so on. The main goal of the OGY approach is to stabilize an unstable orbit embedded into a chaotic attractor by means of a small time-dependent change in some control parameter.

The idea is to vary the control parameter in the neighborhood of the value corresponding to a fixed point, when the uncontrolled trajectory comes close to this fixed point. Without a control, the system approaches the vicinity of an unstable fixed point along the stable manifold and then moves away along unstable manifold. To prevent leaving the vicinity of a fixed point, the control parameter has to be quickly varied by a small amount in such a way that the state of system will land on a stable manifold and consequently the evolution will approach it as $t \to \infty$.

We illustrate this technique by one of improvements of the OGY scheme, suggested by Friedel and co-authors in [147], that uses the center manifold approach. The goal is to stabilize the fixed point (stationary state)

$$x_{st}(a) = \frac{a-1}{a} \tag{2.81}$$

of the logistic map

$$x_{n+1} = ax_n(1-x_n). \tag{2.82}$$

Introducing $a = a_n$, the authors derived the following control law

$$a_n = \frac{a^2(a-2+K+au_n)u_n}{(a-1+au_n)(1-au_n)} + a, \tag{2.83}$$

that guarantee the desired stabilization, where K is coefficient and $u_n = x_n - x_{st}(a)$ the deviation from the fixed point. Bifurcation diagram of the system (2.82) with and without control is shown in Fig. 2.14. Advantage of the scheme, and one of reasons why it has

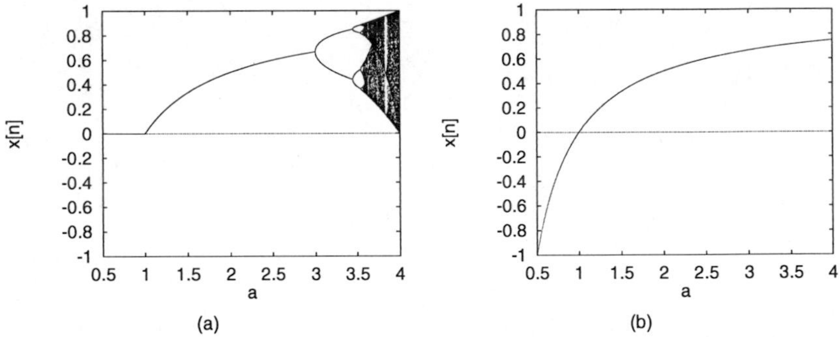

Figure 2.14: (a) Bifurcation diagram of system (2.82) without control; **(b)** Bifurcation diagram of the same system where the control (2.83) is on; $x_0 = 0.4$ in all figures.

gained so much interest, is that it can be easily realized in experimental situations [148], e.g. the necessary information can be extracted from time series data.

2.5.4 Time delayed feedback

The time delayed feedback is the last control approach discussed here. This scheme was first suggested by Pyragas [141] and is not less popular than OGY scheme. Generally, the time delay approach does not require a priori analytical knowledge of the system dynamics and does not require computer analysis. There are some differences between applications of this technique to time continuous and time discrete systems; here the last from them will be presented.

Idea of delayed feedback is to stabilize the unstable periodic orbit by introducing a small continuous perturbation Δx into a dynamics. The value and sign of this perturbation has to be so chosen that an amplitude of motion due to $F(x_n) - \Delta x$ would be decreasing until the system lands into a fixed point where $\Delta x = 0$. Since $F(x_n)$ is periodical with the defined phase, the perturbation has to behave with an inverse phase to $F(x_n)$. The most simple way to obtain such an inverse phase is to select Δ_x as a difference between the last and current states of $F(x_n)$ itself, i.e. $\Delta x = x_{n-1} - x_n$. Specific property of this delay scheme is that it exerts a dedicated influence on the dynamics, e.g. it does not modify non-periodical stationary states, because $x_{n-1} = x_n$ and Δx becomes equal zero. Considering the following logistic map with delayed feedback

$$x_{n+1} = ax_n(1 - x_n) + k(x_{n-1} - x_n), \tag{2.84}$$

we can find that stationary states $x_{st_{1,2}} = \{0, \frac{a-1}{a}\}$ do not depend on k. Since k influences the eigenvalues, modifying this coefficient, we can modify the stability of these stationary states by shifting the corresponding bifurcations. Bifurcation diagram of system (2.84) with the positive and negative k is shown in Fig. 2.15(a),(b). Concluding the overview of

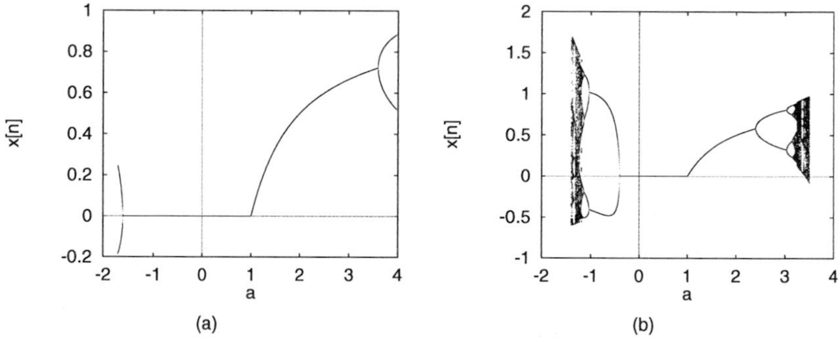

Figure 2.15: (a) Bifurcation diagram of system (2.84) at $k = -0.3$; **(b)** Bifurcation diagram of the same system with $k = 0.3$; $x_0 = 0.4$ in all figures.

this scheme, we remark that it generally increases a dimension of the considered system. Thus, the “one-dimensional” unstable fixed point of (2.84) turns into the “two-dimensional” stable fixed point.

Finally, summarizing this section, we refer the demonstrated techniques to the basic tools of applied nonlinear dynamics. Application of these methods for achieving the desired behavior still remains nontrivial. In the following chapters we propose using the synergetic concept that can guide us through diverse techniques with the goal of obtaining such a desired collective behavior.

Chapter 3

Analytical agent

When you have eliminated all which
is impossible, then whatever remains,
however improbable, must be the truth.
Sir Arthur Conan Doyle
Sherlock Holmes.

The order parameters (OPs), as stated by synergetics, are central to the analytical treatment of collective phenomena. In this chapter, we demonstrate that OPs can be derived for a single (analytical) agent. In this case, they describe macroscopic changes in an agent's individual behavior. In the next chapter, we will demonstrate that OPs of many interacting (analytical) agents consist of these "individual" OPs, fused in such a way that allows the description of macroscopic changes in collective behavior.

Before we start to consider collective phenomena in this manner, we need to consider some mathematical subtleties in deriving OP. Collective systems must be represented in the form of dynamical systems to which mathematical methods can be applied. The difficulty in performing further analysis is that mathematical techniques are primarily developed for time-continuous systems. Therefore, in this chapter we first represent some properties of local analysis for time-discrete dynamical systems; the mathematical framework is given by the adiabatic approximation procedure, different center manifold approaches and the normal form reduction. The methods are compared from the viewpoint of their accuracy of approximation, numerical verification and computational complexity.

3.1 Analysis of elementary agent

Our discussion about emergence in collective systems starts with the description and analysis of the individual elements forming the collective phenomena. In this section, we consider a time-discrete model that demonstrates the functional properties of a simple analytical agent. We focus our attention on deriving order parameters for this discrete model. In Chap. 4 we intend to reveal the mechanisms that changes these OPs when agents start to interact.

We first consider some mathematical subtleties in treating time-discrete systems. Such systems are often used in nonlinear dynamics and generally represent models of real discrete systems existing primarily in technical, physical, and biological areas, as well as in computer science. The best-known examples of these models are given by such unsophisticated systems as the logistic map, the Hénon map, and so on (see [149], [153], [154]). Recently, the high-dimensional coupled maps lattices (see [59], [119]) have been used in studying various complex spatio-temporal phenomena. Analytical and numerical [111] investigations of these discrete systems consist often of the local analysis [155] of a non-periodical or low-periodical reference states, to which the synergetic methodology [43], [44], [104], can be successfully applied.

The details of synergetic methodology, as applied to discrete systems, are discussed below. Specifically, we point out the accuracy of approximation, numerical verification, and the computational complexity of such techniques as the adiabatic elimination procedure, different approaches of the center-manifold- and normal-form-reductions, with explicit calculated coefficients. These methods are known in the analysis of continuous systems [42], [45], [156], but in contrast to that, their application to discrete systems is slightly different and has some specific features.

In this work, the well-known terminology of synergetics [13], [43] is used, in particular, the slaving principle and the order parameters. The mathematical formulation of the slaving principle is given by the center manifold (CM) theorem [157]. In the calculation procedure of the center manifold, described by Carr [106], a control parameter is included as a new dependent variable, whereby the two-dimensional center manifold $h(\xi_n^u, \varepsilon_n)$ is obtained, where ξ_n^u - is a dependent variable and ε_n - a smallness parameter. Note that the explicit form of the function $h(\xi_n^u, \varepsilon_n)$ is not unique. Further, using a priori knowledge about local system dynamics, this method is simplified so that a one-parameter dependent CM $h(\xi_n^u)$ can be obtained. After that, the adiabatic elimination procedure [13] is considered, which can be viewed as a low-order approximation of the center manifold approach.

The "slaving" function allows the derivation of the required low-dimensional order parameter equations. These equations can also be derived on the basis of another approach [158], namely the normal form reduction with explicit calculated coefficients [113]. The normal forms, obtained from the mode equations and the order parameter equations, can then be compared, as shown in Chap. 4. Thus, on the one hand, all results are verified and on the other hand, it allows comparison of the approximation accuracy and computational complexity of these methods. The derived order parameters are later required in Chap. 4, where we demonstrate how new collective behavior emerges from interactions among agents.

3.2 Formalism

Consider the following nonlinear autonomous time discrete system

$$\underline{\mathbf{q}}_{n+1} = \underline{\mathbf{N}}(\underline{\mathbf{q}}_n, \{\alpha\}), \tag{3.1}$$

where $\underline{\mathbf{q}}_n$ is a state vector, $\underline{\mathbf{N}}$ is nonlinear vector function of $\underline{\mathbf{q}}_n$ and $\{\alpha\}$ is a set of control parameters. In the following, we select in the set $\{\alpha\}$ only one relevant control parameter α, assuming others to be fixed. As already mentioned, the main goal of local analysis consists in deriving low-dimensional discrete equations of motion that describe a behaviour of the system (3.1) in a vicinity of bifurcations. In the case of maps this can be achieved, for instance, either by the reduction procedure based on the center manifold theorem or by the normal form reduction with the explicit calculated coefficients [42], [45].

In both cases the coordinates origin is first shifted to the stationary state $\underline{\mathbf{q}}_{st}$ of (3.1) and then the coordinate transformation, which diagonalizes the linear part of (3.1), is performed. The system (3.1) receives the following general form of mode amplitude equations

$$\underline{\boldsymbol{\xi}}^u_{n+1} = \underline{\underline{\boldsymbol{\Lambda}}}_u(\{\alpha\})\underline{\boldsymbol{\xi}}^u_n + \underline{\tilde{\mathbf{N}}}_u(\underline{\boldsymbol{\xi}}^u_n, \underline{\boldsymbol{\xi}}^s_n, \{\alpha\}), \tag{3.2a}$$

$$\underline{\boldsymbol{\xi}}^s_{n+1} = \underline{\underline{\boldsymbol{\Lambda}}}_s(\{\alpha\})\underline{\boldsymbol{\xi}}^s_n + \underline{\tilde{\mathbf{N}}}_s(\underline{\boldsymbol{\xi}}^u_n, \underline{\boldsymbol{\xi}}^s_n, \{\alpha\}), \tag{3.2b}$$

where $\underline{\underline{\boldsymbol{\Lambda}}}_u$, $\underline{\underline{\boldsymbol{\Lambda}}}_s$ are the matrices of eigenvalues of Jacobian, evaluated at the stationary state $\underline{\mathbf{q}}_{st}$ of the system (3.1) and $\underline{\tilde{\mathbf{N}}}_u$, $\underline{\tilde{\mathbf{N}}}_s$ are nonlinear functions of state vectors $\underline{\boldsymbol{\xi}}^u_n$, $\underline{\boldsymbol{\xi}}^s_n$ called the mode amplitudes.

Eqs. (3.2), denoted simply as the mode equations [43], have the decoupled linear parts. From the time scale hierarchy

$$|\lambda_s| < 1, |\lambda_u| > 1, \tag{3.3}$$

$$1 - |\lambda_s| \gg 1 - |\lambda_u|, \tag{3.4}$$

the diagonal matrices of eigenvalues are denoted as the unstable $\underline{\underline{\boldsymbol{\Lambda}}}_u$ and stable $\underline{\underline{\boldsymbol{\Lambda}}}_s$. They correspond to the unstable $\underline{\xi} n^u$ and stable$\underline{\xi} n^s$ mode amplitudes (or simply modes). In accordance with the synergetic concept the unstable mode equations play a key role in a dynamics of the systems in the vicinity of instability [104]. The unstable modes because of $|\lambda_u| \approx 1$ possess the slow intrinsic dynamics in contrast to the stable modes that vary much faster. As shown in [43], the complete dynamics of the system near the bifurcation is governed by the slow components, in other words, by the unstable modes. In consequence, the stable modes are expressed as functions of the unstable modes

$$\underline{\boldsymbol{\xi}}^s_n = \underline{\mathbf{h}}(\underline{\boldsymbol{\xi}}^u_n), \tag{3.5}$$

or more exactly

$$\underline{\boldsymbol{\xi}}^s_n = \underline{\mathbf{h}}(\underline{\boldsymbol{\xi}}^u_n) + O(\underline{\mathbf{h}}^{(r+1)}), \tag{3.6}$$

i.e. they are "enslaved" by the unstable modes. This functional dependence can be derived e.g., by means of the center manifold reduction [42], [106]. The explicit calculation of the required function $\underline{\mathbf{h}}(\underline{\boldsymbol{\xi}}^u_n)$ presents just one of objectives of the next sections.

Now substituting Eq. (3.5) in Eq. (3.2b), we obtain implicit equations for the center manifold $\underline{\mathbf{h}}(\underline{\boldsymbol{\xi}}^u_n)$

$$\begin{aligned} \underline{\mathbf{h}}(\underline{\boldsymbol{\xi}}^u_{n+1}) &= \underline{\mathbf{h}}\left(\underline{\underline{\boldsymbol{\Lambda}}}_u(\{\alpha\})\underline{\boldsymbol{\xi}}^u_n + \underline{\tilde{\mathbf{N}}}_u(\underline{\boldsymbol{\xi}}^u_n, \underline{\mathbf{h}}(\underline{\boldsymbol{\xi}}^u_n), \{\alpha\})\right) = \\ &= \underline{\underline{\boldsymbol{\Lambda}}}_s(\{\alpha\})\underline{\mathbf{h}}(\underline{\boldsymbol{\xi}}^u_n) + \underline{\tilde{\mathbf{N}}}_s(\underline{\boldsymbol{\xi}}^u_n, \underline{\mathbf{h}}(\underline{\boldsymbol{\xi}}^u_n), \{\alpha\}). \end{aligned} \tag{3.7}$$

Substituting the functions $\underline{\mathbf{h}}$ into the unstable mode equations (3.2a), we are able to reduce the dimension of the initial system (3.1) and to derive so-called order parameter equations

$$\underline{\boldsymbol{\xi}}^u_{n+1} = \underline{\underline{\boldsymbol{\Lambda}}}_u(\{\alpha\})\underline{\boldsymbol{\xi}}^u_n + \underline{\tilde{\mathbf{N}}}_u(\underline{\boldsymbol{\xi}}^u_n, \underline{\mathbf{h}}(\underline{\boldsymbol{\xi}}^u_n), \{\alpha\}) + ... + O(\underline{\tilde{\mathbf{N}}}_u^{(r+1)}). \tag{3.8}$$

The exact form of the order parameter equations (3.8) is not unique defined [159].

Accordingly, the aim of reduction can be viewed as decoupling the stable and unstable modes in the mode equations (3.2). This goal can be also achieved by performing the nonlinear coordinate transformation (often called as near identity transformation)

$$\underline{\boldsymbol{\xi}}_n = \underline{\boldsymbol{\varphi}}_n + \underline{\mathbf{p}}^{(2)}(\underline{\boldsymbol{\varphi}}_n) + ... + O(\underline{\mathbf{p}}^{(r+1)}), \tag{3.9}$$

which simplifies the structure of nonlinear functions $\tilde{\underline{\mathbf{N}}}_u$, $\tilde{\underline{\mathbf{N}}}_s$ in Eqs. (3.2). Here $\underline{\mathbf{p}}^{(r)}$ are nonlinear functions of the state vector $\underline{\varphi}_n$ of the order r. This method is known as the normal form reduction [45], [113]. Removing the nonlinear terms, contained couplings between $\underline{\xi}^u_n$, $\underline{\xi}^s_n$, we yield the mode equations (3.2) in the following form

$$\underline{\varphi}^u_{n+1} = \underline{\underline{\mathbf{\Lambda}}}_u(\{\alpha\})\underline{\varphi}^u_n + \hat{\underline{\mathbf{N}}}_u(\underline{\varphi}^u_n, \{\alpha\}) + O(\hat{\underline{\mathbf{N}}}^{(r+1)}_u), \tag{3.10a}$$

$$\underline{\varphi}^s_{n+1} = \underline{\underline{\mathbf{\Lambda}}}_s(\{\alpha\})\underline{\varphi}^s_n + \hat{\underline{\mathbf{N}}}_s(\underline{\varphi}^u_n, \underline{\varphi}^s_n, \{\alpha\}) + O(\hat{\underline{\mathbf{N}}}^{(r+1)}_s), \tag{3.10b}$$

where $\hat{\underline{\mathbf{N}}}_u$, $\hat{\underline{\mathbf{N}}}_s$ are the simplified nonlinear functions $\tilde{\underline{\mathbf{N}}}_u$, $\tilde{\underline{\mathbf{N}}}_s$, whose structure is determined only by eigenvalues $\underline{\underline{\mathbf{\Lambda}}}$ [113] [1].

The normal form (3.10a) possesses the simplest structure, that describes the corresponding local bifurcation. The equivalent to (3.10a) equation can be obtained by applying the normal form reduction to the order parameter equation (3.8). In this case the Eq. (3.10a) and Eq. (3.8) are equal as germs in sense of singularity theory [114], [115]. Since the function (3.5) is implicitly contained in (3.10b), we argue that both equations (3.8) and (3.10a) present the synergetic order parameters (see [158]). Moreover, a comparison between these normal forms with exactly calculated coefficients may appear as a "sui generis" accuracy criterion of reduction methods on the center manifold.

Now, finishing this formalism, we begin a detailed consideration of the mentioned objects. For simplification we assume there is only one stable mode amplitude ξ^s_n and one unstable mode amplitude ξ^u_n in the equations (3.2).

3.3 Parameter dependent center manifold approach for maps

The center manifold reduction is well-know mathematical procedure, that allows reducing a dimension of the original system (3.1) in the vicinity of a bifurcation. This theory was developed in detail for the case of time-continuous systems (in the form of differential equations)[42], [106], [160]. However, the computation procedure for the time discrete dynamical systems for the center manifold depending upon parameters is not researched enough. Further in this section we consider two computational techniques for maps, one of which is based on the conventional approach for time continuous systems, another one is a simplified version of the first procedure with usage of a priori knowledge about system's dynamics.

3.3.1 CM using assumption $h(\xi^u_n, \varepsilon_n)$=$A(\xi^u_n)^2 + B\xi^u_n\varepsilon_n + C\varepsilon^2_n + ...$

The general form of mode amplitude equations (3.2) depends on the bifurcation parameter α in the linear and nonlinear parts. For an application of the center manifold theorem the system (3.2) has to be transformed in such a form, where the linear parts will be independent of the bifurcation parameter α. Such a method is described by Carr [106] for the case of time-continuous systems, and below we demonstrate an application of this idea for the discrete case.

The parameter independent center manifold can be expressed in the general form [42], [45]

$$\xi^s_n = h(\xi^u_n) = \sum_{l=2}^{r-1} g^l(\xi^u_n) + O(r), \tag{3.11}$$

[1]This dimension reduction by means of the NIT transformations can be performed when the stable and unstable modes are decoupled from each other in the linear parts of Eqs. (3.2).

where r is a maximal order of the CM, $g^l(\xi_n^u)$ is the term of the order l for which the following conditions should be fulfilled:

$$h(0) = 0, \quad \left.\frac{dh(\xi_n^u)}{d\xi_n^u}\right|_{\xi_n^u=0} = 0 \ . \tag{3.12}$$

Substituting $\alpha = \alpha_{cr} + \varepsilon$ in the (3.2), where ε is the smallness parameter and α_{cr} is the critical value of the control parameter α, we expand all ε-dependent terms of (3.2) into the Taylor series with respect to ε up to the order r

$$\xi_{n+1}^u = (\lambda_u^{(0)} + \lambda_u^{(1)}\varepsilon + \lambda_u^{(2)}\varepsilon^2 + ...)\xi_n^u + \tilde{N}_u^{(r-1)}(\xi_n^u, \xi_n^s, \varepsilon) + O(r) \tag{3.13a}$$

$$\xi_{n+1}^s = (\lambda_s^{(0)} + \lambda_s^(1)\varepsilon + \lambda_s^{(2)}\varepsilon^2 + ...)\xi_n^s + \tilde{N}_s^{(r-1)}(\xi_n^u, \xi_n^s, \varepsilon) + O(r) \tag{3.13b}$$

where $\lambda_u^{(0)}$, $\lambda_s^{(0)}$ are functions of the constant α_{cr}, namely $\lambda_u^{(0)}(\alpha_{cr})$, $\lambda_s^{(0)}(\alpha_{cr})$ and $|\lambda_u^{(0)}(\alpha_{cr})| = 1$. From now on the upper index in brackets means the expansion order of this expression in the Taylor series.

Now, as suggested by Carr [106], we include the smallness parameter ε as a new dependent variable ε_n and combine all nonlinear terms with respect to ξ_n^u, ξ_n^s, ε_n

$$\xi_{n+1}^u = \lambda_u^{(0)}\xi_n^u + \hat{N}_u^{(r-1)}(\xi_n^u, \xi_n^s, \varepsilon_n) + O(r) \tag{3.14a}$$

$$\xi_{n+1}^s = \lambda_s^{(0)}\xi_n^s + \hat{N}_s^{(r-1)}(\xi_n^u, \xi_n^s, \varepsilon_n) + O(r) \tag{3.14b}$$

$$\varepsilon_{n+1} = \lambda_\varepsilon \varepsilon_n, \tag{3.14c}$$

where $\hat{N}_u, \hat{N}_s$ are nonlinear parts of these equations with respect to $\xi_n^u, \xi_n^s, \varepsilon_n$ up to the order r and λ_ε is constant such as $\lambda_\varepsilon = 1$ or $\lambda_\varepsilon = -1$. The interpretation of λ_ε is that time-discrete dynamical systems, as distinct from time-continuous ones, have two center manifolds corresponding to the eigenvalues $\lambda_\varepsilon = 1$ and $\lambda_\varepsilon = -1$. Concerning the sign of the constant λ_ε, we take into account the following thoughts. The terms of the lowest order with respect to ε_n in the center manifold (accordingly in the order parameter equation too) are independent of the constant λ_ε (as discussed in Sec. 3.3.2), therefore λ_ε doesn't influence the low-order CM.

However, the high-order terms of ε_n contain this constant and our suggestion consist in selecting the values of λ_ε in accordance with the already existed center manifold, namely equal to $\lambda_u^{(0)}(\alpha_{cr})$. The indirect evidence in favour of such suggestion can be obtained by a numerical iteration of the order parameter equation with the high-order CM that is evaluated two times: one with $\lambda_\varepsilon = \lambda_u^{(0)}(\alpha_{cr})$ another one with $\lambda_\varepsilon = -\lambda_u^{(0)}(\alpha_{cr})$ (see Sec. 4). The approximation accuracy (see Sec. 3.3.3, Sec. 3.6.1) in the case $\lambda_\varepsilon = \lambda_u^{(0)}(\alpha_{cr})$ is vastly better than in another case.

As a consequence of the transformations (3.13) and insertion of new state variable ε_n, we obtain the system (3.14) having the two-dimensional center manifold $h(\xi_n^u, \varepsilon_n)$. According to (3.11) the function h can be expressed as the following polynomial up to order r with respect to ξ_n^u and ε_n

$$\xi_n^s = h(\xi_n^u, \varepsilon_n) = A(\xi_n^u)^2 + B\xi_n^u\varepsilon_n + C\varepsilon_n^2 + ... + O(r) \tag{3.15}$$

and accordingly

$$\xi_{n+1}^s = h(\xi_{n+1}^u, \varepsilon_{n+1}) = A(\xi_{n+1}^u)^2 + B\xi_{n+1}^u\varepsilon_{n+1} + C\varepsilon_{n+1}^2 + ... + O(r), \tag{3.16}$$

for which conditions (3.12) should be fulfilled. Substituting the expressions (3.15) and (3.16) into the stable mode equation of the system (3.14), we yield

$$\xi_{n+1}^s = h(\xi_{n+1}^u, \varepsilon_{n+1}) = h(\lambda_u^{(0)}\xi_n^u + \hat{N}_u(\xi_n^u, h(\xi_n^u, \varepsilon_n), \lambda_\varepsilon \varepsilon_n)) =$$
$$= \lambda_s^{(0)} h(\xi_n^u, \varepsilon_n) + \hat{N}_s(\xi_n^u, h(\xi_n^u, \varepsilon_n), \varepsilon_n) \qquad (3.17)$$

Collecting in this equation the terms up to the order r with respect to ε_n and ξ_n^u, we can determine the coefficients $A, B, C, ...$ and accordingly the center manifold $h(\xi_n^u, \varepsilon_n)$. In order to derive of the order parameter equation we substitute $\xi_n^s = h(\xi_n^u, \varepsilon_n)$ into the unstable mode equation of the system (3.14) and finally get

$$\xi_{n+1}^u = \lambda_u(\alpha_{cr})\xi_n^u + \hat{N}_u(\xi_n^u, h(\xi_n^u, \varepsilon_n), \varepsilon_n). \qquad (3.18)$$

Returning first to the systems (3.13) and then to (3.2a) we can rewrite the order parameter as

$$\xi_{n+1}^u = \lambda_u(\alpha)\xi_n^u + \tilde{N}_u(\xi_n^u, h(\xi_n^u, \alpha - \alpha_{cr}), \alpha) \ , \qquad (3.19)$$

where $\varepsilon = \alpha - \alpha_{cr}$ and α_{cr} is already implicated as the explicit parameter into this equation.

3.3.2 CM using assumption $h(\xi_n^u, \varepsilon_n) = [A_2(\varepsilon_n)](\xi_n^u)^2 + [A_3(\varepsilon_n)](\xi_n^u)^3 + ...$

Solving Eq. (3.17) and determining the center manifold $h(\xi_n^u, \varepsilon_n)$, we point to the coefficients corresponding to the zeroth and the first order of ξ_n^u (ε^2, ε^3, $\xi_n^u\varepsilon$, $\xi_n^u\varepsilon^2$, ...) in the function (3.15) that are always equal to zero. In order to demonstrate this, we first determine the coefficients of the center manifold up to the fourth order

$$\xi_n^s = h(\xi_n^u, \varepsilon_n) = A_2(\xi_n^u)^2 + B_2\xi_n^u\varepsilon_n + C_2\varepsilon_n^2 +$$
$$+ A_3(\xi_n^u)^3 + B_3(\xi_n^u)^2\varepsilon_n + C_3\xi_n^u\varepsilon_n^2 + D_3\varepsilon_n^3 + O(4). \qquad (3.20)$$

Then we generalize these calculations to show that the coefficients of the high-order terms with the zeroth and the first order of ξ_n^u also equal zero, independently of the form of nonlinear parts in the mode equations.

Now Eq. (3.17), containing the second order terms with respect to ξ_n^u, ε_n, has the following form

$$A_2(\lambda_u^{(0)}\xi_n^u)^2 + B_2(\lambda_n^{(0)}\xi_n^u)\lambda_\varepsilon\varepsilon_n + C_2(\lambda_\varepsilon\varepsilon_n)^2 =$$
$$= \lambda_s^{(0)}(A_2(\xi_n^u)^2 + B_2\xi_n^u\varepsilon_n + C_2\varepsilon_n^2) + \hat{N}_s^{(2)}(\xi_n^u, h(\xi_n^u, \varepsilon_n), \varepsilon_n). \qquad (3.21)$$

Function $\hat{N}_n^{(2)}$ cannot contain terms with $(\xi_n^u)^0$ or $(\xi_n^u)^1$ therefore ε_n^2 is introduced in the lhs and rhs of (3.21) only by virtue of the function (3.20) i.e. with the same coefficient C_2

$$\varepsilon_n^2 : \ C_2\lambda_\varepsilon^2 = \lambda_s^{(0)}C_2 \qquad (3.22)$$

and the equality (3.22) is fulfilled only when $C_2 = 0$. Equivalent affirmation can be made for the first order of ξ_n^u given by

$$\varepsilon_n\xi_n^u : \ B_2\lambda_\varepsilon\lambda_u^{(0)} = \lambda_s^{(0)}B_2 \qquad (3.23)$$

and correspondingly $B_2 = 0$. Now we consider the third order terms of Eq. (3.17)

$$A_3(\lambda_u^{(0)}\xi_n^u)^3 + B_3(\lambda_n^{(0)}\xi_n^u + \hat{N}_u^{(2)})^2\lambda_\varepsilon\varepsilon_n + C_3(\lambda_u^{(0)}\xi_n^u)(\lambda_\varepsilon\varepsilon_n)^2 +$$
$$+D_3(\lambda_\varepsilon\varepsilon_n)^3 + A_2(2\lambda_u^{(0)}\xi_n^u\hat{N}_u^{(2)}) + A_2(2\lambda_u^{(0)}\lambda_u^{(1)}(\xi_n^u)^2\varepsilon_n) +$$
$$+B_2(\lambda_u^{(1)}\varepsilon_n\xi_n^u)\lambda_\varepsilon\varepsilon_n + B_2(\hat{N}_u^{(2)})\lambda_\varepsilon\varepsilon_n = \lambda_s^{(0)}(A_3(\xi_n^u)^3 + B_3(\xi_n^u)^2\varepsilon_n +$$
$$+C_3\xi_n^u\varepsilon_n^2 + D_3\varepsilon_n^3) + \lambda_s^{(1)}\varepsilon_n(A_2(\xi_n^u)^2 + B_2\xi_n^u\varepsilon_n + C_2\varepsilon_n^2) +$$
$$+\hat{N}_s^{(3)}(\xi_n^u, h(\xi_n^u, \varepsilon_n), \varepsilon_n). \qquad (3.24)$$

The function $\hat{N}_s^{(3)}$ cannot also contain the terms with $(\xi_n^u)^0$, but can contain the terms $(\xi_n^u)^1$ (for example in place of $\xi_n^u \xi_n^s$) which are defined only by the second order of (3.20) and are already equal to zero, i.e. $C_2 \xi_n^u \varepsilon_n^2$. Thus, like previous, the terms of $(\xi_n^u)^0$ depend only on the linear parts of rhs and lhs in (3.24)

$$\varepsilon_n^3 : \; D_3 \lambda_\varepsilon^3 = \lambda_s^{(0)} D_3 + \lambda_s^{(1)} C_2 \tag{3.25}$$

hence it follows that $D_3 = 0$. Carrying out such a calculation for the first order of (ξ_n^u)

$$\xi_n^u \varepsilon_n^2 : \; C_3 \lambda_\varepsilon^2 \lambda_u^{(0)} + B_2 \lambda_u^{(1)} \lambda_\varepsilon = \lambda_s^{(0)} C_3 + \lambda_s^{(1)} B_2 \tag{3.26}$$

we get $C_3 = 0$.

Continuing this calculation further, we permanently ascertain that terms with $(\xi_n^u)^0$ are defined only by the function (3.20) and not by the nonlinear parts of mode equations. In addition, these are always with the same coefficient in rhs and lhs of (3.17) and as a result these coefficients always equal zero.

The terms contained $(\xi_n^u)^1$ in the nonlinear function $\hat{N}^{(l)}(\xi_n^u, h(\xi_n^u, \varepsilon_n), \varepsilon_n)$ are defined by the center manifold function $h^{(l-1)}(\xi_n^u, \varepsilon_n)$, the coefficients, which are already set to zero by previous calculations.

This statement allows selecting the assumption about the center manifold as a polynomial of the variable ξ_n^u whose coefficients depend upon ε_n, i.e the CM-coefficients A, B, C are functions of a smallness parameter ε_n

$$\begin{aligned} h(\xi_n^u, \varepsilon_n) &= [A_2(\varepsilon_n)](\xi_n^u)^2 + [A_3(\varepsilon_n)](\xi_n^u)^3 + ... = \\ &= (a_{20} + a_{21}\varepsilon + a_{22}\varepsilon^2 + ...)(\xi_n^u)^2 + (a_{30} + a_{31}\varepsilon + a_{32}\varepsilon^2 + ...)(\xi_n^u)^3 + ... \end{aligned} \tag{3.27}$$

Further calculations do not diverge from ones described in Sec. 3.3.1.

For an affirmation of the statement (3.27) we have selected the function (3.15) for our test-system (4.1) in Chap. 4. In this example the coefficients of the terms with $(\xi_n^u)^0$, $(\xi_n^u)^1$ up to the sixth order are equal to zero. Generally speaking, the distinction between (3.15) and (3.27) can be considered only as a mathematical subtlety since both assumptions satisfy the center manifold theorem (conditions (3.12)) and both approaches deliver the same coefficients.

3.3.3 CM using assumption $h(\xi^u) = A_2(\xi^u)^2 + A_3(\xi^u)^3 + ...$ and $\xi^u = f(\{\beta\}, \gamma, \varepsilon)$

The above described calculation procedure of the center manifold h is based on polynomials with respect to ξ_n^u, ε_n. But considering an asymptotical dynamics of the unstable mode, we can reach a conclusion that the unstable mode ξ^u itself is a function of ε. Therefore, when we are able to derive the dependence between ξ^u and ε, we can assume that the CM-polynomial depends only upon ξ^u. This assumption simplify the required calculations and, besides, it allows carrying out a numerical verification of the obtained CM. This is shown further in this section.

Asymptotical dynamics of the unstable mode ξ^u

Considering a bifurcation dynamics of the order parameter equation (for example the transcritical bifurcation shown in Fig. 4.6(a) in Chap. 4), we can make one relevant assumption that after a bifurcation an asymptotical dynamics of the unstable mode is a function of the smallness parameter ε.

Indeed, finding the stationary states of the normal form of a transcritical bifurcation

$$\xi^u_{n+1} = (\varepsilon + 1)\xi^u_n - (\xi^u_n)^2 \,, \tag{3.28}$$

we ascertain that the second stationary state is $\xi^u_{st} = \varepsilon$. Taking into account that stable stationary states equal the asymptotically iterated dynamical variable, we write $\xi^u = \varepsilon$. More generally, we can assume

$$\xi^u = \beta_1\varepsilon^\gamma + \beta_2\varepsilon^{2\gamma} + O(\varepsilon^{3\gamma}) = f(\{\beta\},\gamma,\varepsilon) \,, \tag{3.29}$$

where β_1, β_2 and γ are some constants. The assumption (3.29) can numerically and analytically be confirmed. The analytical method consists in obtaining stationary states of the order parameter equation near the bifurcation and then expanding all ε-dependent terms of the stationary state ξ^u_{st} into a Taylor series with respect to ε.

The value γ, β_1 can be also obtained numerically with the least square fit approach by mapping the function $\log(\xi^u) = \gamma\log(\alpha - \alpha_{cr}) + \log(\beta_1)$ from the mode equations or the order parameter equation. Both methods will be demonstrated in the corresponding sections of this work.

Remark, that the highest order of the expression (3.29) defines the order r up to which all terms of the mode equations are expanded. In many cases it is enough to obtain only the first term of the expansion (3.29) that simplifies all further calculation steps.

The function $h(\xi^u)$

Proposition. *Using a priori knowledge about an asymptotical dynamics of the unstable mode $\xi^u = f(\{\beta\},\gamma,\varepsilon)$, the function (3.27) in the vicinity of the bifurcation can be represented in the following form*

$$h(\xi^u) = A_2(\{\beta\},\gamma)(\xi^u)^2 + A_3(\{\beta\},\gamma)(\xi^u)^3 + ... + O(r), \tag{3.30}$$

where $A_2(\{\beta\},\gamma)$, $A_3(\{\beta\},\gamma)$ are coefficients dependent in general upon $\{\beta\}$, γ.

Proof. Inserting

$$\xi^u = \beta_1\varepsilon^\gamma + \beta_2\varepsilon^{2\gamma} + \beta_3\varepsilon^{3\gamma} + O(\varepsilon^{4\gamma}), \tag{3.31}$$

into the function (3.27), we get

$$\begin{aligned} h = (a_{20} + a_{21}\varepsilon + a_{22}\varepsilon^2)(\beta_1\varepsilon^\gamma + \beta_2\varepsilon^{2\gamma})^2 + \\ +(a_{30} + a_{31}\varepsilon + a_{32}\,\varepsilon^2)(\beta_1\varepsilon^\gamma + \beta_2\varepsilon^{2\gamma})^3 + O(\varepsilon^{4\gamma}). \end{aligned} \tag{3.32}$$

Assuming for instance $\gamma = 1$, we rewrite (3.32) up to the fourth order

$$h = a_{21}\beta_1^2\varepsilon^3 + 2a_{20}\beta_1\beta_2\varepsilon^3 + a_{30}\beta_1^3\varepsilon^3 + a_{20}\beta_1^2\varepsilon^2 + O(\varepsilon^4). \tag{3.33}$$

The same result can be obtained if in (3.30) the coefficients will be chosen as follows

$$A_2 = a_{20}, \;\; A_3 = \frac{a_{21}}{\beta_1} + a_{30}. \tag{3.34}$$

Considering the value $\gamma = \frac{1}{2}$, the validity of (3.34) can be also confirmed. Generally speaking, the exact values of coefficient A_3 or other high-order coefficients depend upon the values of γ and β, but they do not affect the form of the function (3.30). □

As already said, the function (3.31) express only the asymptotic dynamics of ξ^u. Consequently, a determination of coefficients in this function can be carried out only on the base of asymptotic dynamics that is not completely correct for the evolutional mode equations. However, this asymptotical limitation is not imposed on the function (3.30), where

$$\xi_n^s = A_2(\xi_n^u)^2 + A_3(\xi_n^u)^3 + ... + O(r), \tag{3.35a}$$
$$\xi_{n+1}^s = A_2(\xi_{n+1}^u)^2 + A_3(\xi_{n+1}^u)^3 + ... + O(r), \tag{3.35b}$$

and accordingly we can consider firstly a non-asymptotic behaviour for deriving the equation (3.17) and then an asymptotic behaviour for defining the desired coefficients.

Now starting from the mode equation (3.13) and substituting the expressions (3.35) in place of stable mode in (3.13), we yield

$$h(\lambda_u^{(0)}\xi_n^u + \hat{N}_u(\xi_n^u, h(\xi_n^u), \varepsilon)) = \lambda_s^{(0)} h(\xi_n^u) + \hat{N}_s(\xi_n^u, h(\xi_n^u), \varepsilon). \tag{3.36}$$

The stable mode appears in both parts of Eq. (3.36) at the same time n, consequently we can apply the asymptotical function (3.31) and rewrite (3.36) as

$$\begin{aligned} &h(\lambda_u^0 f(\{\beta\}, \gamma, \varepsilon) + \tilde{N}_u(f(\{\beta\}, \gamma, \varepsilon), h(f(\{\beta\}, \gamma, \varepsilon), \varepsilon) = \\ &= \lambda_s^0 h(f(\{\beta\}, \gamma, \varepsilon)) + \tilde{N}_s(f(\{\beta\}, \gamma, \varepsilon), h(f(\{\beta\}, \gamma, \varepsilon), \varepsilon). \end{aligned} \tag{3.37}$$

Now inserting the values of $\{\beta\}$, γ and collecting in this equation the terms with respect to ε, we can calculate the coefficients A_2, A_3 and accordingly the function (3.30). Following farther, we derive the order parameter equation from the unstable mode in the form

$$\xi_{n+1}^u = \lambda_u(\alpha)\xi_n^u + \tilde{N}_u(\xi_n^u, h(\xi_n^u), \alpha), \tag{3.38}$$

where $h(\xi_n^u)$ is given by (3.35).

Here we are interested in the essential question of how the obtained center manifold $\xi^s = h(\xi^u)$ does correspond to the existing functional dependence between the stable and unstable modes $\xi^s = h(\xi^u)$ in the mode equations.

Verification of the obtained results

The simplest approach for verification of the obtained function consists in a numerical comparison between both modes and the analytically obtained result. This can be done by a projection of the three-dimensional bifurcation diagram on the plane ξ_n^s, ξ_n^u (see Fig. 3.5 in the example of the Hénon map). The numerical solution from the mode equation must coincide with the analytical curve from (3.30) at least in the vicinity of the $\xi_n^u = 0$. The projection of this three-dimensional bifurcation diagram on the plane ξ_n^u, a must also coincide with the analytically obtained order parameter equation. Furthermore, the form of curve $\xi_n^s(\xi_n^u)$ can give the necessary information about the minimal order of functions (3.15), (3.27) or (3.30).

In addition, we can verify this method in the following way: expanding the asymptotical solution of stable mode ξ^s into a Taylor series with respect to the smallness parameter ε

$$\xi_{st}^s = k_1\varepsilon^2 + k_2\varepsilon^3 + ... + O(r), \tag{3.39}$$

where k_i are coefficients and then comparing coefficients of the same orders in ε given by the lhs and rhs of (3.30)

$$\begin{aligned} &k_1\varepsilon^2 + k_2\varepsilon^3 ... + O(r) = \\ &= A_2(\{\beta\}, \gamma)(f(\{\beta\}, \gamma, \varepsilon))^2 + A_3(\{\beta\}, \gamma)(f(\{\beta\}, \gamma, \varepsilon))^3 + ... + O(r). \end{aligned} \tag{3.40}$$

This method suits also for the verification of the center manifold functions given by (3.15) and (3.27).

3.4 The adiabatic elimination procedure

The adiabatic elimination procedure for time-continuous systems is described e.g. in [13] or [43]. Unfortunately we cannot consider the discrete systems in the same way. Considering the adiabatic elimination, we use another approach, shown in [157]; we compare the slaving principle based on the adiabatic elimination with the center manifold reduction, verifying so the validity of this reduction method.

We restrict our consideration to a low-order approximation. This is motivated by investigations of continuous systems, where the adiabatic elimination is the lowest-order approximation of the "slaving" function [43]. Consequently, we start with the two-dimensional mode equations (3.2) up to the third order

$$\xi^u_{n+1} = \lambda_u \xi^u_n + a_{1u}(\xi^u_n)^2 + a_{2u}\xi^u_n\xi^s_n + a_{3u}(\xi^s_n)^2 + O(3), \tag{3.41a}$$

$$\xi^s_{n+1} = \lambda_s \xi^s_n + a_{1s}(\xi^u_n)^2 + a_{2s}\xi^u_n\xi^s_n + a_{3s}(\xi^s_n)^2 + O(3), \tag{3.41b}$$

where $a_{iu,is}$ are coefficients dependent on the smallness parameter ε. Since the type of instability in this general case is unknown, we use the procedure described in Sec. 3.3.1:

$$\xi^s_n = h(\xi^u_n, \varepsilon_n) = A_2(\xi^u_n)^2 + B_2\xi^u_n\varepsilon_n + C_2\varepsilon^2_n + O(3). \tag{3.42}$$

Carrying out all calculations mentioned in Sec.3.3.1, we derive the coefficients A_2, B_2, C_2:

$$A_2 = \frac{a^{(0)}_{1s}}{(\lambda^{(0)}_u)^2 - \lambda^{(0)}_s}, \quad B_2 = 0, \quad C_2 = 0, \tag{3.43}$$

where $\lambda^{(0)}$, $a^{(0)}$ are the terms of the lowest order in the corresponding Taylor series. Accordingly, the slaving function (3.5) obtained by the center manifold reduction can be written as

$$\xi^s_n = \frac{a^{(0)}_{1s}}{(\lambda^{(0)}_u)^2 - \lambda^{(0)}_s}(\xi^u_n)^2. \tag{3.44}$$

In the following calculations the expression (3.44) will be compared with the result of adiabatic approximation.

As mentioned above, the adiabatic approximation uses an assumption that stable mode can be neglected [13], [161]

$$\xi^s_{n+1} = \xi^s_n. \tag{3.45}$$

Rewriting the equation (3.41b) in the following form

$$a_{3s}(\xi^s_n)^2 + \xi^s_n(\lambda_s + a_{2s}\xi^u_n - 1) + a_{1s}(\xi^u_n)^2 = 0, \tag{3.46}$$

we solve this equation with respect to ξ^s_n

$$\xi^{s_{1,2}}_n = \frac{-a_{2s}\xi^u_n - \lambda_s + 1 \pm \sqrt{(a_{2s}\xi^u_n + \lambda_s - 1)^2 - 4a_{3s}a_{1s}\,(\xi^u_n)^2}}{2a_{3s}}. \tag{3.47}$$

Expanding expression (3.47) into a Taylor series with respect to ξ^u_n up to the third order we get

$$\xi_n^{s1,2} = \frac{-\lambda_s + 1 \pm \sqrt{(\lambda_s - 1)^2}}{2a_{3s}} + \frac{a_{2s}}{2a_{3s}}\left(-1 \pm \frac{\sqrt{(\lambda_s - 1)^2}}{\lambda_s - 1}\right)\xi_n^u \mp$$

$$\mp \frac{a_{1s}}{(\lambda_s - 1)} \frac{\sqrt{(\lambda_s - 1)^2}}{(\lambda_s - 1)} (\xi_n^u)^2 + O(3). \tag{3.48}$$

Per definition $|\lambda_s| < 1$, consequently $(\lambda_s - 1) < 0$ and so the expression (3.48) can be reduced to the form

$$\xi_n^{s1} = -\frac{\lambda_s - 1}{a_{3s}} - \frac{a_{2s}}{a_{3s}}\xi_n^u + \frac{a_{1s}}{\lambda_s - 1}(\xi_n^u)^2 + O(3), \tag{3.49a}$$

$$\xi_n^{s2} = \frac{a_{1s}}{1 - \lambda_s}(\xi_n^u)^2 + O(3). \tag{3.49b}$$

Taking into account properties (3.12) of the center manifold, we conclude that the required solution for the adiabatic elimination is defined by the expression (3.49b).

A comparison between the center manifold (3.44) and the adiabatic approximation (3.49b) shows, that these functions are identical at the bifurcation point where the control parameter a equals a_{cr} and accordingly $(\lambda_u(a_{cr}))^2 = (\lambda_u^{(0)})^2 = 1$. Obviously, the farther away from a bifurcation we take numerical points, the less both solutions will coincide. In general, this procedure can be improved by expanding (3.49b) into a Taylor series with respect to the smallness parameter ε.

Now we attempt to generalize the considered approach for the case of mode equations with several stable modes. For that we rewrite the mode equations in the following form:

$$\xi_{n+1}^u = \lambda_u \xi_n^u + f_u(\xi_n^u, \xi_n^{s1}, \xi_n^{s2}) + O(3), \tag{3.50a}$$

$$\xi_{n+1}^{s1} = \lambda_{s_1}\xi_n^{s1} + a_{1s_1}(\xi_n^u)^2 + a_{2s_1}\xi_n^u\xi_n^{s1} + a_{3s_1}(\xi_n^{s1})^2 + f_{s_1}(\xi_n^u, \xi_n^{s1}, \xi_n^{s2}) + O(3), \tag{3.50b}$$

$$\xi_{n+1}^{s2} = \lambda_{s_2}\xi_n^{s2} + a_{1s_2}(\xi_n^u)^2 + a_{2s_2}\xi_n^u\xi_n^{s2} + a_{3s_2}(\xi_n^{s2})^2 + f_{s_2}(\xi_n^u, \xi_n^{s1}, \xi_n^{s2}) + O(3), \tag{3.50c}$$

where f_u is the second order polynomial of the variables $\xi_n^u, \xi_n^{s1}, \xi_n^{s2}$ and f_{s_1}, f_{s_2} are defined by

$$f_{s_1} = a_{4s_1}(\xi_n^{s2})^2 + a_{5s_1}\xi_n^u\xi_n^{s2} + a_{6s_1}\xi_n^{s1}\xi_n^{s2}, \tag{3.51a}$$

$$f_{s_2} = a_{4s_2}(\xi_n^{s1})^2 + a_{5s_2}\xi_n^u\xi_n^{s1} + a_{6s_2}\xi_n^{s1}\xi_n^{s2}. \tag{3.51b}$$

The system (3.50) has two center manifolds $\xi_n^{s1} = h_1(\xi_n^u, \varepsilon_n)$ and $\xi_n^{s2} = h_2(\xi_n^u, \varepsilon_n)$ which can be calculated by the method described above. Finally, we derive

$$\xi_n^{s1} = \frac{a_{1s_1}^{(0)}}{(\lambda_u^{(0)})^2 - \lambda_{s_1}^{(0)}}(\xi_n^u)^2, \quad \xi_n^{s2} = \frac{a_{1s_2}^{(0)}}{(\lambda_u^{(0)})^2 - \lambda_{s_2}^{(0)}}(\xi_n^u)^2. \tag{3.52}$$

We obtain the same result as (3.52) at $a = a_{cr}$ by the adiabatic approximation, if we consider the stable modes in (3.50) without functions f_{s_1}, f_{s_2} and assuming $\xi_{n+1}^{s1} = \xi_n^{s1}$, $\xi_{n+1}^{s2} = \xi_n^{s2}$

$$\xi_n^{s1} = \lambda_{s_1}\xi_n^{s1} + a_{s_1}(\xi_n^u)^2 + a_{s_2}\xi_n^u\xi_n^s + a_{s_3}(\xi_n^s)^2, \tag{3.53a}$$

$$\xi_n^{s2} = \lambda_{s_2}\xi_n^{s2} + a_{s_1}(\xi_n^u)^2 + a_{s_2}\xi_n^u\xi_n^s + a_{s_3}(\xi_n^s)^2, \tag{3.53b}$$

and then solving these equations with respect to the variables ξ_n^{s1} and ξ_n^{s2}.

Thus, as demonstrated, the adiabatic approximation procedure allows us to carry out the simple calculation of the functions $h(\xi_n^u, \varepsilon_n)$ which consists in solving quadratic equations. The results of this procedure coincide with the low-order center manifold near the bifurcation point.

3.5 The normal form reduction for maps

The normal form (NF) reduction allows transforming the original system to the "simplest possible" form, the so-called normal form, by means of a consecutive coordinate transformation, denoted as "near identity transformations" [113]. Since this technique is in fact well-known concerning a simplification of structure of nonlinear systems [42], [45], [162] we show only calculations of the explicit coefficients for the reduced equation.

Let us start with the mode equations (3.2), written as

$$\underline{\xi}_{n+1} = \underline{\underline{\Lambda}}\, \underline{\xi}_n + \underline{\tilde{N}}^{(2)}(\underline{\xi}_n) + \underline{\tilde{N}}^{(3)}(\underline{\xi}_n) + O(4). \tag{3.54}$$

Here, $\underline{\tilde{N}}^{(l)}$ are nonlinear functions of the order l, $\underline{\xi} = \begin{pmatrix} \xi^u \\ \xi^s \end{pmatrix}$ and $\underline{\underline{\Lambda}}$ is the Jacobian of the system (3.1). We perform all transformations and calculations up to the fourth order with respect to $\underline{\xi}$.

Now using the following near-identity transformation

$$\underline{\xi}_n = \underline{x}_n + \underline{g}^{(2)}(\underline{x}_n) + \underline{g}^{(3)}(\underline{x}_n) + O(4), \tag{3.55}$$

where $\underline{g}^{(l)}$ are a polynomial function of order l, $\underline{x} = \begin{pmatrix} x \\ y \end{pmatrix}$, and accordingly

$$\underline{\xi}_{n+1} = \underline{x}_{n+1} + \underline{g}^{(2)}(\underline{x}_{n+1}) + \underline{g}^{(3)}(\underline{x}_{n+1}) + O(4), \tag{3.56}$$

for small $\underline{x}$, we choose the functions $\underline{g}$ so that it takes the simplest possible form

$$\underline{x}_{n+1} = \underline{\underline{\Lambda}}\, \underline{x}_n + \underline{\tilde{g}}^{(2)}(\underline{x}_n) + \underline{\tilde{g}}^{(3)}(\underline{x}_n) + O(4), \tag{3.57}$$

where the $\underline{\tilde{g}}^{(l)}$ are resonant terms of the order l.

Substituting (3.55) and (3.56) into (3.54), we get

$$\begin{aligned} &\underline{x}_{n+1} + \underline{g}^{(2)}(\underline{x}_{n+1}) + \underline{g}^{(3)}(\underline{x}_{n+1}) + O(4) = \underline{\underline{\Lambda}}(\underline{x}_n + \underline{g}^{(2)}(\underline{x}_n) + \underline{g}^{(3)}(\underline{x}_n)) + \\ &+\underline{\tilde{N}}^{(2)}(\underline{x}_n + \underline{g}^{(2)}(\underline{x}_n) + \underline{g}^{(3)}(\underline{x}_n)) + \underline{\tilde{N}}^{(3)}(\underline{x}_n + \underline{g}^{(2)}(\underline{x}_n) + \underline{g}^{(3)}(\underline{x}_n)) + O(4) \end{aligned} \tag{3.58}$$

or collecting the coefficients with the same powers of $\underline{x}$

$$\begin{aligned} \underline{x}_{n+1} = {} & \underline{\underline{\Lambda}}\, (\underline{x}_n) + [\underline{\underline{\Lambda}}\, \underline{g}^{(2)}(\underline{x}_n) - \underline{g}^{(2)}(\underline{x}_{n+1}) + \underline{F}^{(2)}(\underline{x}_n)] + \\ & + [\underline{\underline{\Lambda}}\, \underline{g}^{(3)}(\underline{x}_n) - \underline{g}^{(3)}(\underline{x}_{n+1}) + \underline{F}^{(3)}(\underline{x}_n)] + O(4), \end{aligned} \tag{3.59}$$

where $\underline{F}^{(l)}$ are functions of only the order l.

The terms $\underline{g}^{(2)}(\underline{x}_{n+1})$ and $\underline{g}^{(3)}(\underline{x}_{n+1})$ can be eliminated by using either the method of operators shown e.g., in [113] or we can follow Wiggins [42], using Eq. (3.57) and taking into account the consecutive transformations up to the order $O(4)$

$$\underline{g}^{(2)}(\underline{x}_{n+1}) = \underline{g}^{(2)}(\underline{\underline{\Lambda}}\, \underline{x}_n + \underline{\tilde{g}}^{(2)}(\underline{x}_n)) + O(4) \tag{3.60}$$

and

$$\underline{g}^{(3)}(\underline{x}_{n+1}) = \underline{g}^{(3)}(\underline{\underline{\Lambda}}\, \underline{x}_n) + O(4). \tag{3.61}$$

For the time-discrete systems the last case is simpler.

As a result of the specified transforms (3.60),(3.61) the Eq. (3.59) is rewritten as

$$\underline{x}_{n+1} = \underline{\underline{\Lambda}}\,(\underline{x}_n) + [\underline{\underline{\Lambda}}\,\underline{g}^{(2)}(\underline{x}_n) - \underline{g}^{(2)}(\underline{\underline{\Lambda}}\,\underline{x}_n) + \underline{F}^{(2)}(\underline{x}_n)] + [\underline{\underline{\Lambda}}\,\underline{g}^{(3)}(\underline{x}_n) -$$
$$-\underline{g}^{(3)}(\underline{\underline{\Lambda}}\,\underline{x}_n)) + \underline{F}^{(3)}(\underline{x}_n) - \underline{g}^{(2)\to O(3)}(\underline{\underline{\Lambda}}\,\underline{x}_n + \underline{\tilde{g}}^{(2)}(\underline{x}_n))] + O(4), \qquad (3.62)$$

where $\underline{g}^{(2)\to O(3)}(\underline{\underline{\Lambda}}\,\underline{x}_n + \underline{\tilde{g}}^{(2)}(\underline{x}_n))$ are nonlinear functions of order $O(3)$ defined by expression (3.60). From the Eq. (3.62) the terms of the order $O(2)$ can be selected and, if it is possible, eliminated

$$O(2) \to \underline{\underline{\Lambda}}\,\underline{g}^{(2)}(\underline{x}_n) - \underline{g}^{(2)}(\underline{\underline{\Lambda}}\,\underline{x}_n) + \underline{F}^{(2)}(\underline{x}_n) = 0. \qquad (3.63)$$

As a consequence of this calculation the resonant terms $\underline{\tilde{g}}^{(2)}$ (3.62) are already determined and we can begin the calculation of the cubical terms using

$$O(3) \to \underline{\underline{\Lambda}}\,\underline{g}^{(3)}(\underline{x}_n) - \underline{g}^{(3)}(\underline{\underline{\Lambda}}\,\underline{x}_n)) + [\underline{F}^{(3)}(\underline{x}_n) - \underline{g}^{(2)\to O(3)}(\underline{\underline{\Lambda}}\,\underline{x}_n + \underline{\tilde{g}}^{(2)}(\underline{x}_n))] = 0. \qquad (3.64)$$

The calculation of coefficients in cubic terms differs from quadratic ones. The cubic nonlinearity contains the terms of quadratic nonlinearity which must be first calculated. This mathematical subtlety has a dominant influence on the coefficients of high-order terms. An example of an explicit calculation of NF-coefficients with cubic and quadratic nonlinearity is shown in Sec. 4.2.4.

3.6 Accuracy and complexity of reduction procedures

In the last part we compare the CM approaches described in Sec. 3.3.1, 3.3.3 and the NF approach from the viewpoint of accuracy and computational complexity.

3.6.1 Approximation accuracy

Investigation of approximation accuracy between different reduction procedures is primarily caused by application-oriented aspects of bifurcation analysis. Performing bifurcation control, the main attention is focused on the type of instability, described by the reduced order parameter(s) or by the normal form. Control of qualitative behaviour, e.g., a shift of bifurcations, does not depend on the approximation accuracy. However, a control of quantitative behaviour near bifurcation is in fact restricted by this accuracy, thereby the less an accuracy is the less is a reliability of control mechanisms. Being motivated by this reason, we intend to look for such a reduction approach that guarantees better approximation accuracy.

Accuracy of reduction procedures can be compared in two ways: numerically and analytically. Results of numerical investigations depend on the system chosen, however, they can serve as sui generis numerical verification of analytically obtained statements. Such a numerical comparison is performed for the OLL map in Chap. 4. Accuracy, from analytical viewpoint, can be estimated by expanding the stationary states of OP, derived by means of different reduction procedures, into Taylor series, like the Eq. (3.39). Comparing the coefficients of equal degrees derived from OPs and mode equations, we can conclude about a reduction method with the best accuracy.

Starting from the mode equations (3.50) at e.g., $\lambda_u(a_{cr}) = 1$, the corresponding local bifurcation can be described by the following normal form

$$\xi^u_{n+1} = \lambda_u \xi^u_n + a_{1u}(\xi^u_n)^2, \qquad (3.65)$$

derived by means of the NF reduction. Applying the CM approach to the same mode equations, we get the following order parameter

$$\xi^u_{n+1} = \lambda_u \xi^u_n + a_{1u}(\xi^u_n)^2 + a_{2u}\Omega_1(\xi^u_n)^3 + a_{3u}\Omega_2(\xi^u_n)^4 + O(2r-1), \tag{3.66}$$

where Ω are coefficients, depending on coefficients of the CM assumptions (3.15), (3.20) or (3.30) up to the order r. Assuming λ_u, a_{iu}, Ω_i to be functions of the smallness parameter ε, we can expand ξ^u_{st} from these equations and modes into Taylor series with respect to ε, namely

$$\xi^u_{st} = k_1\varepsilon + k_2\varepsilon^2 + k_3\varepsilon^3 + ... \tag{3.67}$$

Since all of the (3.65), (3.66) and (3.50) are equal as germs, coefficients k_1 (the first nonzero coefficients) have to coincide at ξ^u_{st} derived from all of them. The NF (3.65) does not have the high-order terms, that the mode equations have, therefore the coefficients k_2 and others obtained from NF (3.65) and from modes (3.50) will be the most different. Thus, the NF reduction is expected to have the worse approximation accuracy. Correspondingly, the CM approach guarantees better accuracy, even it may be improved by increasing the order of CM assumptions, that is numerically confirmed in Chap. 4. The procedure of adiabatic approximation can be viewed as the low-order CM approach. Moreover, independently of chosen unstable eigenvalues, the relation of approximation accuracy between CM and NF reduction remains the same.

3.6.2 Computational complexity

There are different definitions of computational complexity. They are related to e.g., the number of symbolic operations or relative time required for an execution of a program [202]. However, the exact value of the number of operations depends upon implementation details in the symbolic calculation program, and the run-time does not have any importance in our case. In this case we can exploit the fact that a relation of complexities between different methods is preserved the same for different definitions of the computational complexity. Then for a calculation of the relative computational complexity (RCC) the following way can be used: algorithm is divided into the parts with a complexity C_i; they are approximately equal for different reduction methods. Here loops denote the n-time repetition for the computation of terms of the order n. The relative computational complexity of the algorithm can be written as $\sum_i^k C_i n$, where n is the order of the reduction procedure, k is the number of structural operations in the algorithm. This method is similar to ones accepted in the algorithms analysis [163].

The algorithmic structure of different reduction procedures is shown in Fig. 3.1. The parts with approximately equal computational complexity are: the generation of the CM assumption or the near-identity transformation with computational complexity C_1, the calculation of different terms with RCC C_2, and determination of coefficients either of CM reduction or NF transformation with RCC C_3. If n is the order of a reduction procedure, the relative computational complexity of NF and both CM reductions is proportional to $C_1n + 3C_2n + 2C_3n$, $C_1n + 3C_2n + C_3n$ and $C_1n + 2C_2n + C_3n$. In Table 3.1 shows the number of operations in our realization. They are the number of Maple code lines in accordant subroutines.

Thus, we can resume, that computational complexity of NF reduction is maximal in comparison with CM reductions, but its accuracy is minimal. The minimal RCC has CM (ξ^u) but at the cost of high computational complexity of a priori information about $\{\beta\}, \gamma$ (see Sec. 3.3.3). If coefficients $\{\beta\}, \gamma$ are known, an application of CM (ξ^u) is optimal.

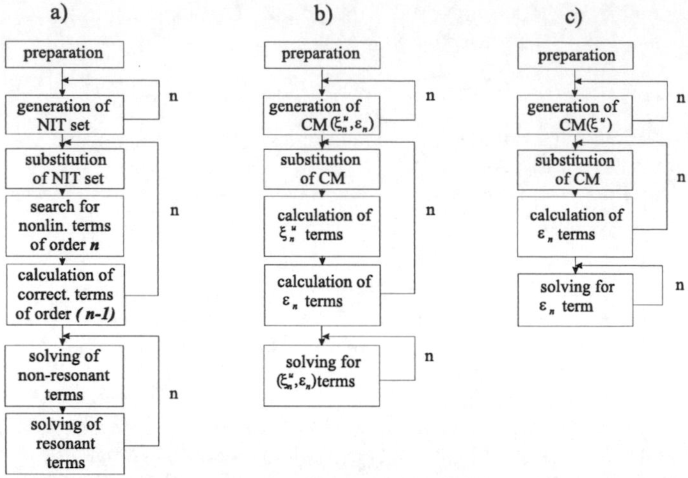

Figure 3.1: Structure of algorithms for **a)** normal form reduction; **b)** CM reduction (ξ^u_n, ε_n); **c)** CM reduction (ξ^u), where n is the order of a reduction procedure, C_i the computational complexity of i-structural operation.

Reduction Procedure	NF	CM (ξ^u_n, ε_n)	CM (ξ^u)
RCC	$C_1n + 3C_2n + 2C_3n$	$C_1n + 3C_2n + C_3n$	$C_1n + 2C_2n + C_3n$
Number of code lines	108	73	24

Table 3.1: Relative computational complexity (RCC) and the number of code lines for NF and both CM reductions.

3.7 Order parameter of a single analytical agent

3.7.1 Motivation

After explaining some details of calculation techniques, we can start to deal with the primary goal of this chapter: derivation of the order parameter from a simple analytical agent. Moreover, we are interested in whether a simple single agent can possess the order parameter(s) and, if yes, what it does mean and could it be used by e.g., control? Answering these questions, let us consider one model whose simulation is performed by $Simulink^{©}$. Structure of this model is shown in Fig. 3.2. Considering this structure, we can ascertain that:

- it possesses the internal feedback drawn by bold lines, thereby its behaviour can be viewed in a sense of being self-regulated, independent of environment. In other words, this is one of simplest forms of autonomy [65];
- this system can perceive and act in environment;
- this model has a defined goal originated from weather forecasting [149], [150], therefore its behaviour is goal-oriented;

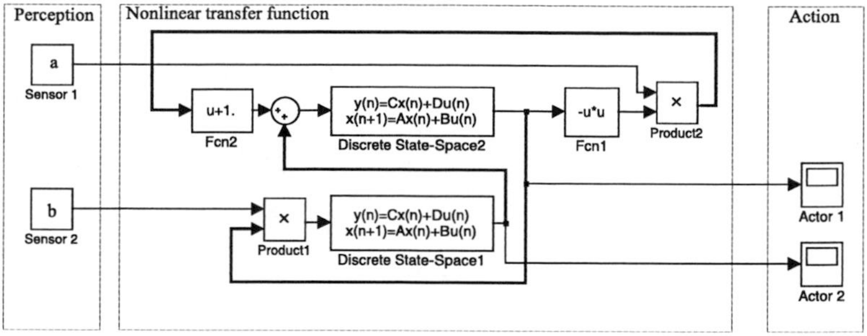

Figure 3.2: *Simulink©* simulation model of a simple single agent. Bold lines represent internal feedback pointing to its autonomy, two oscilloscopes are the actors for this simulation. Environmental feedback is not shown.

- this model is rational (or more exactly pre-rational) in sense that it follows predetermined rules and tries to achieve its goal;
- this model possesses implicitly the processing and planning in the form of nonlinear function incorporated into its structure;
- this model is not intelligent because it does not have any form of initiative in goal achieving, it does not possess also the social ability.

Following the definitions of agent [65], [58], this model can be viewed in sense of autonomous, pre-rational, non-intelligent and non-social agent. Whereas further models will have social ability, this model is destined to demonstrate that a single agent exhibits different complex behaviour, even chaotic one. For those behavioural types we can derive the order parameter. Accordingly, further investigations have to answer the question of how the order parameter of a single agent is influenced by its "social" abilities.

Information, delivered by sensors, can be thought of as parameters varied in some range. Thus, skipping physical actors, we can express the structure, shown in Fig. 3.2, by the following nonlinear equations,

$$x_{n+1} = 1 + y_n - ax_n^2, \tag{3.68a}$$

$$y_{n+1} = bx_n \ , \tag{3.68b}$$

where a, b are parameters, $x_n, y_n \in \mathbb{R}$ are state variables. In the vast literature on the object this mathematical model is well-known and denoted as two-dimensional Hénon map (e.g., [149], [151], [152]) that is just intended for the further analysis within this section. The Hénon map as an analytical agent is chosen with an intension to compare the derived results with already published ones. Moreover it offers such a degree of complexity that allows a demonstration of considered techniques. More complex system, such OLL-map, is considered later.

3.7.2 Linear stability analysis

[2]The represented investigations of Hénon map are generally restricted to the parameter $b = 0.3$ and focused on the first bifurcation of the period-doubling cascade (see bifurcation

[2]Subsections 3.7.2-3.7.5 represent results that are parts of the work done in collaboration between author and S. Kornienko, M. Schanz, P. Levi, see [161].

diagram in Fig. 3.3) which occur at the critical parameter value $a_{cr1} = a_{cr} = 0.3675$. For

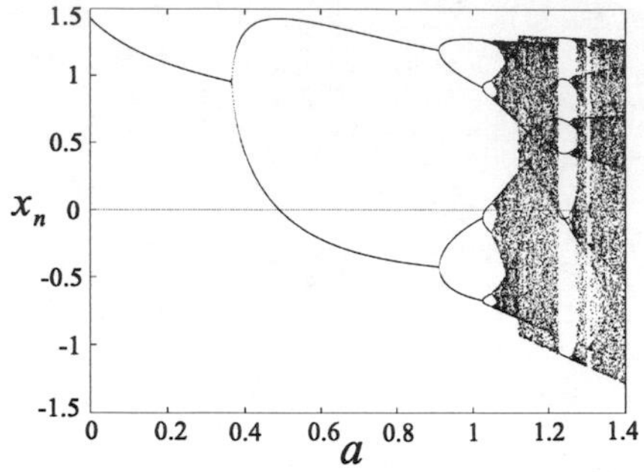

Figure 3.3: Bifurcation diagram of the original Hénon map (3.68), $x_0 = y_0 = 0.5$, $b = 0.3$

this bifurcation we derive the order parameter equation in the above formulated straight forward way and analyze its properties. The state vector of this system is defined by $\underline{\mathbf{q}}_n = (q_n^1, q_n^2)^T = (x_n, y_n)^T$. The stationary states $\underline{\mathbf{q}}_{st}$ of Eq. (3.68) are given by

$$x_{st_{(1,2)}} = \frac{b - 1 \pm \sqrt{1 - 2b + b^2 + 4a}}{2a}, \quad y_{st_{(1,2)}} = \frac{b(b - 1 \pm \sqrt{1 - 2b + b^2 + 4a})}{2a}, \tag{3.69}$$

and the Jacobian of the system (3.68) evaluated at the stationary states (3.69) is then

$$\underline{\underline{\mathbf{J}}} = \begin{pmatrix} -2ax_{st} & 1 \\ b & 0 \end{pmatrix}. \tag{3.70}$$

For the eigenvalues of the matrix (3.70) we get

$$\lambda_{1,2} = -ax_{st} \pm \sqrt{a^2 x_{st}^2 + b}\,, \tag{3.71}$$

and the corresponding eigenvectors are:

$$\underline{\mathbf{v}}_1 = \begin{pmatrix} \frac{\lambda_1}{b} \\ 1 \end{pmatrix}, \ \underline{\mathbf{v}}_2 = \begin{pmatrix} \frac{\lambda_2}{b} \\ 1 \end{pmatrix}. \tag{3.72}$$

From the linear stability analysis of the system it follows that the stationary state (x_{st_2}, y_{st_2}) is unstable because the condition for a contracting linear map $|\lambda| < 1$ is violated in the whole region of interest of the parameter a, i.e., the interval [0, 1.4]. The stationary state (x_{st_1}, y_{st_1}) loses its stability at the parameter value $a_{cr1} = a_{cr} = \frac{147}{400} = 0.3675$ (see Fig. 3.4), because at this value of the parameter a the condition for a contracting linear map $|\lambda| < 1$ is violated by the eigenvalue λ_2.

As a consequence we focus in the following on the stable stationary state (x_{st_1}, y_{st_1}) and denote the corresponding eigenvalue λ_2 from now on as λ_u and the other eigenvalue λ_1 as λ_s.

3.7.3 The mode amplitude equations

In the following we use the compact tensor notation for our investigations

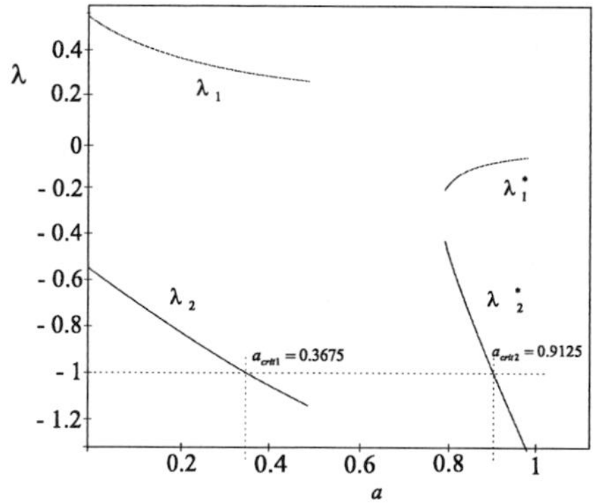

Figure 3.4: Eigenvalues (3.71) in dependence on the bifurcation parameter a, $\lambda^*_{1,2}$ are the eigenvalues of the second iterated Hénon map, presenting the second period-doubling bifurcation at $a_{cr2} = 0.9125, b = 0.3$.

$$\underline{\mathbf{q}}_{n+1} = \Gamma_{(1)} + \Gamma_{(2)}(:\underline{\mathbf{q}}_n) + \Gamma_{(3)}(:\underline{\mathbf{q}}_n)^2 + ... = \sum_{r=0}^{p} \Gamma_{(r+1)}(:\underline{\mathbf{q}}_n)^r, \tag{3.73}$$

Expression (3.73) reads in components:

$$q^i_{n+1} = (\Gamma_{(1)})_i + \sum_{j_1}(\Gamma_{(2)})_{ij_1} q^{j_1}_n + \sum_{j_1 j_2}(\Gamma_{(3)})_{ij_1 j_2} q^{j_1}_n q^{j_2}_n + ... = \tag{3.74}$$

$$= \sum_{r=0}^{p} \sum_{ij_1 ... j_r} (\Gamma_{(r+1)})_{ij_1...j_r} q^{j_1}_n ... q^{j_r}_n .$$

Herein p is the maximal order of the equation of motion and the indices in brackets denote the rank of the corresponding tensors. The non-vanishing components of the tensors for the Hénon map (3.68) are shown in Table 3.2.

$\Gamma_{(1)}$	$\Gamma_{(2)}$	$\Gamma_{(3)}$
$(\Gamma_{(1)})_1 = 1$	$(\Gamma_{(2)})_{12} = 1$	$(\Gamma_{(3)})_{111} = -a$
$(\Gamma_{(1)})_2 = 0$	$(\Gamma_{(2)})_{21} = b$	

Table 3.2: Non-vanishing tensors components $\Gamma_{(1)}, \Gamma_{(2)}, \Gamma_{(3)}$ of the Hénon map.

For the further analysis we investigate the behaviour of small deviations from the stationary states. Therefore we insert $\underline{\mathbf{q}}_n = \underline{\mathbf{q}}_{st} + \Delta\underline{\mathbf{q}}_n = (x_{st} + \Delta x_n, y_{st} + \Delta y_n)$ in the evolution Eq. (3.73) and derive the equation of motion for the deviation $\Delta\underline{\mathbf{q}}_n$

$$\Delta\underline{\mathbf{q}}_{n+1} = \Gamma^L_{(2)}(:\Delta\underline{\mathbf{q}}_n) + \sum_{r=2}^{p} \Gamma^N_{(r+1)}(:\Delta\underline{\mathbf{q}}_n)^r, \tag{3.75}$$

where Γ^L represents the linear part which is given by the Jacobian and Γ^N represent the nonlinear parts. The non-vanishing components of the tensors in Eq. (3.75) for the systems (3.68) are shown in Table 3.3.

$\Gamma^L_{(2)}$	$\Gamma^N_{(3)}$
$(\Gamma^L_{(2)})_{11} = -2ax_{st}$	$(\Gamma^N_{(3)})_{111} = -a$
$(\Gamma^L_{(2)})_{12} = 1$	
$(\Gamma^L_{(2)})_{21} = b$	

Table 3.3: Non-vanishing tensors components $\Gamma^L_{(2)}, \Gamma^N_{(3)}$ of the Hénon map.

To proceed further we apply the following coordinate transformation introducing hereby the modes ξ^s_n, ξ^u_n and $\underline{\boldsymbol{\xi}}_n = (\xi^s_n, \xi^u_n)^T$

$$\Delta\underline{\mathbf{q}}_n = \sum_k \xi^k_n \underline{\mathbf{v}}_k = \underline{\underline{\mathbf{V}}}\underline{\boldsymbol{\xi}}_n = \Gamma^V_{(2)}(:\underline{\boldsymbol{\xi}}_n), \tag{3.76}$$

where the columns of the matrix $\underline{\underline{\mathbf{V}}}$ are given by the eigenvectors of the Jacobian, i.e., $\underline{\underline{\mathbf{V}}} = (\underline{\mathbf{v}}^1, \underline{\mathbf{v}}^2)$ and $\Gamma^V_{(2)}$ is the corresponding tensor. Inserting Eq. (3.76) in Eq. (3.75) we get then

$$\begin{aligned}\Gamma^V_{(2)}(:\underline{\boldsymbol{\xi}}_{n+1}) &= \Gamma^L_{(2)}\Gamma^V_{(2)}(:\underline{\boldsymbol{\xi}}_n) + \sum_{r=2}^{p}\Gamma^N_{(r+1)}(\Gamma^V_{(2)}(:\underline{\boldsymbol{\xi}}_n))^r = \\ &= \Gamma^L_{(2)}\Gamma^V_{(2)}(:\underline{\boldsymbol{\xi}}_n) + \Gamma^N_{(3)}(\Gamma^V_{(2)}(:\underline{\boldsymbol{\xi}}_n))^2.\end{aligned} \tag{3.77}$$

Multiplying from the left side with the inverse tensor $\Gamma^{V^{-1}}_{(2)}$, we obtain the mode equations

$$\underline{\boldsymbol{\xi}}_{n+1} = \Gamma^\Lambda_{(2)}(:\underline{\boldsymbol{\xi}}_n) + \sum_{r=2}^{p}\Gamma^{\tilde{N}}_{(r+1)}(:\underline{\boldsymbol{\xi}}_n)^r = \Gamma^\Lambda_{(2)}(:\underline{\boldsymbol{\xi}}_n) + \Gamma^{\tilde{N}}_{(3)}(:\underline{\boldsymbol{\xi}}_n)^2. \tag{3.78}$$

with the tensor products $\Gamma^\Lambda_{(2)} = \Gamma^{V^{-1}}_{(2)}\Gamma^L_{(2)}\Gamma^V_{(2)}$ and $\Gamma^{\tilde{N}}_{(3)} = \Gamma^{V^{-1}}_{(2)}\Gamma^N_{(3)}\Gamma^V_{(2)}\Gamma^V_{(2)}$. We thereby made use of the identity

$$(\Gamma^{V^{-1}}_{(2)}\Gamma^V_{(2)})_{ij} = \delta_{ij}, \tag{3.79}$$

where δ_{ij} is the well-known Kronecker-Symbol. As an example we show the explicit calculation of the tensor

$$\Gamma^{\tilde{N}}_{(3)} = \Gamma^{V^{-1}}_{(2)}\Gamma^N_{(3)}\Gamma^V_{(2)}\Gamma^V_{(2)}\,, \tag{3.80}$$

which reads in components:

$$(\Gamma^{\tilde{N}}_{(3)})_{ijk} = \sum_{lmn}(\Gamma^{V^{-1}}_{(2)})_{il}(\Gamma^N_{(3)})_{lmn}(\Gamma^V_{(2)})_{mj}(\Gamma^V_{(2)})_{nk}\,. \tag{3.81}$$

The non-vanishing components of the tensor $\Gamma^\Lambda_{(2)}$ are $(\Gamma^\Lambda_{(2)})_{11} = \lambda_s$ and $(\Gamma^\Lambda_{(2)})_{22} = \lambda_u$. With respect to the tensor components, summarized in Table 3.3, Eq. (3.78) reads in detail (we denote ξ^u_n the mode with eigenvalue λ_u and ξ^s_n the mode with the eigenvalue λ_s)

$$\xi^s_{n+1} = \lambda_s\xi^s_n - \frac{a(\lambda_s\xi^s_n + \lambda_u\xi^u_n)^2}{b(\lambda_s - \lambda_u)}, \tag{3.82a}$$

$$\xi^u_{n+1} = \lambda_u\xi^u_n + \frac{a(\lambda_s\xi^s_n + \lambda_u\xi^u_n)^2}{b(\lambda_s - \lambda_u)}\,, \tag{3.82b}$$

where the first equation is the one for the stable mode and the second one that for the unstable mode. The Eqs. (3.82) are the mode equations of the system (3.68). They describe the behaviour of the stable and unstable modes of the Hénon map at the first bifurcation of the period-doubling cascade (see Fig. 3.3, and Fig. 3.5). In terms of syner-

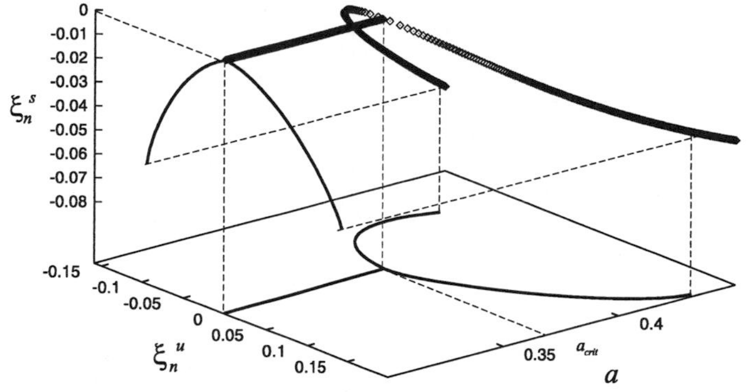

Figure 3.5: Bifurcation diagram of the mode equations (3.82) at the first bifurcation in a 3D representation (from [161]).

getics the unstable mode ξ_n^u corresponds to the order parameter, while the stable mode ξ_n^s corresponds to the enslaved mode, as was already mentioned in Sec. 3.2.

3.7.4 Order parameter using the adiabatic approximation

Near an instability the stable mode ξ_n^s shows an interesting characteristic behaviour. The dynamics is mainly influenced by two effects. Firstly, an intrinsic dynamics which is governed by the eigenvalue λ_s and, secondly, an external dynamics which is governed by the nonlinear coupling with the unstable mode. Due to the time scale hierarchy expressed in (3.3) the intrinsic dynamic of the stable mode ξ_n^s is much faster than that of the unstable mode ξ_n^u. Therefore, we can say that the intrinsic dynamics decays very fast and only the dynamics which is caused by the unstable mode remains. This characteristic behaviour is mathematically expressed by the center manifold theorem. To derive an approximation of the center manifold $h(\xi_n^u)$ we can exploit this behaviour using the adiabatic approximation procedure for the stable mode

$$\xi_{n+1}^s = \xi_n^s \,. \tag{3.83}$$

After this calculation we can determine, whether the properties:

$$h(0) = 0 \,, \tag{3.84a}$$

$$\left.\frac{dh(\xi_n^u)}{d\xi_n^u}\right|_{\xi_n^u=0} = 0 \tag{3.84b}$$

are fulfilled. From Eq. (3.82a) we derive the following equation for the stable mode ξ_n^s:

$$\xi_n^s = \lambda_s \xi_n^s - \frac{a(\lambda_s \xi_n^s + \lambda_u \xi_n^u)^2}{b(\lambda_s - \lambda_u)} \,. \tag{3.85}$$

Solving this quadratic equation we obtain finally the following approximation of the center manifold:

$$\xi_n^{s_1,s_2} = \frac{2ab\xi_n^u - K_a \pm \sqrt{K_a(K_a - 4ab\xi_n^u)}}{2a\lambda_s^2} \tag{3.86}$$

where

$$K_a = b(\lambda_s - 1)(\lambda_u - \lambda_s) \tag{3.87}$$

and $\lambda_s\lambda_u = -b$. In Eq. (3.86) we select the sign of the square root so that the order parameter equation for values $a < 0.3675$ lead to the already known stationary state (x_{st_1}, y_{st_1}). This condition is satisfied by the plus sign in Eq. (3.86). Substituting the value $\xi_n^u = 0$ in the expression for $\xi_n^{s_1} = h(\xi_n^u)$ we immediately obtain $\xi_n^s = 0$, which means that requirement (3.84a) is fulfilled. Similarly calculating the derivative of $\xi_n^{s_1}$ with respect to ξ_n^u and substituting again $\xi_n^u = 0$ we see that (3.84b) is also fulfilled. The dependence of $\xi_n^{s_1} = \xi_n^s = h(\xi_n^u)$ on the unstable mode ξ_n^u is shown in Fig. 3.6. Substituting the ex-

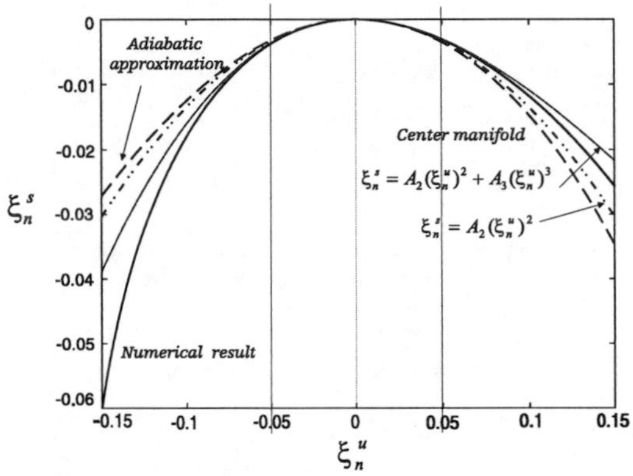

Figure 3.6: Dependence between the variables ξ_n^u, ξ_n^s, obtained by a simulation and analytically, using the adiabatic approximation (3.86) and the center manifold (3.89)(from [161]).

pression for the center manifold in that part of Eqs. (3.82b) which represents the unstable mode we yield the order parameter equation for the first bifurcation via the adiabatic approximation procedure

$$\xi_n^u = \lambda_u\xi_n^u + \frac{a}{b(\lambda_s - \lambda_u)}\left(\frac{2ab\xi_n^u - K_a + \sqrt{K_a(K_a - 4ab\xi_n^u)}}{2a\lambda_s} + \lambda_u\xi_n^u\right)^2. \tag{3.88}$$

where K_a is defined by the expression (3.87).

3.7.5 Order parameter using the center manifold theorem

In this section we show the calculation of the order parameter equation, using the well-known center manifold theorem [42], [106], [157]. Considering Eq. (3.5), we use the following hypothesis about a center manifold:

$$\xi_n^s = h(\xi_n^u) = A_2(\xi_n^u)^2 + A_3(\xi_n^u)^3 + O(4), \tag{3.89a}$$
$$\xi_{n+1}^s = h(\xi_{n+1}^u) = A_2(\xi_{n+1}^u)^2 + A_3(\xi_{n+1}^u)^3 + O(4), \tag{3.89b}$$

where the coefficients A_2, A_3 are determined from the solution of Eq. (3.36). For the calculations, we use the following assumption:

$$\xi_n^u \sim \varepsilon^\alpha, \tag{3.90}$$

which will be confirmed numerically (see Table 5.1 in Chap. 5), where $\varepsilon = a - a_{crit}$ is the so-called smallness parameter, $\alpha = \frac{1}{2}$ and a_{crit} is the critical value of the bifurcation parameter a. Our further approach is the following: Using $a = \varepsilon + a_{crit}$ we expand all ε-dependent expressions $a, \lambda_u(a), \lambda_s(a)$ in the Eqs. (3.82) into a Taylor series with respect to ε and take into account only terms up to the order of $\varepsilon^{3\alpha}$. Because $\varepsilon^{3\alpha} = \varepsilon^{\frac{3}{2}}$ terms with the order $\varepsilon^{2\alpha+1} = \varepsilon^{\frac{4}{2}}$ we do not consider.

First we rewrite the Eqs. (3.82) in the following form:

$$\xi_{n+1}^s = \lambda_s \xi_n^s - K\Phi, \tag{3.91a}$$
$$\xi_{n+1}^u = \lambda_u \xi_n^u + K\Phi, \tag{3.91b}$$

with the abbreviations:

$$K = \frac{a}{b(\lambda_s - \lambda_u)}, \quad \Phi = (\lambda_s \xi_n^s + \lambda_u \xi_n^u)^2. \tag{3.92}$$

Using this approach the resulting terms of Eqs. (3.91) are summarized in Table 3.4.

Terms	Taylor series
$\lambda_u(a_{crit} + \varepsilon)$	$\lambda_u^0 + \lambda_u^1 \varepsilon + ...$
$\lambda_s(a_{crit} + \varepsilon)$	$\lambda_s^0 + \lambda_s^1 \varepsilon + ...$
$K = \dfrac{(a_{crit} + \varepsilon)}{(b(\lambda_s(a_{crit} + \varepsilon) - \lambda_u(a_{crit} + \varepsilon))}$	$K^0 + ...$
$\Phi = (\lambda_s(a_{crit} + \varepsilon)(A_2\varepsilon^{2\alpha} + A_3\varepsilon^{3\alpha}) + \lambda_u(a_{crit} + \varepsilon)\varepsilon^\alpha)^2$	$2A_2\lambda_s^0\lambda_u^0\varepsilon^{3\alpha} + (\lambda_u^0)^2\varepsilon^{2\alpha} + ...$

Table 3.4: Expansion of the terms of Eqs. (3.91) into a Taylor series with respect to ε up to the relevant order.

From now on whenever in an expression an upper index exist which is equal to zero we denote with it the lowest order of this expression. Therefore, the expressions in table 3.4 are:

$$K^0 = \frac{a_{crit}}{b(\lambda_s^0 - \lambda_u^0)}, \ \lambda_u^0 = \lambda_u(a_{crit}), \ \lambda_s^0 = \lambda_s(a_{crit}). \tag{3.93}$$

Using Eq. (3.36) for the calculation of the center manifold we yield in this case

$$h(\xi_{n+1}^u) = \lambda_s h(\xi_n^u) - K\Phi. \tag{3.94}$$

where K, Φ are given by (3.92). Obeying further Eqs. (3.88)-(3.92) and keeping only terms up to the order $\varepsilon^{3\alpha}$ we end up with

$$\begin{aligned} &A_2\left(\lambda_u^0\varepsilon^\alpha + \frac{a_{crit}}{b(\lambda_s^0 - \lambda_u^0)}(\lambda_u^0)^2\varepsilon^{2\alpha}\right)^2 + A_3(\lambda_u^0\varepsilon^\alpha)^3 = \\ &= \lambda_s^0(A_2\varepsilon^{2\alpha} + A_3\varepsilon^{3\alpha}) - \frac{a_{crit}}{b(\lambda_s^0 - \lambda_u^0)}(2A_2\lambda_s^0\lambda_u^0\varepsilon^{3\alpha} + (\lambda_u^0)^2\varepsilon^{2\alpha}). \end{aligned} \tag{3.95}$$

From this equation we determine the coefficients A_2, A_3:

$$A_2 = \frac{a_{crit}}{b(\lambda_s^0 - \lambda_u^0)} \frac{(\lambda_u^0)^2}{(\lambda_s^0 - (\lambda_u^0)^2)}, \tag{3.96a}$$

$$A_3 = 2A_2 \frac{a_{crit}}{b(\lambda_s^0 - \lambda_u^0)} \frac{\lambda_u^0(\lambda_s^0 + (\lambda_u^0)^2)}{(\lambda_s^0 - (\lambda_u^0)^3)}. \tag{3.96b}$$

Thus we yield the cubic center manifold (3.89a) in the following form

$$\xi_n^s = \frac{a_{crit}}{b(\lambda_s^0 - \lambda_u^0)} \frac{(\lambda_u^0)^2}{(\lambda_s^0 - (\lambda_u^0)^2)} \left((\xi_n^u)^2 + 2\frac{a_{crit}}{b(\lambda_s^0 - \lambda_u^0)} \frac{\lambda_u^0(\lambda_s^0 + (\lambda_u^0)^2)}{(\lambda_s^0 - (\lambda_u^0)^3)} (\xi_n^u)^3 \right). \tag{3.97}$$

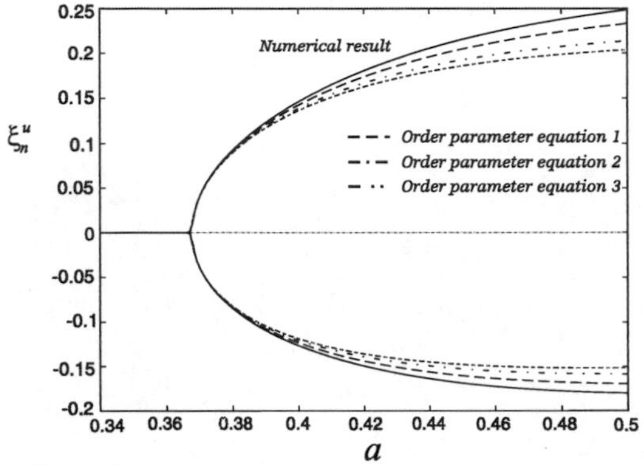

Figure 3.7: Bifurcation diagram of unstable mode of the mode equation (3.82) and the order parameter equations (3.98) with a quadratic and cubic center manifold. The order parameter 1 is obtained with the cubic center manifold, the order parameter 2 - with the quadratic center manifold and the order parameter 3 is based on the adiabatic elimination procedure (from [161]).

The projection of the numerically obtained bifurcation diagram of the mode Eqs. (3.82) (see Fig. 3.5) on the plane ξ_n^u, ξ_n^s show the dependence between the variables ξ_n^u and ξ_n^s. This dependence should coincide with that of Eq. (3.97) at least in the vicinity of $\xi_n^u = 0$. The dependence between the variables ξ_n^u, ξ_n^s, obtained numerically and analytically are shown in Fig. 3.6. Substituting Eq. (3.97) in the second part of Eqs. (3.82) we yield the order parameter equation

$$\xi_{n+1}^u = \lambda_u \xi_n^u + \frac{a}{b(\lambda_s - \lambda_u)} \left(\lambda_s A_2 (\xi_n^u)^2 + \lambda_s A_3 (\xi_n^u)^3 + \lambda_u \xi_n^u \right)^2, \tag{3.98}$$

where the coefficients A_2, A_3 are given by the expressions (3.96). The bifurcation diagrams of ξ_n^u with respect to the parameter a of the mode equations, the adiabatic approximation (3.88) and the order parameter equation with a quadratic and a cubic center manifold are shown in Fig. 3.7. Numerical results of Eq. (3.98) show, that our assumption $\xi_n^u \sim \varepsilon^{\frac{1}{2}}$ in the vicinity of the bifurcation point is satisfied (see Table 5.1 in Chap. 5).

3.8 Summary

In this chapter, we considered a derivation of the order parameter from a simple non-interacting analytical agent. A mathematical model of such an agent has been represented by the well-known Hénon map. Based on this map, two important methodological points have been clearly demonstrated. First, even a simple single agent undergoes qualitative changes in its own dynamics. Derived OPs describe these changes. The phenomenon of collective behavior can be thought of as a dedicated synchronization between different OPs when all agents coherently undergo behavioral changes. The goal of further investigations is to show how the OPs of a single agent are changed during interactions by the agent's social abilities. Second, the structure shown in Fig. 3.2 and its mathematical model (3.68) belong to different levels of system simplification. The difference between them is that actors are completely eliminated at the mathematical level, whereas sensors are involved simply as parameters in the model (3.68), without considering their nature. In this way, an algorithmic simplification is performed, whereby we obtain a mathematical model. A detailed treatment of this topic is considered in Chap. 6.

In this chapter, we also considered several methods of a local analysis for time-discrete dynamical systems. These methods present a systematic method for the dimension reduction and to a structural simplification of an initial nonlinear system and the derivation of the order parameter equation(s). Subsequently, the approximation accuracy and computational complexity of these methods are compared.

The CM approach and NF reduction in the discrete case appreciably diverges from the continuous. That is primarily caused by critical values of the unstable eigenvalues taking both $\lambda_u = 1$ and $\lambda_u = -1$. As turned out, the CM $h(\xi_n^u, \varepsilon_n)$ can be represented by a polynomial function of either ξ_n^u, ε_n or only ξ_n^u, where polynomial coefficients depend generally on ε_n. Considering the asymptotic dependence $\xi^u = f(\{\beta\}, \gamma)$, where $\{\beta\}$, γ are a priori known coefficients, the third polynomial function for the CM reduction in the form $h(\xi^u)$ has been obtained. As shown, the computational complexity is minimal in the last case. In contrast to the CM approach, the NF reduction is more formalized in a computational sense, but its accuracy remains minimal, compared with CM approaches.

Chapter 4

Interacting agents

All perception of truth is the detection of an analogy...
Henry David Thoreau
The Journal of Henry D. Thoreau
(Volume II September 5, 1851 (p. 463)).

This chapter addresses interacting agents. Through interactions, they simultaneously undergo changes in qualitative behavior so that the whole group turns over into a new state. From the viewpoint of synergetics such a simultaneous change of qualitative dynamics is described by the group's order parameter(s). These "collective" OPs are in fact the OPs contained in single agents that are modified by interactions. Investigation of the group's OPs reveals the influence of cooperative activity caused by interactions in a coalition of agents. In particular, there are four mechanisms of such influence treated in this chapter.

4.1 Two interacting agents

Interaction is a kind of action that occurs as two or more objects have an effect on one another [164]. As demonstrated in Chap. 3, a single agent can possess complex dynamics, for which the order parameters can be derived. We are interested in the case in which a few simple agents interact, while still retaining their autonomy. These agents, as a common system, behave in an ordered way and exhibit so-called "collective behavior". The order parameters of overall system can be thought of as OPs from a single agent that is modified by the agent's social capabilities[1]. Therefore, the order parameters are worth investigating in order to understand how the interactions influence the collective properties of such systems.

We start with the simplest case, in which only two agents are involved, as shown in Fig. 4.1. There are two methods of interaction to be considered. In the first, both agents have equal rights and each can influence the internal processes of its neighbor. In the second, agents are not always equal within the group, therefore one can exert a small influence or control over another. Each of these interaction schemes therefore has a regulator, and it must be emphasized that the interaction pattern is not predetermined: that is, it can be influenced by for example, the environment or another agent. By means of an appropriate regulator we can set for example which of the interaction schemes (or both simultaneously) is to be activated. The internal structure of the interaction scheme has

[1] This assumption is valid for homogeneous systems and for so-called kernels in the heterogeneous case.

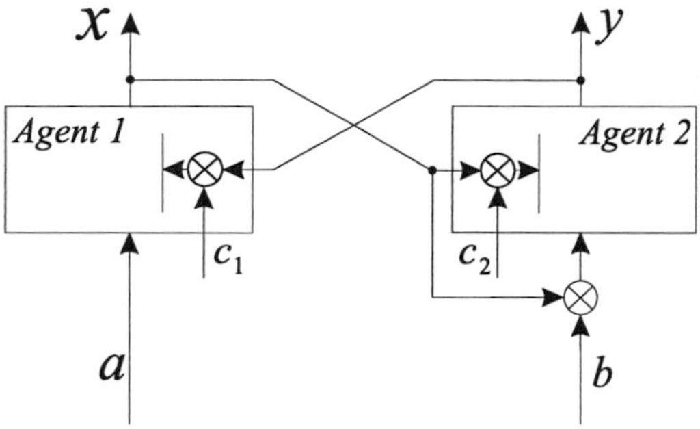

Figure 4.1: Two autonomous systems incorporated into a common structure. There are two ways of interaction when each agent can influence some internal process of neighbor and when one of them can exert a small influence on a control of other.

to be chosen such that first, the behavior of both agents is independent of the environment, and second, that if the first interaction scheme is terminated, the autonomy of both agents is not disrupted. In this way, both agents remain autonomous, even though they are integrated into a common system.

An essential point of this structure is a concrete model used for agent modeling, see Fig. 4.1. The question the investigations undertaken seek to answer is how the interactions modify the OP of the overall system. The concrete agent's model is not so important regarding this question, because the results obtained are expected to be valid for whatever model we use. Therefore, it makes sense to choose as simple a model as possible, in order to demonstrate the methodology of the results. The $Simulink^{©}$ simulation model of both agents is shown in Fig. 4.2. The bold lines depict the feedback that provides autonomy for the agents, and the shadowed quadrants show elements of the interaction scheme. Sensors 1, and 4 can be viewed in the sense of the perceiving elements, whereas sensors 2, and 3 are primarily destined for modification of the interactions. Certainly, all can be considered as control elements for modules A1, and A2. Two oscilloscopes are the actors in this simulation. Following the definition of an agent and taking into account the similar case shown in Chap. 3, both agents in Fig. 4.2 can be denoted as autonomous, rational (pre-rational), social, and non-intelligent.

In seeking an appropriate mathematical agent model, our attention was attracted to numerous publications (see [135], [165]) having as their subject coupled logistic maps (see [166], [167]). Possessing diverse dynamics and being simple, these systems can be seen as *sui generis* test systems [142]. Therefore, wishing to use the simplest model for the agents, we assume the logistic map to be such a model. First, as already mentioned, this map is well investigated, and second, we achieve a two-dimensional coupled system, to which the computational techniques shown in Chap. 3 can be applied. Moreover, we would like to demonstrate the accuracy of transformation, explained theoretically in Chap. 3, by increasing the order of CM approximation and NF reduction.

Thus, a mathematical model of the functional structure shown in Fig. 4.1 and whose simulation model shown in Fig. 4.2 is a pair of logistic map fused multiplicatively and additively in the following way:

$$x_{n+1} = c_1 y_n + a x_n (1 - x_n), \tag{4.1a}$$

$$y_{n+1} = c_2 x_n + b x_n y_n (1 - y_n), \tag{4.1b}$$

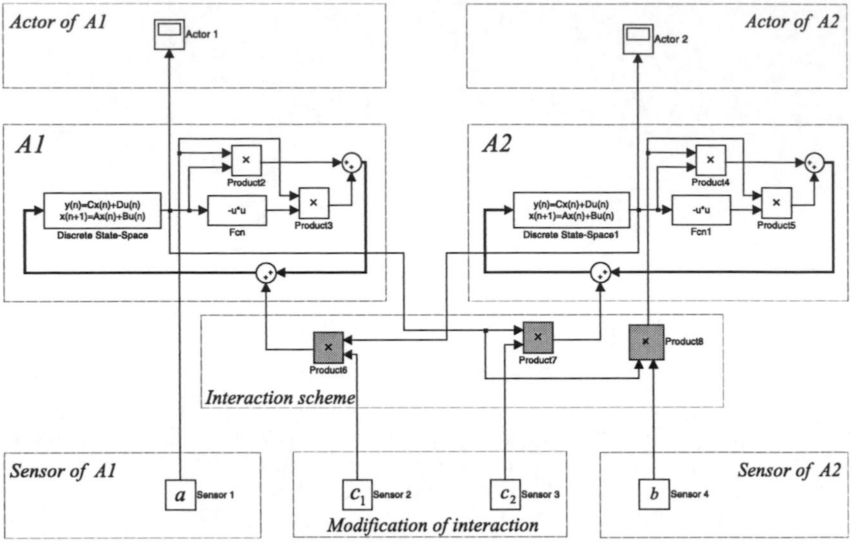

Figure 4.2: The $Simulink^{©}$ simulation model of the structure shown in Fig. 4.1 (see description in text).

where $x_n \in \mathbb{R}$, $y_n \in \mathbb{R}$, c_1, c_2 are the coefficients of linear coupling, namely $c_1 = c_2 = c$ for further simplification, b is the coefficient of nonlinear coupling, and a is the general bifurcation parameter. Looking at the system (4.1), we ensure it satisfies the requirement for autonomy, since at $c = 0$ the equation (4.1a) is uncoupled from (4.1b), whereas term bx_n in (4.1b) can be viewed as a control parameter for this equation.

4.2 Synergetic analysis

In this section, all stages of the above-mentioned synergetic analysis will be applied to the investigation of the system (4.1). At different values of coefficients c, b this system exhibits different types of dynamical behavior, including the well-know period-doubling scenario and quasi-periodic route to chaos, some bifurcations and their unfolding with critical eigenvalue $+1$, -1, a complex-conjugate, and even an appearance of two critical eigenvalues. In this context, the system (4.1) presents itself as a successful object for an application and appraisal of various theoretical methods analyzing local and global dynamics. Further, the system (4.1) is denoted as an ordinary logistic-logistic map (OLL) and can appear as a *sui generis* test system.

The synergetic investigations of the OLL map are intended to show modifications of the order parameter caused by changes of couplings. Moreover, we intend to demonstrate a mechanism of this modification independently from the current system and couplings. Following this strategy, the common OP of (4.1) can be thought of as the OP of the single system (4.1a) modified by the couplings. When coefficients $c = 0$ and (4.1a) are uncoupled from (4.1b), such an initial order parameter may easily be derived. More difficult cases arise when these coefficients are so small that they in fact do not perturb the dynamics of (4.1) and therefore the dynamics are not distinguishable from the dynamics of the initial logistic maps, although the systems (4.1a) and (4.1b) remain coupled. In this case, we assume that the OP derived from the coupled system (4.1) at small c presents such an initial OP. The desired behavior of (4.1) with transcritical and period-doubling bifurcations intended for a derivation of OP is obtained at values $c = 0.1$, $b = 1$ and is shown in Fig. 4.5(a).

After this OP is obtained, its modification can be shown in two ways. First, we can vary numerical the values of the coefficients b and c and then look at the bifurcation diagrams arising. Second, we can try to explain these modifications analytically, deriving the mechanism of the OP's modification. We follow the second method, which is shown in Sec. 4.3, and return to the derived OP of the OLL map in Sec. 4.3.2.3 to verify the theoretical explanations.

4.2.1 The linear stability analysis

At different values of coefficients c, b the system (4.1) exhibits qualitatively different types of behavior including the period-doubling scenario, quasi-periodic route to chaos, transcritical, period-four, Hopf, shifted saddle-node bifurcations and so on. Therefore we have first to restrict such areas of parameters values that behavior of (4.1) will be further separately investigated in. In Fig. 4.3 the region of parameters $b = [-2...2]$ and $c = [-1...1]$ at two-dimensional plane is divided into circa one hundred points where the qualitative behavior of (4.1) is tentatively categorized. In the central area of the parameters space

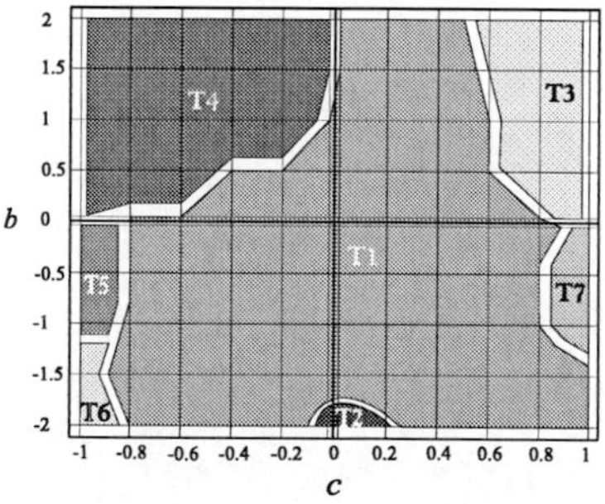

Figure 4.3: The approx. regions of qualitatively different behavior of system (4.1) in dependence of the parameters b, c. The behavior types T1, T2 correspond to the bifurcation diagrams shown in Fig. (4.4) (a)-(b), types T3,T4 - Fig. (4.4) (c)-(d) and finally T5,T6,(T7) - Fig. (4.4) (e)-(f).

in Fig. 4.3, the dynamics of system (4.1) has only small perturbations. Therefore, the behavior in this area is similar to the qualitative behavior, containing in original logistic maps. Further towards boundaries of Fig. 4.3 the perturbations of NF are growing and the system (4.1) begins to show completely new behavior types.

Generally speaking, the coupling modifies in fact all elements of the dynamics of (4.1), but taking into account only a local bifurcation, we can distinguish six types of behavior shown in Fig. 4.4(a)-(f). In order to characterize these types of dynamics we first perform the linear stability analysis.

The stationary states $\underline{q}_{st} = (x_{st}, y_{st})$ for the Eqs. (4.1) is given by the trivial values

$$x_{st_{(0)}} = 0, \; y_{st_{(0)}} = 0 \,, \tag{4.2}$$

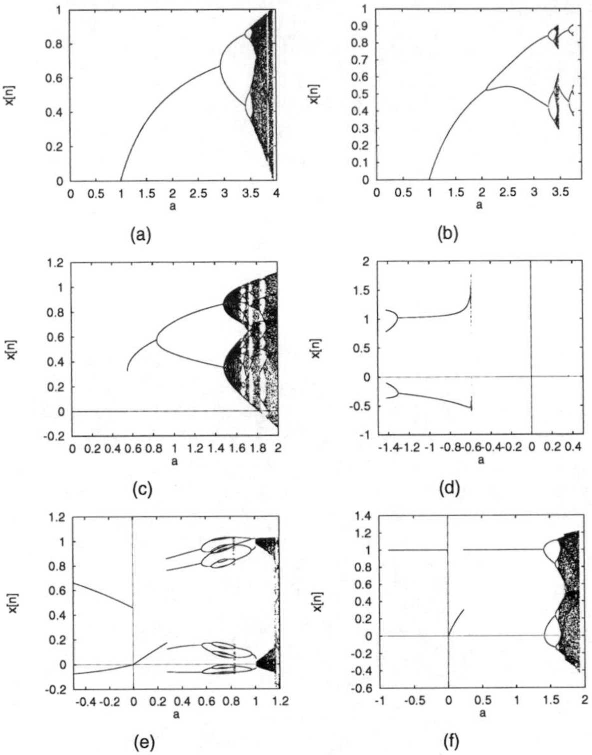

Figure 4.4: (a) The bifurcation diagram of the OLL map (4.1) at parameters $b = 1$, $c = 0.1$, $x_0 = 0.1$. It contains the transcritical (at $a_{cr} = 0.99$) and the period-doubling bifurcation with period-doubling scenario, occurring in the logistic map; **(b)** The bifurcation diagram of the OLL map at parameters $b = -2$, $c = 0.1$, $x_0 = 0.4$. It contains the same bifurcation types, but in this case the perturbation of global dynamics is observed; **(c)** The bifurcation diagram of the OLL map at parameters $b = 2$, $c = 0.6$, $x_0 = 0.4$. In this case there are the perturbed bifurcation with NF (4.116a) up to the sixth order (at $a_{cr} = 0.64$), the Neimark-Sacker bifurcation with the quasiperiodicity and then period-doubling scenario; **(d)** The bifurcation diagram of the OLL map at parameters $b = 0.5$, $c = -0.8$, $x_0 = 0.01$. The same bifurcation type as in (c), but in this case the nonlocal stationary state is unstable, causing hereby the loss of dissipation in system (4.1) after the bifurcation point $a_{cr} = 0.36$; **(e)** The bifurcation diagram of the OLL map at parameters $b = -1.5$, $c = -1$, $x_0 = 0.1$. Here at $a = 0$ one can observe a period-two motion, that is abruptly jumped on a nonperiodical motion and then on period-four motion. The bifurcation at $a_{cr} = 0$ is caused by two unstable eigenvalues $\lambda_{u_{1,2}} = \pm 1$ with two-dimensional normal form (4.118); **(f)** The bifurcation diagram of the OLL map at parameters $b = -0.5$, $c = -1$, $x_0 = 0.1$. At $a_{cr} = 0$ the same bifurcation as in (e), but one can observe the stabilization of periodical motion on the values ± 1. These Figures are published in [168].

and nontrivial values that for $x_{st_{(1,2,3,4)}}$ are determined by roots of the following equation

$$\begin{aligned} x_{st_{(1,2,3,4)}} &= a^2bx^4 + (2a^2b - 2ab)x^3 + (b - abc + a^2b - 2ab)x^2 + \\ &+ (abc - bc + ac)x - c^3 - ac + c), \\ y_{st_{(1,2,3,4)}} &= x_{st_{(1,2,3,4)}}(ax_{st_{(1,2,3,4)}} - a + 1)/c. \end{aligned} \tag{4.3}$$

Forasmuch as the symbolic solution of the polynomials (4.3) is very lengthy, the stationary states $x_{st_{(1,2,3,4)}}$, $y_{st_{(1,2,3,4)}}$ and the corresponding eigenvalues $\lambda_{1_{(1,2,3,4)}}$, $\lambda_{2_{(1,2,3,4)}}$ will be calculated numerically with the fixed parameters b and c. Then, this numerical solutions will be substituted everywhere, where there are the symbols x_{st}, y_{st} and λ_1, λ_2. Thus, on the one hand we show in fact always the analytical expressions for the above-mentioned approach, on the other hand for these analytical expressions are received only numerical results.

The Jacobian $\underline{\underline{\mathbf{L}}}$ of system (4.1) evaluated at the stationary states x_{st}, y_{st} is given by

$$\underline{\underline{\mathbf{L}}} = \begin{bmatrix} a\,(1 - 2x_{st}) & c \\ c + by_{st}\,(1 - y_{st}) & bx_{st}\,(1 - 2y_{st}) \end{bmatrix}, \tag{4.4}$$

and the corresponding eigenvectors are defined by the following expressions:

$$\begin{aligned} \lambda_{1,2} = -ax_{st} - bx_{st}y_{st} + \frac{1}{2}(a + bx_{st}) \pm \frac{1}{2}(4a^2x_{st}^2 - 4a^2x_{st} + a^2 - 8abx_{st}^2y_{st} + 4abx_{st}^2 + \\ + 4abx_{st}y_{st} - 2abx_{st} + 4c^2 - 4b^2x_{st}^2y_{st} + 4b^2x_{st}^2y_{st}^2 - 4cby_{st}^2 + b^2x_{st}^2 + 4cby_{st})^{\frac{1}{2}}, \end{aligned} \tag{4.5}$$

and the corresponding eigenvectors are

$$\underline{\mathbf{v}}_1 = \begin{pmatrix} -\dfrac{\lambda_1 + 2bx_{st}y_{st} - bx_{st}}{-c - by_{st} + b(y_{st})^2} \\ 1 \end{pmatrix}, \underline{\mathbf{v}}_2 = \begin{pmatrix} -\dfrac{\lambda_2 + 2bx_{st}y_{st} - bx_{st}}{-c - by_{st} + b(y_{st})^2} \\ 1 \end{pmatrix}. \tag{4.6}$$

Now we separately investigate the cases shown in Fig. 4.4(a)-(f).

Case I, Fig. 4.4 (a).
From the linear stability analysis follows, that in the first area of qualitative behavior($b = 1$, $c = 0.1$) the system (4.1) undergoes the transcritical bifurcation ($a = a_{cr}^{I} = 0.99$) and the period-doubling one ($a = a_{cr}^{II} = 2.938298$), that we are interested in. The numerical values of the eigenvalues (4.5) together with the stable states (4.3) are shown in Fig. 4.5(b)-(d).

In the case of the transcritical bifurcation, the first eigenvalue λ_1 is stable and the second eigenvalue λ_2 is unstable; in the case of the period-doubling bifurcation, the situation is vice versa, namely λ_1 is stable and accordingly λ_2 is unstable as shown in Fig. 4.5(d).

This is the area where the OP of coupled system (4.1) is similar to the OP containing in the initial logistic map. Therefore, the desired in the first phase "initial" OP will be obtained for this type of behavior. In the cases III and IV this "initial" OP is already influenced by couplings, therefore these cases are intended to show the modification of OP and to exemplify the made in Sec. 4.3 theoretical proposals about a mechanism of modifications.

Case II, Fig. 4.4 (b).
This is a similar case to Fig. 4.4(a), where the local bifurcations are not changed, but the coupling modified the global bifurcations.

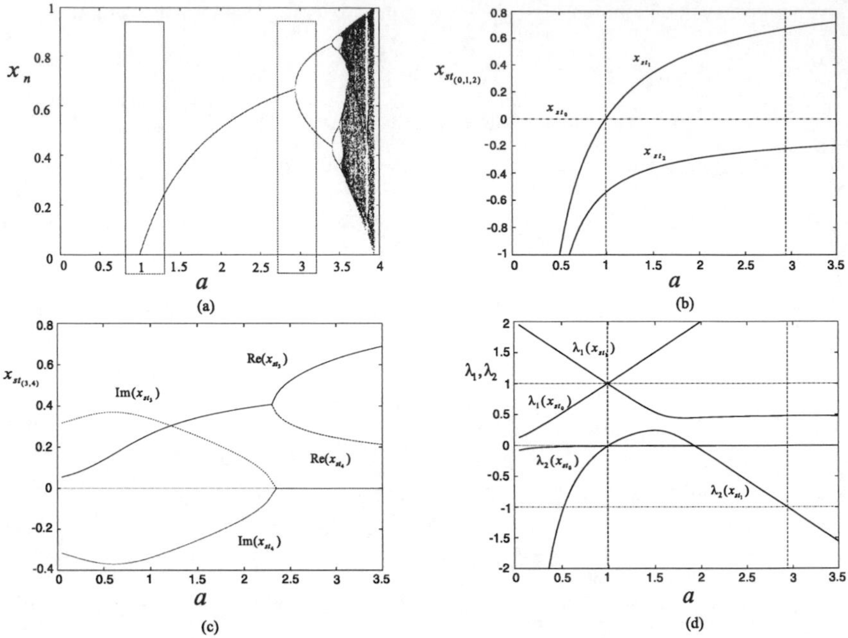

Figure 4.5: Relevant stationary states and eigenvalues of system (4.1); **(a)** The bifurcation diagram of the OLL map (4.1) (variable x_n) at values $c = 0.1$, $b = 1$; **(b)** Real stable states $x_{st_{(0,1,2)}}$; **(c)** Imaginary stable states $x_{st_{(3,4)}}$; **(d)** Relevant eigenvalues λ_1, λ_2 of system (4.1)

Case III, Fig. 4.4 (c),(d).
Next type of behavior, intended to a further investigation, arises at parameters $b = 2$, $c = 0.6$ and shown in Fig. 4.4(c). The numerical values of the eigenvalues (4.5) together with the stationary states (4.3) in dependence on the bifurcation parameter a are shown in Fig. 4.14(a). As we can see, here arises the similar to discussed in Sec. 4.3.2.1 case, when at unstable eigenvalues $\lambda_u = 1$ ($a_{cr} = 0.64$) the system possesses the complex stationary states $x_{st_{3,4}}$ connected to the origin and is in fact forced to jump on the nonlocal real state x_{st_3}.

The similar case is shown in Fig. 4.4(d) at values $b = 0.5$, $c = -0.8$. Here at unstable eigenvalues $\lambda_u = 1$ ($a_{cr} = 0.36$) the system undergoes the same type of bifurcation, but the real stationary state x_{st_3} is unstable. As a result the system (4.1) immediately after the bifurcation loses dissipation. Both instabilities are described by the same normal form shown in Sec. 4.3.2.3.

Case IV, Fig. 4.4 (e),(f).
The third case is less trivial and illustrates the last possibility of a modification of normal forms, caused by the couplings, namely, a change of the dimension in NFs. This case is shown in Fig. 4.4(e),(f). Here the eigenvalues at the zero stationary states are given by

$$\lambda_{1,2} = \frac{a \pm \sqrt{(a^2 + 4c^2)}}{2}. \tag{4.7}$$

Because of zero stationary states, the coupling coefficient b except from the expression (4.5). Choosing the coupling coefficient $c = 1$ or $c = -1$, we get hereby at the critical value of the bifurcation parameter $a_{cr} = 0$ two unstable eigenvalues of the OLL map (4.1)

$$\lambda_{u1,2} = \pm 1. \tag{4.8}$$

The eigenvalues $\lambda_{1,2}$ (4.5) evaluated numerically on x_{st_1}, y_{st_1} at $b = -1.5$, $c = -1$ are also equal to ± 1 at $a_{cr} = 0$. They, together with the relevant stationary states x_{st_0}, x_{st_1} are shown in Fig. 4.14(b). Thus we get here two unstable eigenvalues that cause two-dimensional normal form. The derivation of this NF is shown in Sec. 4.3.2.3.

4.2.2 The mode amplitude equations

For the further analysis, the original system (4.1) should be transformed in the form of the mode amplitude equations [43]. By means of this transformation the system (4.1) is decomposed into "fast and slow components", the so-called stable and unstable modes [13]. Further, using the slaving principle (3.5), it is possible to replace, in the vicinity of the instability, the "fast" stable modes through the "slow" unstable modes and to receive the order parameter equation, as it is pointed out in Sec. 3.2.

For this transformation the compact tensor notation is used

$$\underline{\mathbf{q}}_{n+1} = \Gamma_{(2)}(:\underline{\mathbf{q}}_n) + \Gamma_{(3)}(:\underline{\mathbf{q}}_n)^2 + \Gamma_{(4)}(:\underline{\mathbf{q}}_n)^3, \tag{4.9}$$

where $\Gamma_{(n)}$ is the tensor of rank n. The non-vanishing tensors components for the system (4.1) are shown in Table 4.1.

$\Gamma_{(2)}$	$\Gamma_{(3)}$	$\Gamma_{(4)}$
$(\Gamma_{(2)})_{11} = a$	$(\Gamma_{(3)})_{111} = -a$	$(\Gamma_{(4)})_{2122} = -b$
$(\Gamma_{(2)})_{12} = c$	$(\Gamma_{(3)})_{212} = b$	
$(\Gamma_{(2)})_{21} = c$		

Table 4.1: The non-vanishing tensors components of the system (4.1)

Inserting $\underline{\mathbf{q}}_n = \underline{\mathbf{q}}_{st} + \Delta\underline{\mathbf{q}}_n$, Eq. (4.9) is rewritten concerning the small derivation from the stationary states

$$\Delta\underline{\mathbf{q}}_{n+1} = \Gamma^L_{(2)}(:\Delta\underline{\mathbf{q}}_n) + \Gamma^N_{(3)}(:\Delta\underline{\mathbf{q}}_n)^2 + \Gamma^N_{(4)}(:\Delta\underline{\mathbf{q}}_n)^3, \tag{4.10}$$

where $\Gamma^L_{(2)}$ is given by the Jacobian of the system (4.1) and $\Gamma^N_{(i)}$ represent the nonlinear part. The non-vanishing components of equation (4.10) are shown in Table 4.2. The next

$\Gamma^L_{(2)}$	$\Gamma^N_{(3)}$	$\Gamma^N_{(4)}$
$(\Gamma^L_{(2)})_{11} = a(1-2x_{st})$	$(\Gamma^N_{(3)})_{111} = -a$	$(\Gamma_{(4)})_{2122} = -b$
$(\Gamma^L_{(2)})_{12} = c$	$(\Gamma^N_{(3)})_{212} = b(1-2y_{st})$	
$(\Gamma^L_{(2)})_{21} = c + by_{st}(1-y_{st})$	$(\Gamma^N_{(3)})_{222} = -bx_{st}$	
$(\Gamma^L_{(2)})_{22} = bx_{st}(1-2y_{st})$		

Table 4.2: Non-vanishing tensors components of Eq. (4.10)

step consist of performing of the following coordinate transformation

$$\Delta\underline{\mathbf{q}}_n = \sum_{k=1}^{2} \xi^k_n \underline{\mathbf{v}}_k = \Gamma^V_{(2)}(:\underline{\boldsymbol{\xi}}_n), \tag{4.11}$$

where $\Gamma^V_{(2)}$ is the tensor of eigenvectors $\underline{\mathbf{v}}_k$ defined by (4.6). Substitution of the expression (4.11) into Eq. (4.10) gives the following equation

$$\Gamma^V_{(2)}(:\underline{\boldsymbol{\xi}}_{n+1}) = \Gamma^L_{(2)}\Gamma^V_{(2)}(:\underline{\boldsymbol{\xi}}_n) + \Gamma^N_{(3)}(\Gamma^V_{(2)}(:\underline{\boldsymbol{\xi}}_n))^2 + \Gamma^N_{(4)}(\Gamma^V_{(2)}(:\underline{\boldsymbol{\xi}}_n))^3. \tag{4.12}$$

Multiplying now from the left side with the inverse tensor $\Gamma^{V^{-1}}_{(2)}$, the mode equations have finally the following form

$$\underline{\boldsymbol{\xi}}_{n+1} = \Gamma^\Lambda_{(2)}(:\underline{\boldsymbol{\xi}}_n) + \Gamma^{\tilde{N}}_{(3)}(:\underline{\boldsymbol{\xi}}_n)^2 + \Gamma^{\tilde{N}}_{(4)}(:\underline{\boldsymbol{\xi}}_n)^3, \tag{4.13}$$

where $\Gamma^\Lambda_{(2)}$ is the tensor of the eigenvalues [2] (4.5) (the non-vanishing components are $(\Gamma^\Lambda_{(2)})_{11} = \lambda_1$, $(\Gamma^\Lambda_{(2)})_{22} = \lambda_2$). Hereby is used

$$(\Gamma^{V^{-1}}_{(2)}\Gamma^V_{(2)})_{ij} = \delta_{ij}, \tag{4.14}$$

where δ is the well-known Kronecker-Symbol. With respect to the tensor components, summarized in Table 4.2, the equation (4.13) reads in detail

$$\xi^1_{n+1} = \lambda_1\xi^1_n + \frac{P\mathcal{W}^2 a - N_1\mathcal{L}b[-x_{st}\mathcal{L} + \mathcal{W}(1-2y_{st}-\mathcal{L})]}{\lambda_1-\lambda_2}, \tag{4.15a}$$

$$\xi^2_{n+1} = \lambda_2\xi^2_n - \frac{P\mathcal{W}^2 a - N_2\mathcal{L}b[-x_{st}\mathcal{L} + \mathcal{W}(1-2y_{st}-\mathcal{L})]}{\lambda_1-\lambda_2}, \tag{4.15b}$$

where

$$\mathcal{W} = -\frac{N_2\xi^1_n + N_1\xi^2_n}{P}, \quad \mathcal{L} = \xi^1_n + \xi^2_n, \quad P = -c - by_{st}(1-y_{st}),$$
$$N_1 = \lambda_2 + bx_{st}(2y_{st}-1), \quad N_2 = \lambda_1 + bx_{st}(2y_{st}-1).$$

In the equations (4.15) the modes ξ^1_n, ξ^2_n will be denoted as the stable ξ^s_n and unstable modes ξ^u_n in accordance with the stable λ_s and unstable eigenvalues λ_u standing in the linear part; namely $\xi^u_n = \xi^1_n$, $\xi^s_n = \xi^2_n$ for the transcritical bifurcation and $\xi^u_n = \xi^2_n$, $\xi^s_n = \xi^1_n$ for period-doubling one.

Both modes $\xi^{s,u}_n$ have the intrinsic dynamics which is governed by the eigenvalues and the external dynamics which is defined by the nonlinear coupling with the other mode. Because the unstable eigenvalue is $|\lambda_u| \approx 1$ near the bifurcation, the intrinsic dynamics of the unstable mode ξ^u_n varies slowly and therefore ξ^u_n is called the slow mode and correspondingly ξ^s_n the fast mode.

There is a correlation between both modes, which is mathematically expressed by the center manifold theorem. Further, according to the approach explicated in Sec. 3.2, the equation of the stable mode ξ^s_n will be used to obtain the center manifold and the equation of the unstable mode ξ^u_n for the derivation of the order parameter equation.

As already mentioned in Sec. 3.3.3 (see Fig. 3.5) the curves, which approximate the center manifold and the asymptotic behavior of the order parameter, can be obtained numerically from the mode equations as dependencies $\xi^s_n(\xi^u_n)$ and $\xi^u_n(a)$. Thus, these curves will be used further for the verification of the analytical results in the following sections.

[2] In general tensor $\Gamma^\Lambda_{(2)}$ represents the eigenvalues in Jordan normal form, see [43].

4.2.3 The center manifold approach and derivation of the order parameter equation

As it is shown by the linear analysis in 4.2.1, depending on the parameter a, the unstable eigenvalue λ_1 and accordingly the mode ξ_n^1 correspond to the transcritical bifurcation and the unstable λ_2, ξ_n^2 correspond to period-doubling one. Further, for the both instabilities we get two order parameter equations derived by means of the center manifold's reduction described in Sec. 3.3.1 and Sec. 3.3.3. For the coefficients of the center manifold up to the third order analytical expressions are given. The analytical expressions for coefficients up to sixth order are very lengthy therefore we show only numerical values computed from the analytically derived expressions.

4.2.3.1 The center manifold and order parameter equation for the transcritical bifurcation (ξ_n^1 is the unstable mode)

The center manifold using $h(\xi_n^1, \varepsilon_n)$

As the first calculation procedure, we apply the parameter dependent center manifold reduction $h(\xi_n^1, \varepsilon_n)$ described in Sec. 3.3.1, supposing that ξ_n^1 is the unstable mode, which corresponds to the transcritical bifurcation. As already mentioned in Sec. 3.3, for a bifurcation with a normal form up to third order the center manifold based on the second-order polynomial is minimally needed

$$\xi_n^2 = h(\xi_n^1, \varepsilon_n) = A_2(\xi_n^1)^2 + B_2(\xi_n^1)\varepsilon_n + C_2\varepsilon_n^2 + O(3). \tag{4.16}$$

Substituting the right side of expression (4.16) accordingly into Eqs. (4.15), we obtain equation (3.17) up to order $O(3)$ from which the coefficients A_2, B_2, C_2 of the center manifold (4.16) can be determined:

$$\begin{aligned}
&C_2(\lambda_\varepsilon)^2\varepsilon_n^2 + B_2\lambda_\varepsilon\varepsilon_n\xi_n^1\lambda_1^{(0)} + A_2(\xi_n^1)^2(\lambda_1^{(0)})^2 = \lambda_2^{(0)}A_2(\xi_n^1)^2 - \\
&-S^{(0)}a_{cr}P^{(0)}(\xi_n^1)^2(N_2^{(0)})^2(M^{(0)})^2 - S^{(0)}b(\xi_n^1)^2N_2^{(0)}x_{st}^{(0)} + \lambda_2^{(0)}B_2\xi_n^1\varepsilon_n - \\
&-S^{(0)}b(\xi_n^1)^2(N_2^{(0)})^2M^{(0)} + 2S^{(0)}b(\xi_n^1)^2(N_2^{(0)})^2M^{(0)}y_{st}^{(0)} + \lambda_2^{(0)}C_2\varepsilon_n^2.
\end{aligned} \tag{4.17}$$

Collecting in this equation coefficients with the same power with respect to ξ_n^1, ε_n we end up finally with

$$(\xi_n^1)^2: \quad A_2 = \frac{N_2^{(0)}(bN_2^{(0)}(2y_{st}^{(0)} - 1) - bP^{(0)}x_{st}^{(0)} - a_{cr}N_2^{(0)})}{((\lambda_1^{(0)})^2 - \lambda_2^{(0)})(\lambda_1^{(0)} - \lambda_2^{(0)})P^{(0)}}, \tag{4.18a}$$

$$\xi_n^1, \varepsilon_n: \quad B_2 = 0, \tag{4.18b}$$

$$\varepsilon_n^2: \quad C_2 = 0. \tag{4.18c}$$

Note, that λ_ε (see Eq. (3.14)) is not represented in the expression for coefficient A_2. On this ground it can be asserted that the second-order center manifold (4.16) is unsusceptible to a choice of λ_ε.

As already pointed out in Sec. 3.3 we purpose to show the accuracy of the approximation of the center manifold. To this effect the center manifold is calculated using the polynomial of the fifth order

$$\begin{aligned}
\xi_n^2 = h(\xi_n^1, \varepsilon_n) &= A_2(\xi_n^1)^2 + B_2(\xi_n^1)\varepsilon_n + C_2\varepsilon_n^2 + A_3(\xi_n^1)^3 + B_3(\xi_n^1)^2\varepsilon_n + C_3(\xi_n^1)\varepsilon_n^2 + D_3\varepsilon_n^3 + \\
&+A_4(\xi_n^1)^4 + B_4(\xi_n^1)^3\varepsilon_n + C_4(\xi_n^1)^2\varepsilon_n^2 + D_4(\xi_n^1)\varepsilon_n^3 + E_4\varepsilon_n^4 + A_5(\xi_n^1)^5 + B_5(\xi_n^1)^4\varepsilon_n + \\
&+C_5(\xi_n^1)^3\varepsilon_n^2 + D_5(\xi_n^1)^2\varepsilon_n^3 + E_5(\xi_n^1)\varepsilon_n^4 + F_5\varepsilon_n^5 + O(6).
\end{aligned} \tag{4.19}$$

The numerical calculated coefficients of (4.19) are summarized in Table 4.5, considering both cases $\lambda_\varepsilon = 1$ and $\lambda_\varepsilon = -1$. We can see, an influence of $\lambda_\varepsilon = -1$ exists only in coefficients B and C which correspond to the variable ε_n. Consequently, in the vicinity of the bifurcation λ_ε is incidental, but further away from the bifurcation point the dominance of this term gradually grows.

Now, using the expression (4.16) with coefficients (4.18), the required order parameter equation can be derived from the equation of the unstable mode (4.15a), replacing the stable variable ξ_n^2 as it is shown in end of this subsection.

The center manifold using $h(\xi^1)$

Here for the calculation of the center manifold we use the approach based on $h(\xi^1)$ and $\xi^1 = f(\{\beta\}, \gamma, \varepsilon)$ described in Sec. 3.3.3. In this case we need a priori information about the coefficients $\{\beta\}, \gamma$. These coefficients can be obtained by two different ways. It is possible either to calculate this coefficients numerically, using the original system (4.1) or to use the mode equations (4.15) and to get these values analytically. We will show the both ways and compare the results.

The numerical calculated coefficients $\{\beta\}, \gamma$ using assumption

$$x = \beta_x \varepsilon^{\gamma_x} + O(r), \quad y = \beta_y \varepsilon^{\gamma_y} + O(r) \tag{4.20}$$

in the vicinity after the transcritical bifurcation are shown in Table 4.3.

N	$a = 0.9901 - 0.991$ $step = 0.00001$	$a = 0.991 - 1$ $step = 0.0001$	$a = 0.991 - 1.1$ $step = 0.001$
$5 \cdot 10^4$	γ_x =0.99838 β_x =1.00740 γ_y =0.99875 β_y =0.10108	γ_x =0.99636 β_x =0.99533 γ_y =1.00003 β_y =0.10206	γ_x =0.98274 β_x =0.92168 γ_y =1.00002 β_y =0.10206
$2 \cdot 10^5$	γ_x =0.99963 β_x =1.01700 γ_y =1.00000 β_y =0.10204	γ_x =0.99636 β_x =0.99533 γ_y =1.00003 β_y =0.10206	γ_x =0.98274 β_x =0.92168 γ_y =1.00002 β_y =0.10206
$5 \cdot 10^5$	γ_x =0.99963 β_x =1.01700 γ_y =1.00000 β_y =0.10204	γ_x =0.99636 β_x =0.99533 γ_y =1.00003 β_y =0.10206	γ_x =0.98274 β_x =0.92168 γ_y =1.00002 β_y =0.10206

Table 4.3: The numerically calculated coefficients $\beta_{x,y}, \gamma_{x,y}$ for (4.20) in the transcritical bifurcation, occurring at $a = a_{cr}^I = \frac{99}{100} = 0.99$.

In Table 4.3, N is the number of iterations. In the intervals of the parameter a one hundred logarithmically equidistant points are chosen and the values of γ and β were obtained by a least squares fit assuming the following functional dependence $\log(x) = \gamma \log(\varepsilon) + \log(\beta)$. Considering Table 4.3, we can ascertain that the best approximation is obtained for the interval which is the closest to the bifurcation point. Therefore, we select data from the first column of Table 4.3 and from the row with the maximal amount of performed iterations. Numerically obtained data approve our assumption $\gamma = 1$ for the transcritical bifurcation. In order to obtain the value of β for ξ^u, taking into account the transformation (4.10), (4.11), we have to multiply the values (β_x, β_y) on the matrix $\underline{\underline{V}}^{-1}$ and as result we get $(\beta_{\xi^u}, \beta_{\xi^s})$=(0.101704, 0.000338).

For the analytical calculation of $\{\beta\}, \gamma$ we substitute $a = a_{cr}^I + \varepsilon$ into equations (4.10), (4.11) and then expand the nonzero stationary states (which correspond to the transcritical bifurcation) into a Taylor series with respect to ε. The values of $\{\beta\}$ calculated in this way are shown in Table 4.4. Similar calculations can be also performed for the original system (4.1), but in this case the obtained results should be transformed to the mode amplitude form with provision for expressions (4.10), (4.11).

	ε	ε^2	ε^3	ε^4	ε^5
β_{ξ^u}	0.102040	- 0.202166	0.298171	-0.388514	0.471714
β_{ξ^s}	0	0.203122	-0.308994	0.388117	0.469401

Table 4.4: The analytically calculated coefficients $\{\beta_{\xi^s}\}, \{\beta_{\xi^u}\}, \gamma$ for the transcritical bifurcation.

As we can see, the numerically and analytically obtained values of $\{\beta\}$ in the lowest order of ε coincide in three digits after the decimal point.

Continuing the calculation of the center manifold in analogy to the first part of this subsection we use the following polynomial

$$\xi^2 = h(\xi^1) = A_2(\xi^1)^2 + O(3) = A_2(\beta\varepsilon^\gamma)^2 + O(3), \tag{4.21}$$

where β ,γ are already defined. Following the described method, the equation of the stable mode (4.15b) is used for the calculation of the center manifold (4.21). At $\gamma = 1$ the equation (3.17) up to order $O(3)$ has the following form

$$\begin{aligned} A_2\varepsilon^2\beta^2(\lambda_1^{(0)})^2 = (\lambda_2^{(0)}A_2 - a_{cr}P^{(0)}(N_2^{(0)})^2(M^{(0)})^2S^{(0)} - b(N_2^{(0)})^2M^{(0)}S^{(0)} + \\ + 2b(N_2^{(0)})^2M^{(0)}y_{st}^{(0)}S^{(0)} - N_2^{(0)}bx_{st}^{(0)}S^{(0)})\varepsilon^2\beta^2. \end{aligned} \tag{4.22}$$

From this we get for the coefficient A_2

$$\varepsilon^2: \; A_2 = \frac{N_2^{(0)}(bN_2^{(0)}(2y_{st}^{(0)} - 1) - bP^{(0)}x_{st}^{(0)} - a_{cr}N_2^{(0)})}{((\lambda_1^{(0)})^2 - \lambda_2^{(0)})(\lambda_1^{(0)} - \lambda_2^{(0)})P^{(0)}}. \tag{4.23}$$

As expected, the expressions (4.23) and (4.18a) are identical. Similarly to (4.19), now we use the polynomial up to the sixth order for a numerical calculation of the center manifold's coefficients

$$\xi^2 = h(\xi^1) = A_2(\xi^1)^2 + A_3(\xi^1)^3 + A_4(\xi^1)^4 + A_5(\xi^1)^5 + O(6). \tag{4.24}$$

The numerical values of these coefficients are summarized in Table 4.5.

The CM	A_2	A_3	A_4	A_5	B_3
$h(\xi_n^1, \varepsilon_n)$, $\lambda_\varepsilon = 1$	19.50789	550.55409	18649.87982	710422.15124	-18.35856
$h(\xi_n^1, \varepsilon_n)$, $\lambda_\varepsilon = -1$	19.50789	550.55409	18649.87982	710422.15124	-18.72944
$h(\xi^1)$	19.507891	370.64012	5011.888917	-17232.88606	—

The CM	B_4	B_5	C_4	C_5	D_5
$h(\xi_n^1, \varepsilon_n)$, $\lambda_\varepsilon = 1$	-1195.19883	-66944.48165	16.32837	1825.93652	-13.34891
$h(\xi_n^1, \varepsilon_n)$, $\lambda_\varepsilon = -1$	866.75031	39904.23395	17.41909	1274.77076	15.77842
$h(\xi^1)$	—	—	—	—	—

Table 4.5: The numerical values of coefficients of the center manifold (4.19) and (4.24).

In the first and second rows of Table 4.5 the coefficients of the center manifold (4.19) are represented for the cases $\lambda_\varepsilon = 1$ and $\lambda_\varepsilon = -1$ (see Eqs. 3.14). The third row shows the coefficients of the center manifold (4.24). The coefficients A_2 of both methods are equal, this is shown analytically.

Different orders of the center manifold (4.24), together with the numerical dependence $\xi^2 = h(\xi^1)$ obtained by simulations of the mode equations (4.15), are shown in Fig. 4.6.

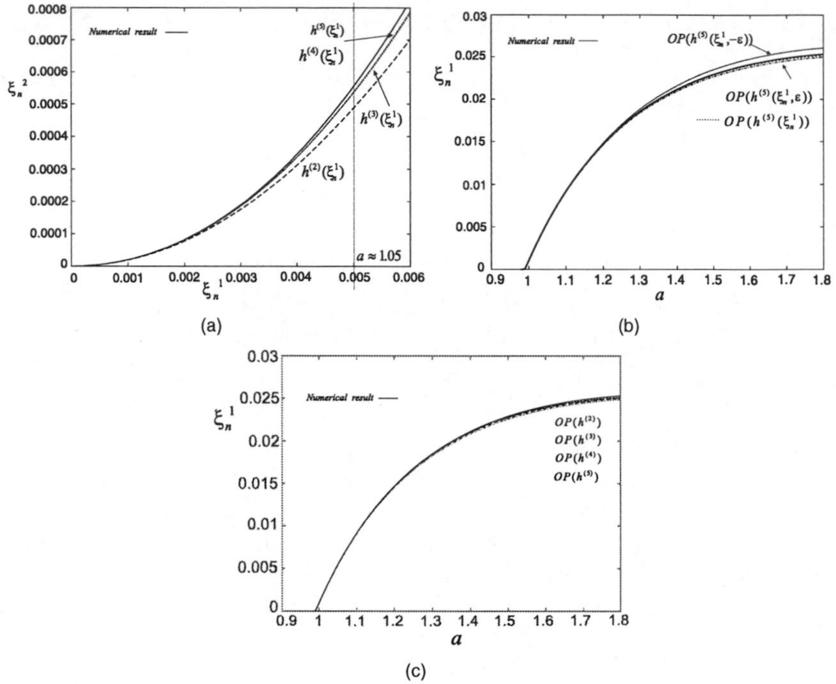

Figure 4.6: (a) The different orders of the center manifold (4.24) and the numerical obtained dependence $\xi^2 = h(\xi^1)$ from the mode equations (4.15); **(b)** The order parameter equation (4.26) using the center manifold (4.19) and the numerical dependence $\xi^1(a)$ from the mode equations (4.15). **(c)** The order parameter equation (4.26) using the center manifold (4.24) and the numerical dependence $\xi^1(a)$ from the mode equations (4.15).

Now, as it is described in Sec. 3.3.3, we are able to compare the results of both center manifold approaches obtained on the base of (4.19) and (4.24) comparing the coefficients of Taylor series of rhs and lhs of

$$\xi^s = h(\xi^u), \tag{4.25}$$

where the values of ξ^u can be calculated from (3.31) using the Table 4.4. The numerical values of Taylor series's coefficients of $h(\xi^u)$ are shown in Table 4.6 and the numerical values of the coefficients of ξ^s in Table 4.4. As we can see, between both approaches $h(\xi^u, \varepsilon)$ and $h(\xi^u)$ in fact don't exist essential distinctions.

	ε^2	ε^3	ε^4	ε^5
$h(\xi^u, \varepsilon)$	0.203122	-0.411066	0.187145	-0.497238
$h(\xi^u)$	0.203122	-0.411066	0.187145	-0.306026

Table 4.6: The coefficients of the Taylor series expansion of rhs (4.25) for the both center manifold approaches (4.19), (4.24) in the transcritical bifurcation.

The order parameter equation

Now using the determined center manifolds h, the corresponding order parameter equation can be derived from the equation (4.15a):

$$\xi^1_{n+1} = \lambda_1 \xi^1_n + \frac{P(\mathcal{W}^*)^2 a - N_1 \mathcal{L}^* b(-x_{st}\mathcal{L}^* + \mathcal{W}^*(1 - 2y_{st} - \mathcal{L}^*))}{\lambda_1 - \lambda_2}, \tag{4.26}$$

where

$$\mathcal{W}^* = -N_2 \xi^1_n - N_1 h(\xi^1_n), \quad \mathcal{L}^* = \xi^1_n + h, \quad P = -c^2 - b y_{st}(1 + b y_{st})$$
$$N_1 = \lambda_2 + 2b x_{st} y_{st} - b x_{st}, \quad N_2 = \lambda_1 + 2b x_{st} y_{st} - b x_{st}$$

and $h = h(\xi^1_n, \varepsilon_n)$ (or $h = h(\xi^1)$) is given on the basis of the expressions (4.16),(4.19) or (4.21), (4.24) accordingly. In Fig. 4.6 the numerical solution of the order parameter equation (4.26), using the center manifolds (4.16) and (4.24), and the numerical dependence $\xi^1(\alpha)$ obtained by simulations of the mode equations (4.15) are shown.

4.2.3.2 The center manifold and order parameter equation for the period-doubling bifurcation (ξ^2_n is the unstable mode)

In this subsection, the second case is considered when ξ^2_n is unstable. All calculations are performed similarly to these, described above, namely firstly the calculation of the center manifold based on $h(\xi^2_n, \varepsilon_n)$, secondly, the calculation of $h(\xi^2)$ and, finally, the derivation of the order parameter equation.

The polynomial for the center manifold analogous to (4.19) has the following form:

$$\begin{aligned}\xi^1_n = h(\xi^2_n, \varepsilon_n) = {} & A_2(\xi^2_n)^2 + B_2(\xi^2_n)\varepsilon_n + C_2\varepsilon^2_n + A_3(\xi^2_n)^3 + B_3(\xi^2_n)^2\varepsilon_n + C_3(\xi^2_n)\varepsilon^2_n + D_3\varepsilon^3_n + \\ & + A_4(\xi^2_n)^4 + B_4(\xi^2_n)^3\varepsilon_n + C_4(\xi^2_n)^2\varepsilon^2_n + D_4(\xi^2_n)\varepsilon^3_n + E_4\varepsilon^4_n + A_5(\xi^2_n)^5 + B_5(\xi^2_n)^4\varepsilon_n + \\ & + C_5(\xi^2_n)^3\varepsilon^2_n + D_5(\xi^2_n)^2\varepsilon^3_n + E_5(\xi^2_n)\varepsilon^4_n + F_5\varepsilon^5_n + O(6).\end{aligned} \tag{4.27}$$

For the second-order coefficients A_2, B_2, C_2 we obtain:

$$(\xi^2_n)^2: \quad A_2 = \frac{N_1^{(0)}(bN_1^{(0)}(2y_{st}^{(0)} - 1) - bP^{(0)}x_{st}^{(0)} - a_{cr}N_1^{(0)})}{((\lambda_1^{(0)})^2 - \lambda_2^{(0)})(\lambda_1^{(0)} - \lambda_2^{(0)})P^{(0)}}, \tag{4.28a}$$

$$\xi^2_n, \varepsilon_n: \quad B_2 = 0, \tag{4.28b}$$

$$\varepsilon^2_n: \quad C_2 = 0. \tag{4.28c}$$

The remain coefficients of (4.27) calculated numerically are represented in Table 4.8.

Accordingly, for the second approach based on Eq. (4.24) the following polynomial is used:

$$\xi^1 = h(\xi^2_n) = A_2(\xi^2)^2 + A_3(\xi^2)^3 + A_4(\xi^2)^4 + A_5(\xi^2)^5 + O(6) \tag{4.29}$$

For the analytical calculation of $\{\beta\}$, γ in the case of period-doubling bifurcation it is necessary to get the stationary states of the second iterated system of (4.15). However it is not possible because of a high complexity of these equations, therefore the values of $\{\beta\}$ and γ will be obtained only numerically by a least squares fit approach, like that presented in Sec. 4.2.3.1. Because in the period-doubling bifurcation there are lower and upper branches, we obtain correspondingly for x and y both $\gamma_{u,l}$, $\beta_{u,l}$ which are represented in Table 4.7. Since in the previous case the best approximation are found in

N	$a = 2.9383 - 2.9400$ $step = 0.00001$	$a = 2.940 - 2.950$ $step = 0.0001$
$2 \cdot 10^5$	γ_{x_u} =0.495691 β_{x_u} =0.320422	γ_{x_u} =0.480875 β_{x_u} =0.290721
	γ_{x_l} =0.501489 β_{x_l} =0.341859	γ_{x_l} =0.515767 β_{x_l} =0.376949
	γ_{y_u} =0.486993 β_{y_u} =0.045721	γ_{y_u} =0.423335 β_{y_u} =0.030097
	γ_{y_l} =0.509469 β_{y_l} =0.058762	γ_{y_l} =0.554972 β_{y_l} =0.079913
$1 \cdot 10^6$	γ_{x_u} =0.496098 β_{x_u} =0.321471	γ_{x_u} =0.480875 β_{x_u} =0.290721
	γ_{x_l} =0.501898 β_{x_l} =0.342983	γ_{x_l} =0.515767 β_{x_l} =0.376949
	γ_{y_u} =0.487399 β_{y_u} =0.045869	γ_{y_u} =0.423335 β_{y_u} =0.030097
	γ_{y_l} =0.509880 β_{y_l} =0.058956	γ_{y_l} =0.554972 β_{y_l} =0.079913

Table 4.7: The numerically calculated coefficients $\beta_{x,y}, \gamma_{x,y}$ in the period-doubling bifurcation.

the vicinity of the bifurcation, we conclude that $\gamma = 0.5$ for the period-doubling bifurcation as pointed out earlier. For the further calculations we use one of the two values $\beta_{u,l}$, for instance β_u, in the case of $N = 10^6$ iterations.

Carrying out the required calculation to obtain (4.29), we get the following second-order coefficient A_2

$$(\xi_n^2)^2 : \quad A_2 = \frac{N_1^{(0)}(a_{cr}N_1^{(0)} + b(P^{(0)}x_{st}^{(0)} + N_1 - 2N_1^{(0)}y_{st}^{(0)})}{(\lambda_s^{(0)} - (\lambda_u^{(0)})^2)(\lambda_u^{(0)} - \lambda_s^{(0)})P^{(0)}} \tag{4.30}$$

and numerical value for A_3 which are shown together in Table 4.8.

The CM	A_2	A_3	A_4	A_5	B_3
$h(\xi_n^2, \varepsilon_n)$, $\lambda_\varepsilon = -1$	-45.07891	960.55532	-89184.31778	3407491.54046	-52.42483
$h(\xi_n^2, \varepsilon_n)$, $\lambda_\varepsilon = 1$	-45.07891	960.55532	-89184.31778	3407491.54046	148.56628
$h(\xi^2)$	-45.07891	960.55532	—	—	—

The CM	B_4	B_5	C_4	C_5	D_5
$h(\xi_n^2, \varepsilon_n)$, $\lambda_\varepsilon = -1$	4996.41389	-312703.50616	-110.477286	9002.80751	-112.99060
$h(\xi_n^2, \varepsilon_n)$, $\lambda_\varepsilon = 1$	-3449.24714	794007.47405	-476.34543	10310.48731	1525.56488
$h(\xi^2)$	—	—	—	—	—

Table 4.8: The numerical values of coefficient of the center manifolds (4.27) and (4.29)

Using the equation (4.15b) of the unstable mode ξ_n^2, we get the following order parameter equation:

$$\xi_{n+1}^2 = \lambda_2 \xi_n^2 + \frac{P(\mathcal{W}^{**})^2 a - N_2 \mathcal{L}^{**} b(-x_{st}\mathcal{L}^{**} + \mathcal{W}^{**}(1 - 2y_{st} - \mathcal{L}^{**}))}{\lambda_1 - \lambda_2} \tag{4.31}$$

where

$$\mathcal{W}^{**} = -N_1 \xi_n^2 - N_2 h(\xi_n^2), \mathcal{L}^{**} = \xi_n^2 + h, P = -c^2 - by_{st}(1 + by_{st})$$
$$N_1 = \lambda_2 + 2bx_{st}y_{st} - bx_{st}, N_2 = \lambda_1 + 2bx_{st}y_{st} - bx_{st}$$

and $h = h(\xi_n^2, \varepsilon_n)$ (or $h = h(\xi^2)$) are defined on the basis of the expressions (4.27) or (4.29) accordingly. In Fig. 4.7(a) different orders of the center manifold (4.29) with the numerical obtained dependence $\xi^1 = f(\xi^2)$ from the mode equations (4.15) are shown. Bifurcation diagram of the order parameter equation (4.31) based on the center manifold (4.27) and (4.29) together with numerical dependence $\xi^1(a)$ from the mode equations is shown in Fig. 4.7(b).

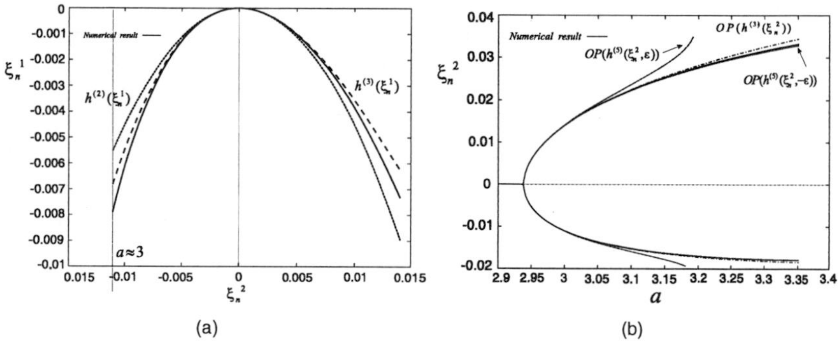

Figure 4.7: (a) Different orders of the center manifold (4.29) and the numerical obtained dependence $\xi^1 = f(\xi^2)$ from the mode equations (4.15); **(b)** Bifurcation diagram of the order parameter equation (4.31) using the center manifold (4.27) and the numerical dependence $\xi^1(a)$ from the mode equations (4.15).

4.2.4 The normal form reduction

All required theoretical outlines of the NF reduction are described in Sec. 3.5. Our idea is that by means of a structural simplification we can get a decoupled system. The explicit calculated coefficients obtained from the mode equations must be equal to the ones obtained from the order parameter equation at least at the second order terms. The high order terms will be different due to the influence of the center manifold reduction.

Being guided by this idea, we first calculate the structure of normal forms from mode equations and the order parameter equation which are determined by resonant terms. Then we separately calculate the coefficients for both normal forms.

The calculations of the second and third order resonant terms based on the expressions, (3.63), (3.64) are explicitly shown in Appendix B.1 (see Tables B.1 and B.2). The coefficients in those tables are calculated for the cases $\lambda_1 = 1$ (transcritical bifurcation) and $\lambda_2 = -1$ (period-doubling bifurcation). The normal forms obtained from the order parameter equation and mode equations are presented in Table 4.9, where $\mu, \tilde{\mu}$ are co-

Eigenvalues	Normal forms of the mode Eqs. (4.15)	Order parameter equations (4.26),(4.31)
$\lambda_1 = 1$ up to $O(3)$	$x_{n+1} = \lambda_1 x_n + \mu^{x^2} x_n^2 + O(3)$ $y_{n+1} = \lambda_2 y_n + \mu^{xy} x_n y_n + O(3)$	$x_{n+1} = \lambda_1 x_n + \tilde{\mu}^{x^2} x_n^2 + O(3)$
$\lambda_2 = -1$ up to $O(4)$	$x_{n+1} = \lambda_1 x_n + \mu^{xy^2} x_n y_n^2 + O(4)$ $y_{n+1} = \lambda_2 y_n + \mu^{y^3} y_n^3 + O(4)$	$x_{n+1} = \lambda_2 x_n + \tilde{\mu}^{x^3} x_n^3 + O(4)$

Table 4.9: The normal forms of the mode equations (4.15) and of the OP equations (4.26),(4.31) with $\lambda_1 = 1, \lambda_2 = -1$.

$\lambda_1 = 1$	Mode equations (4.15)
x^2	$\mu^{x^2} = S(N_1 b x_{st} + Pa_{cr} M^2 (N_2)^2 - N_1 N_2 M b(2y_{st} - 1))$
xy	$\mu^{xy} = SN_2(bM(N_2 + N_1)(2y_{st} - 1)) + 2bx_{st} - 2N_1 Pa_{cr} M^2)$
$\lambda_2 = -1$	
xy^2	$\mu^{xy^2} = S(B_2^x((N_1 N_2 bM + N_1^2 bM(1 - 2y_{st}) + 2x_{st} bN_1 +$ $+2Pa_{cr} M^2 N_1 N_2)) + 2C_2^x(N_1 N_2 bM(1 - 2y_{st}) + x_{st} bN_1 +$ $+Pa_{cr} M^2 N_2^2) + 2B_2^y(N_1^2 bM(1 - 2y_{st}) + x_{st} bN_1 + Pa_{cr} M^2 N_1^2) +$ $+C_2^y((N_1 N_2 bM + N_1^2 bM)(1 - 2y_{st}) + 2x_{st} bN_1 + 2Pa_{cr} M^2 N_1 N_2) -$ $-bN_1 M(2N_1 + N_2))$
y^3	$\mu^{y^3} = S(N_1 bMN_2 - 2C^y(N_1 N_2 bM(1 - 2y_{st}) + PaM^2 N_1^2 +$ $+2bN_2 x_{st}) - C_2^x((N_1 N_2 bM + N_2^2 bM)(1 - 2yst) + 2x_{st} bN_2 +$ $+2PaM^2 N_1 N_2)$

Table 4.10: The coefficients of the normal forms obtained from the mode equations (4.15) with $\lambda_1 = 1, \lambda_2 = -1$

efficients shown in Tables 4.10 and 4.11. In Table 4.10 are shown the coefficients $\tilde{\mu}$ calculated for the case $\lambda_1 = 1$ using Eq. (4.26) and for case $\lambda_2 = -1$ using Eq. (4.31). The coefficient A_2 in Table 4.11 is the center manifold coefficients defined by (4.28) or

$\lambda_1 = 1$	Order parameter equation (4.26)
x^2	$\tilde{\mu}^{x^2} = -S(N_1 N_2 bM - N_1 b x_{st} + 2N_1 N_2 bM y_{st} - Pa_{cr} M^2 (N_2^2)$
$\lambda_2 = -1$	Order parameter equation (4.31)
x^3	$\tilde{\mu}^{y^3} = SN_1 N_2 bM - A_2 SN_2(bM(1 - 2y_{st})(N_1 + N_2) + 2bx_{st} +$ $2Pa_{cr} M^2 N_1) - 2A_2^x(MbN_1 N_2(1 - 2y_{st}) + x_{st} bN_2 + Pa_{cr} M^2 N_1^2$

Table 4.11: The coefficients of the normal forms obtained from the order parameter equation (4.26), (4.31) with $\lambda_1 = 1, \lambda_2 = -1$

(4.30). The normal form's coefficients μ^{x^2} from the Table 4.10 and $\tilde{\mu}^{x^2}$ from the Table 4.11 are identical. As mentioned in the beginning of this section, by means of a structural simplification we are able to get a decoupled system. Furthermore, a dimension reduction occurs when the dimension of the normal form is less than dimension of the mode equations. In this case in fact there aren't essential distinctions between the order parameter equation obtained on the base of the CM reduction and NF reduction in the vicinity of a bifurcation.

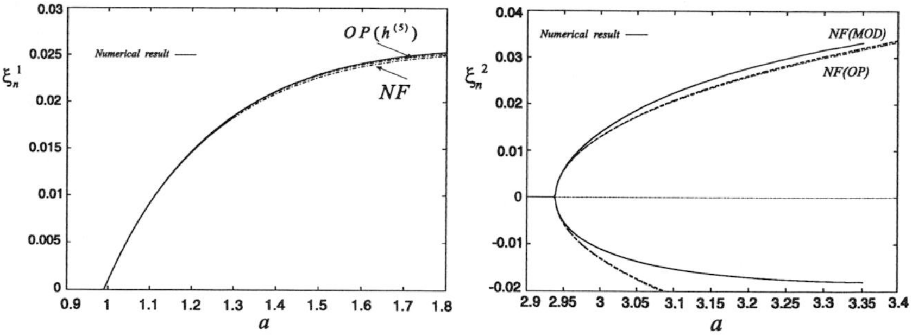

Figure 4.8: (a) Bifurcations diagram of the normal form obtained from the mode equations (4.15) and order parameter equation (4.26) (Table 4.11 and 4.10) with the center manifold (4.19) and the numerical obtained dependence $\xi_n^1(a)$ from the mode equations (4.15) for the transcritical bifurcation; **(b)** Bifurcations diagram of the normal form obtained from the mode equations (4.15) and order parameter equation (4.31) (Table 4.11 and 4.10) with the center manifold (4.27) and the numerical obtained dependence $\xi_n^2(a)$ from the mode equations (4.15) for the period-doubling bifurcation.

4.2.5 Numerical comparison of accuracy

The results of all reduction procedures coincide in a small neighbourhood of bifurcations. In other words when the task is to investigate the type of bifurcation, the accuracy of reduction procedures hasn't any importance. However, for a behavioral investigation, the approximation accuracy of a reduced system is very important. In this connection we show in Fig. 4.9 the numerical values of absolute error $\Delta\xi_n^u$ in dependence upon $\frac{a-a_{cr}}{a_{cr}}$ for different reduction procedures in the case of a transcritical bifurcation.

This dependence of the error can't be generalized for the case of other bifurcations, but we would like to make two essential remarks, which are illustrated by this figure. Firstly, the approximation accuracy of the normal form reduction is less than the CM reduction, even less than the accuracy of the second order CM reduction. This can be explained by the absence of high order terms in the normal form which influence the system dynamics away a bifurcation. We suppose that this observation will also be valid in a general case.

Secondly, increasing the approximation order, the accuracy of CM reduction will also be increased but noticeably less than the increase of computational complexity for the calculation of high order terms. Generally speaking, we can draw a conclusion that a calculation of high-order terms of CM reduction hasn't a significance forasmuch it doesn't lead to an essential improvement of the approximation accuracy.

4.2.6 Appraisal of results

The derived OPs (4.26), (4.31) or the NF shown in Table 4.9 contain the coupling coefficients b and c. Varying these coefficients, we can establish a dependence between the couplings and order parameters. However, there are two problems that are encountered by following this way. Firstly, modifying b and c, we in fact modify all the elements of dynamics, even those which are not affected by a local analysis, e.g., making the stationary

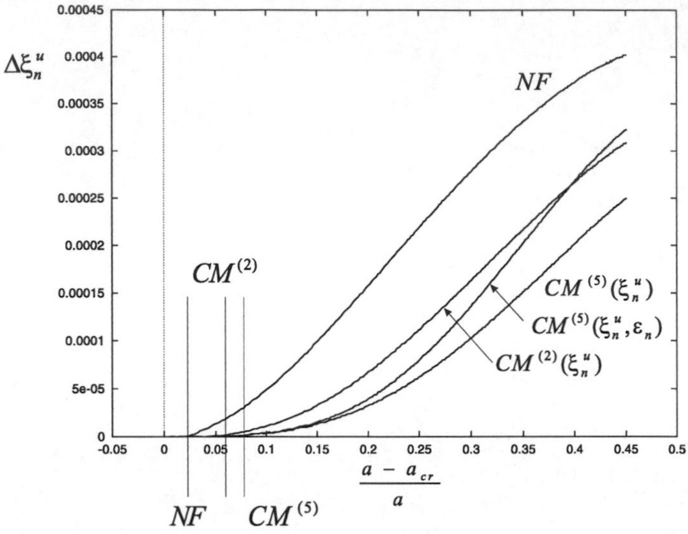

Figure 4.9: Numerical values of the absolute error $\Delta\xi_n^u = |\xi_n^u - \tilde{\xi}_n^u|$, where ξ_n^u is obtained from the mode equations and $\tilde{\xi}_n^u$ from the reduced system, in dependence upon $\frac{a-a_{cr}}{a_{cr}}$ for normal form reduction and different CM approaches in the case of the transcritical bifurcation of system 4.1.

states complex-conjugate or causing some other global effects. In this case the OP has to be re-derived when these new conditions will be taken into account. Secondly, small numerical changes of coupling coefficients do not allow understanding the mechanism of how the couplings influence the OP; we will be getting only numberless bifurcation diagrams that do not tell anything about their origin. Therefore, in the next section we investigate such a mechanism. After that, this consideration will be continued for the cases III and IV mentioned by the linear analysis in Sec. 4.2.1.

4.3 Mechanism of order parameter modification

An influence of coupling methods on a qualitative behavior of interacting systems, described by the normal forms of local instabilities, is investigated in this section. The modification of eigenvalues, being the most intense influence, the increase of dimension and determinancy order, and, finally, an unfolding of normal forms compose four main effects, caused by a coupling's change. Usage of the synergetic methodology allows the dimension scaling, where the results obtained first for low-dimensional systems can then be expanded for an arbitrary high-dimensional case. The undertaken investigation is illustrated by examples, featuring different scaling possibilities as well as coupling methods.

Let us consider the following nonlinear autonomous system discrete in time

$$q_{n+1} = N(q_n, \{\alpha\}), \tag{4.32}$$

that presents the initial or so-called "basic" system. In general case the system (4.32) can be multidimensional. Coupling m such equations by means of nonlinear functions $\underline{\mathbf{F}}$, that consist of multiplicative $\underline{\mathbf{F}}^M$ and additive $\underline{\mathbf{F}}^A$ components, we get the following coupled system

$$\underline{\mathbf{q}}_{n+1} = \underline{\mathbf{N}}(\underline{\mathbf{q}}_n, \{\alpha\})\underline{\mathbf{F}}_n^M(\underline{\mathbf{q}}_n, \{\beta_M\}) + \underline{\mathbf{F}}_n^A(\underline{\mathbf{q}}_n, \{\beta_A\}), \tag{4.33}$$

where $\underline{\mathbf{q}}_n = (q_n^1, q_n^2, ...q_n^m)^T \in \mathbb{R}^m$ is a state vector in the m-dimensional space, $\underline{\mathbf{N}}$, $\underline{\mathbf{F}}$ are in general nonlinear functions of $\underline{\mathbf{q}}_n$ with the control parameters $\{\alpha\}$, $(\{\beta_M\}, \{\beta_A\} \in \{\beta\})$(see Sec. (2.4.2)). Evolution of the coupling functions $\underline{\mathbf{F}}^M$, $\underline{\mathbf{F}}^A$ can be written as follows

$$\begin{aligned} \underline{\mathbf{F}}_{n+1}^M(\underline{\mathbf{q}}_n, \{\beta_M\}) &= \underline{\mathbf{F}}_n^M(\underline{\mathbf{q}}_n, \{\beta_M\}) + \underline{\tilde{\mathbf{F}}}_n^M(\underline{\mathbf{q}}_n, \{\beta_M\}), \\ \underline{\mathbf{F}}_{n+1}^A(\underline{\mathbf{q}}_n, \{\beta_A\}) &= \underline{\mathbf{F}}_n^A(\underline{\mathbf{q}}_n, \{\beta_A\}) + \underline{\tilde{\mathbf{F}}}_n^A(\underline{\mathbf{q}}_n, \{\beta_A\}), \end{aligned} \tag{4.34}$$

where $\underline{\tilde{\mathbf{F}}}$ denotes the part that determines modifications of couplings. The function $\underline{\tilde{\mathbf{F}}}$ plays in fact the role of a "program" governing an evolution of the system (4.33), for instance, in [169] the self-referenced term $\underline{\tilde{\mathbf{F}}}_n = \underline{\mathbf{F}}_n \circ \underline{\mathbf{F}}_n$ has been suggested and investigated. Complete dynamics of the system (4.33) is determined by own dynamics of this system at fixed couplings and by functional dynamics (4.34) of couplings. However, due to nonlinear dynamical dependence, every change of functions $\underline{\mathbf{F}}$ modifies a dynamics of (4.33) in an "unexpected" way. The required in practical applications purposeful changes of (4.33) by means of (4.34) seem problematical, if the mechanism of this modification is not understood. Thus there are three components of self-modification and we intend to bring out the "bridge" between functional dynamics (4.34) and the system (4.33).

An appearance of new functional structures points to qualitative changes in a collective dynamics. Therefore, being guided by the synergetic concept, we concentrate on the question of how the couplings $\underline{\mathbf{F}}$ between initial systems (4.32) exert an influence on the qualitative behavior of the system (4.33). Here, the qualitative behavior of the system (4.33) implies a local dynamics given by normal forms (NF) of the appropriate local bifurcations

$$\underline{\mathbf{q}}_{n+1} = \underline{\underline{\Lambda}}_u(\{\alpha\}, \{\beta\})\underline{\mathbf{q}}_n + \underline{\mathbf{g}}^{(2)}(\underline{\mathbf{q}}_n, \{\alpha\}, \{\beta\}) + ... + O(\underline{\mathbf{g}}^{(r+1)}). \tag{4.35}$$

In (4.35) the term $\underline{\underline{\Lambda}}_u$ presents the diagonal matrix of eigenvalues, $\underline{\mathbf{g}}$ are the resonant terms, dependent on both $\{\alpha\}$ and $\{\beta\}$ and r is the determinancy order of this NF. Remark, due to a performing of the coordinate transformation, the initial functions $\underline{\mathbf{N}}$ and $\underline{\mathbf{F}}$ are fused into one function, whose linear part is given by $\underline{\Lambda}$ and the nonlinear part by $\underline{\tilde{\mathbf{N}}}$ (see Sec. 3.2). Thus the linear part as well as nonlinear part of Eqs. (4.35) are dependent on set of both parameters $\{\alpha\}$ and coefficients $\{\beta\}$.

Our idea consists in the following assumption that will be further discussed for homogeneous and nonhomogeneous couplings. The initial system (4.32), being uncoupled, demonstrates local behavior described by an "initial" NF. Local behavior of the coupled system (4.33) differs from the behavior of the initial system (4.32), that can be viewed as a perturbation of this initial NF caused by the coupling $\underline{\mathbf{F}}$. Although this coupling mechanism simultaneously influences all elements of NF, we can at least distinguish between four main components of such an influence:

- change of the unstable eigenvalue $\underline{\underline{\Lambda}}_u$. Generally speaking, the derived in this case normal form completely differs from the initial one;

- appearance of multiple unstable eigenvalues that increase the dimension of normal form (4.35). Forasmuch as there are two critical values ± 1 of unstable eigenvalues, the derived multidimensional normal forms possess specific properties;

- change of the determinancy order r. This includes into consideration additional high-order terms that often lead to an appearance of non-local solutions of the bifurcation problem (4.35);

- perturbation of different resonant terms $\underline{g}$ of NF and an unfolding of the normal form.

These four topics make the main focus of the given section. The essential question concerns the dimension m of the considered system (4.33). As turned out, homogeneous couplings do not modify the local properties of coupled maps. Therefore, exemplifying the suggested approach, low-dimension systems will be first considered. After that the dimension scaling will be performed to prove whether the obtained results are also valid for high-dimensional systems.

4.3.1 Modification of linear terms: change of the eigenvalue

According to NF theory [113], the resonant terms of one-dimensional Eq. (4.35) are defined by the unstable eigenvalue. The couplings, changing this eigenvalue, have the most intense influence on a qualitative behaviour of the whole system. The unstable eigenvalue can undergo either a change of sign, an appearance of complex-conjugate values or a shift of critical values. The case of multiple unstable eigenvalues is considered in Sec. 4.3.2.3.

The shift of critical values refers to the problem of bifurcation control and has been already studied in detail (see [161,]. Consequently here we deal only with a change of sign at the unstable eigenvalue

$$\lambda_u^{init}(\{\alpha\}) = -\lambda_u(\{\alpha\}, \{\beta\}), \tag{4.36}$$

and an appearance of complex-conjugate eigenvalues $\lambda_u^{\pm}(\{\alpha\}, \{\beta\})$ where

$$\lambda_u^{init}(\{\alpha\}) = \lambda_u^{\pm}(\{\alpha\}, 0) \tag{4.37}$$

and $\lambda_u^{init}(\{\alpha\})$ is the initially unstable eigenvalue of (4.32). Before we start treating this matter, it is necessary to make some preliminary remarks. Firstly, we impose restrictions on the coupling method. Now let us consider the system (4.33), linearized at the stationary state $\underline{\mathbf{q}}_{st} = (q_{st}^1, ..., q_{st}^m)$ with the eigenvalues

$$\underline{\lambda}(\{\alpha\}, \{\beta\}, \underline{\mathbf{q}}_{st}(\{\alpha\}, \{\beta\})), \quad i = 1, ..., m. \tag{4.38}$$

As can see,the eigenvalues (4.38) are in general dependent on the $\underline{\mathbf{q}}_{st}(\{\alpha\}, \{\beta\})$, that are the nonlinear functions themselves. Forasmuch as a qualitative behavior is the focus of the work, performing linear coordinate change, the dynamic system can be transformed, so that the stationary states get $\underline{\mathbf{q}}_{st}(\{\alpha\}, \{\beta\}) = 0$. Therefore, without loss of generality, the further consideration may be restricted by the cases when $\underline{\mathbf{q}}_{st}(\{\alpha\}, \{\beta\}) = 0$ or $\underline{\mathbf{q}}_{st}(\{\alpha\})$ are independent of the coefficients $\{\beta\}$. The last condition can be achieved by using the coupling methods that do not modify stationary states of the initial system

(4.32). For this effect we propose applying the coupling methods based on the generalized time delay function (taking into account the limitations of delayed feedbacks)

$$\underline{\mathbf{F}} = k(\{\beta\})(k_1(\{\beta\})\underline{\mathbf{q}}_{n-(2\tau-1)} - k_2(\{\beta\})\underline{\mathbf{q}}_{n-(\tau-1)}), \tag{4.39}$$

which in original form was first suggested by Pyragas [141].

As shown in [161], the parameter τ is the period of period-doubling cascade and $k(\{\beta\})$ is some function. Being aimed to change the local behavior of the system (4.33), we use the control function (4.39) in the coupling methods in order to perform the bifurcation control and hereby to modify the behavior of the system (4.33). In the further consideration we set $\tau = 1$, meaning hereby that only the first local bifurcation undergoes the changes. But generally, setting τ on some other values, we are able to influence the τ-first local bifurcations.

Hereby, inserting the function (4.39) at $k_{1,2}(\{\beta\}) = 1$, $\tau = 1$ either into the unstable mode equations (3.2a) (i.e. into the order parameter equations (3.8)) or into all mode equations (3.2a), (3.2b), we obtain by means of linear transformation the local feedback

$$F^i = k(\{\beta\})(q^i_{n-1} - q^i_n), \tag{4.40}$$

or the global feedback

$$F^i = k(\{\beta\})\sum_{j=1}^{m} c^j(q^j_{n-1} - q^j_n), \tag{4.41}$$

where c^j are coefficients. Generally, the utilized couplings are not restricted only by delayed functions, however, the global feedback (4.41) is the most simple coupling scheme, that does not only combine the initial equations (4.32) into the coupled system (4.33), but also changes their local behavior. Remark, that the case $k_{1,2}(\{\beta\}) \neq 1$ pertains to $\underline{\mathbf{q}}_{st} = 0$ for non-periodical behavior. The second remark pertains to the bifurcation parameter α. It can appear either in all basic equations or only in a few of them. The last case leads finally to the effect of unfolding and it is considered further in Sec. 4.3.2.3. Here we assume that the bifurcation parameter appears in all basic equations.

The last remark relates to basic systems (4.32). They can be already coupled together and we start the investigations with the coupled system (4.33). In contrast to that, we can connect the basic systems ourselves, changing hereby their qualitative behavior. These cases lead to different coupling mechanisms and accordingly they have to be considered separately.

4.3.1.1 Change of sign at the eigenvalues in the initially uncoupled systems

In order to give a small introductory example for demonstration the main points of mentioned approach, let the initial system (4.32) be given by the normal form of a transcritical bifurcation so that

$$q^i_{n+1} = aq^i_n - (q^i_n)^2 + F^i_{coupl}, \quad i = 1, ...m, \tag{4.42}$$

where m is the dimension. The system (4.42) exhibits the transcritical bifurcation at $a = 1$ with $\lambda_u = 1$ (see Fig. 4.10(a)) and a period-doubling bifurcation at $a = 3$ with $\lambda_u = -1$.

Intending to change the local dynamics of coupled systems, we can influence both bifurcations. Inverting a sign at the eigenvalue, the transcritical bifurcation will be transformed into period-doubling and period-doubling into a transcritical one. But forasmuch

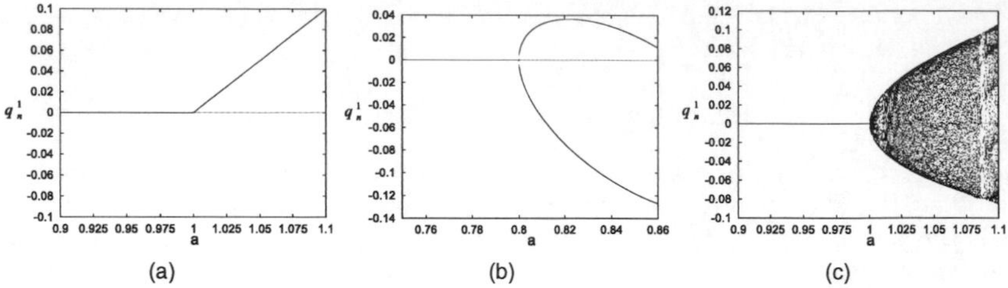

Figure 4.10: (a) The bifurcation diagram of the uncoupled system (4.42) in the neighbourhood of $a_{cr} = 1$; **(b)** The bifurcation diagram of the coupled system (4.42) in the neighbourhood of $a_{cr} = 1 - \beta$ for $\beta = 0.2$; **(c)** The bifurcation diagram of the system (4.42) with the control part (4.92) in the neighbourhood of $a_{cr} = 1$.

as the system 4.35 does not possess the second stable stationary state in the neighbourhood of $a = 3$, it makes no sense to transform the period-doubling bifurcation into the trancritical one. In this relation, the task is to find such a coupling function F^i_{coupl} that to transform the transcritical bifurcation into the period-doubling one in the neighbourhood of $a = 1$. For demonstration only two basic systems are chosen ($m = 2$) and coupled by means of the function (4.41), where all coefficients $c^j = 1/2$ (see Eq. 4.74 in Sec. 4.3.1.2).

$$x_{n+1} = ax_n - (x_n)^2 + k(A(x_{n-1} - x_n) + B(y_{n-1} - y_n)), \tag{4.43a}$$

$$y_{n+1} = ay_n - (y_n)^2 + k(C(x_{n-1} - x_n) + D(y_{n-1} - y_n)). \tag{4.43b}$$

Now introducing two new state variables, representing the time delay of x_n and y_n (as shown in [161]), we calculate the Jacobian $\underline{\underline{\mathrm{J}}}$ of this four-dimensional system evaluated at $x_{st} = 0$, $y_{st} = 0$

$$\underline{\underline{\mathrm{J}}} = \frac{1}{2}\begin{pmatrix} 2a-k & -k & k & k \\ -k & 2a-k & k & k \\ 2 & 0 & 0 & 0 \\ 0 & 2 & 0 & 0 \end{pmatrix}. \tag{4.44}$$

The eigenvalues are given by

$$\lambda_{1,2} = \frac{a - k \pm \sqrt{(k-a)^2 + 4k}}{2}, \; \lambda_3 = 0, \; \lambda_4 = a. \tag{4.45}$$

The system (4.43) without coupling has the unstable eigenvalues that are equal to a. The coupled system (4.43) possesses the unstable eigenvalue λ_u given by λ_1 (4.45) that is dependent on k. In order to create the period-doubling bifurcation we require for the unstable eigenvalue in accordance with (4.36) λ_1 that

$$\lambda_1 = \frac{a - k + \sqrt{(k-a)^2 + 4k}}{2} = -a - \beta \tag{4.46}$$

The coefficient β is introduced because otherwise the extended system (4.43) has three unstable eigenvalues a, a and $-a$, that leads to the three-dimensional NF at $a_{cr} = 1$. Introducing β, we shift backward the instability with the negative eigenvalue, causing hereby only a period-doubling bifurcation with one-dimensional NF. Solving (4.46), we obtain

$$k = \frac{2a^2 + 3a\beta + \beta^2}{a + 1 + \beta}. \tag{4.47}$$

With function k, calculated above, the system (4.43) undergoes the period-doubling bifurcation at the critical values of the bifurcation parameter $a_{cr} = 0.8$ and $\beta = 0.2$. The bifurcation diagram is shown in Fig. 4.10(b).

Dimension scaling

The two-dimensional example (4.43) is destined to show the simplest possibility of the sign inverting at the unstable eigenvalue. However the principal question still remains open: can such mechanism be expanded to high-dimensional systems ? We suppose that certain classes of couplings, calculated first for the low-dimensional systems, can be scaled up to arbitrary high-dimensional systems.

In this case the theoretical ground for the dimension scaling consists in preserving unstable eigenvalues derived by the linear stability analysis. The high-dimensional system, being transformed into the form of the mode equations (3.2), has some unstable mode equations (3.2a) and a plenty of stable mode equations (3.2b). Our observation is that an increase of the number of initial equations leads to an increase of stable mode equations; the number of unstable mode equations remains the same. Inserting functions, similar to (4.39), into these unstable mode equations and performing the back-transformations, we get couplings in the original system, which are invariant to the dimension of a system.

The mathematical formulation of this matter will be shown in Sec. 4.3.1.2. For the system (4.43) one of possible scaling schemes has the following form

$$q_{n+1}^i = aq_n^i - (q_n^i)^2 + k\sum_{j=1}^{m}\frac{1}{m}(q_{n-1}^j - q_n^j), \qquad (4.48)$$

where m is the number of equations and k is defined by (4.47).

Here we can point to the questions about application of the time-delay control as a coupling method. When one of equations is influenced from outside and the control is on, this perturbation reaches other equations. During some time, when the control is stabilizing the behavior, all equations are interacting. After the perturbation is decayed, this control is turned off. In this sense the time-delay control can be also viewed as a specific coupling method. However further we try to separate the coupling part from a control part, providing hereby more “clear” interactions.

4.3.1.2 Change of sign at the eigenvalues in the initially coupled systems

Since in Sec. 4.3.1.1 we presented one example of how the coupling can influence the unstable eigenvalues, we would like here to introduce a mathematical formalism that generalizes our approach. We use for this the tensor notation that simplifies a notation of required calculations.

Considering the case of initially coupled systems, it is assumed that they can be coupled in different ways. We impose only one condition on these “initial” couplings, namely they have to preserve the unstable eigenvalues independently of the dimension of an original system. This dimension scaling property will be used at a transition from low-dimensional systems to high-dimensional ones.

Assume, the basic systems (4.42) are additively coupled as the two-way ring (see [131]) (taking into account the boundary conditions for $i+1$ and $i-1$)

$$q_{n+1}^i = aq_n^i - (q_n^i)^2 + \varepsilon(q_n^{i-1} + q_n^{i+1}), \quad i = 1, ..., m, \qquad (4.49)$$

where ε is a small coupling parameter.

To verify the dimension scaling property of two-way ring coupling, we have to consider the special square matrix

$$\begin{pmatrix} a & \varepsilon & 0 & 0 & \ldots & 0 & \varepsilon \\ \varepsilon & a & \varepsilon & 0 & \ldots & 0 & 0 \\ 0 & \varepsilon & a & \varepsilon & \ldots & 0 & 0 \\ 0 & 0 & \varepsilon & a & \ldots & 0 & 0 \\ \vdots & \vdots & \vdots & \vdots & \ddots & \vdots & \vdots \\ 0 & 0 & 0 & 0 & \ldots & a & \varepsilon \\ \varepsilon & 0 & 0 & 0 & \ldots & \varepsilon & a \end{pmatrix}, \tag{4.50}$$

which is the Jacobian of the system (4.49) evaluated at $q^i_{st} = 0$. The maximal eigenvalue of the matrix (4.50) in absolute magnitude is equal to $a + 2\varepsilon$ independently of the size of this matrix (see [131], [129]).

The system (4.49) can be written in the following generalized form (see also [161], [111])

$$\underline{\mathbf{q}}_{n+1} = \Gamma_{(2)}(:\underline{\mathbf{q}}_n) + \Gamma_{(3)}(:\underline{\mathbf{q}}_n)^2 = \sum_{r=1}^{2} \Gamma_{(r+1)}(:\underline{\mathbf{q}}_n)^r, \tag{4.51}$$

where $\Gamma_{(r)}$ are the tensors of rank r. The non-vanishing components of the tensors for system (4.51) are shown in Table 4.12.

$\Gamma_{(2)}$	$\Gamma_{(3)}$
$(\Gamma_{(2)})_{ii} = a$	$(\Gamma_{(3)})_{iii} = -1$
$(\Gamma_{(2)})_{j,j+1} = \varepsilon$	
$(\Gamma_{(2)})_{k,k-1} = \varepsilon$	
$(\Gamma_{(2)})_{1,m} = \varepsilon$	
$(\Gamma_{(2)})_{m,1} = \varepsilon$	
$j = 1, ..., m-1,$	$k = 2, ..., m$
$i = 1, ..., m,$	$m \geq 3$

Table 4.12: Non-vanishing tensors components $\Gamma_{(2)}, \Gamma_{(3)}$ of system (4.51).

Forasmuch as $\underline{\mathbf{q}}_n = \underline{\mathbf{q}}_{st} + \Delta\underline{\mathbf{q}}_n$ and $\underline{\mathbf{q}}_{st} = 0$, the following coordinate transformation can be performed

$$\underline{\mathbf{q}}_n = \sum_{k=1}^{m} \xi^k_n \underline{\mathbf{v}}_k = \Gamma^V_{(2)}(:\underline{\boldsymbol{\xi}}_n), \tag{4.52}$$

where $\underline{\boldsymbol{\xi}}_n$ are amplitudes and $\underline{\mathbf{v}}_k$ eigenvectors of the matrix (4.50). For the special case $m = 3$ the eigenvectors $\underline{\mathbf{v}}_k$ are given by

$$\underline{\mathbf{v}}_1 = \begin{pmatrix} 1 \\ 1 \\ 1 \end{pmatrix}, \ \underline{\mathbf{v}}_2 = \begin{pmatrix} -1 \\ 1 \\ 0 \end{pmatrix}, \ \underline{\mathbf{v}}_3 = \begin{pmatrix} -1 \\ 0 \\ 1 \end{pmatrix}. \tag{4.53}$$

Substituting (4.52) into Eq. (4.51), we get the following equation

$$\Gamma^V_{(2)}(:\underline{\boldsymbol{\xi}}_{n+1}) = \Gamma_{(2)}\Gamma^V_{(2)}(:\underline{\boldsymbol{\xi}}_n) + \Gamma_{(3)}(\Gamma^V_{(2)}(:\underline{\boldsymbol{\xi}}_n))^2. \tag{4.54}$$

$\Gamma^{V}_{(2)}$	$\Gamma^{\Lambda}_{(2)}$	$\Gamma^{\tilde{N}}_{(3)}$		
$(\Gamma^{V}_{(2)})_{ii,21,31} = 1$	$(\Gamma^{\Lambda}_{(2)})_{11} = a + 2\varepsilon$	$(\Gamma^{\tilde{N}}_{(3)})_{111} = -1$	$(\Gamma^{\tilde{N}}_{(3)})_{212} = -2$	$(\Gamma^{\tilde{N}}_{(3)})_{313} = -2$
$(\Gamma^{V}_{(2)})_{32,23} = 0$	$(\Gamma^{\Lambda}_{(2)})_{22} = a - \varepsilon$	$(\Gamma^{\tilde{N}}_{(3)})_{122} = -2/3$	$(\Gamma^{\tilde{N}}_{(3)})_{222} = -1/3$	$(\Gamma^{\tilde{N}}_{(3)})_{322} = 2/3$
$(\Gamma^{V}_{(2)})_{12,13} = -1$	$(\Gamma^{\Lambda}_{(2)})_{33} = a - \varepsilon$	$(\Gamma^{\tilde{N}}_{(3)})_{123} = -2/3$	$(\Gamma^{\tilde{N}}_{(3)})_{223} = 2/3$	$(\Gamma^{\tilde{N}}_{(3)})_{323} = 2/3$
	$(\Gamma^{\tilde{N}}_{(3)})_{133} = -2/3$	$(\Gamma^{\tilde{N}}_{(3)})_{233} = 2/3$	$(\Gamma^{\tilde{N}}_{(3)})_{333} = -1/3$	

Table 4.13: Non-vanishing tensors components $\Gamma^{V}_{(2)}$ of Eq. (4.54) and $\Gamma^{\Lambda}_{(2)}$, $\Gamma^{\tilde{N}}_{(3)}$ of Eq. (4.55) for the special case $m = 3$.

Multiplying from the left side with the inverse tensor $\Gamma^{V^{-1}}_{(2)}$, we finally obtain the mode equation

$$\underline{\boldsymbol{\xi}}_{n+1} = \Gamma^{\Lambda}_{(2)}(:\underline{\boldsymbol{\xi}}_n) + \Gamma^{\tilde{N}}_{(3)}(:\underline{\boldsymbol{\xi}}_n)^2, \tag{4.55}$$

where $\Gamma^{\Lambda}_{(2)} = \Gamma^{V^{-1}}_{(2)}\Gamma_{(2)}\Gamma^{V}_{(2)}$, $\Gamma^{\tilde{N}}_{(3)} = \Gamma^{V^{-1}}_{(2)}\Gamma_{(3)}\Gamma^{V}_{(2)}\Gamma^{V}_{(2)}$ and $(\Gamma^{V^{-1}}_{(2)}\Gamma^{V}_{(2)})_{ij} = \delta_{ij}$ (δ_{ij} is the Kronecker-Symbol). The non-vanishing components of the tensors for system (4.55) are shown in Table 4.13. The unstable eigenvalue λ_u, given by $(\Gamma^{\Lambda}_{(2)})_{11} = a + 2\varepsilon$, is invariant to the dimension m of the coupled system (4.49). Now using the approach, similar to that presented in Sec. 4.3.1.1, we insert a feedback into the equation with the unstable eigenvalue so that the statement (4.36) will be satisfied. Here, for the delayed feedback mechanism we use the modified scheme (4.39), where $\tau = 1$, $k_1(\{\beta\}) \neq 1$ and $k_2(\{\beta\}) = 1$.

$$\underline{\mathbf{F}} = k(\{\beta\})\left(k_1(\{\beta\})\underline{\mathbf{q}}_{(n-1)} - \underline{\mathbf{q}}_n\right). \tag{4.56}$$

The feature of the function (4.56) consists in possessing two coefficients. By virtue of these we are able to transform the unstable eigenvalue separately within the radical expression and outside the radical as shown further. Remark that the above mentioned property of the function (4.56) pertains to the cases when $\underline{\mathbf{q}}_{st} = 0$.

Further we introduce the tensor $(\Gamma^{C}_{(2)})$ showing mode equations with the inserted function (4.56). As already said, if the expression (4.56) is introduced only into the unstable mode equation, then

$$(\Gamma^{C}_{(2)})_{11} = 1, \quad (\Gamma^{C}_{(2)})_{ij} = 0, \quad i \neq 1, \quad j \neq 1. \tag{4.57}$$

Substituting (4.56) into (4.55) and taking into account (4.57), we rewrite the mode equations in the following form

$$\underline{\boldsymbol{\xi}}_{n+1} = \Gamma^{\Lambda}_{(2)}(:\underline{\boldsymbol{\xi}}_n) + \Gamma^{\tilde{N}}_{(3)}(:\underline{\boldsymbol{\xi}}_n)^2 + k\Gamma^{C}_{(2)}(:k_1\underline{\boldsymbol{\xi}}_{n-1} - \underline{\boldsymbol{\xi}}_n). \tag{4.58}$$

Carrying out the linear stability analysis for (4.58), the four eigenvalues are obtained

$$\lambda_{1,2} = 0.5a + \varepsilon - 0.5k \pm 0.5\sqrt{(a + 2\varepsilon - k)^2 + 4kk_1}, \quad \lambda_{3,4} = a - \varepsilon. \tag{4.59}$$

In accordance with (4.36) we require that λ_1 becomes unstable, getting hereby $\lambda_1 = -(a + 2\varepsilon)$. This can be achieved setting

$$k_1 = -\frac{(a + 2\varepsilon - k)^2}{4k}, \quad k = 3a + 6\varepsilon. \tag{4.60}$$

Then the eigenvalues (4.59) get the following form

$$\lambda_{1,2} = -(a + 2\varepsilon), \quad \lambda_{3,4} = a - \varepsilon. \tag{4.61}$$

As we can see from the linear stability analysis, the system (4.58) has the bifurcation at $a_{cr} = 1 - 2\varepsilon$ (ε is small) with the unstable eigenvalue $\lambda_u = -1$ that corresponds to the period-doubling bifurcation.

Thus, the coefficients k, k_1, being calculated, enable us to transform the system (4.58) backwards to the initial form (4.49). Multiplying (4.58) with the tensor $\Gamma^V_{(2)}$ and taking into account (4.52), we finally get system (4.51) with included coupling terms

$$\underline{\mathbf{q}}_{n+1} = \Gamma_{(2)}(:\underline{\mathbf{q}}_n) + \Gamma_{(3)}(:\underline{\mathbf{q}}_n)^2 + k[\Gamma^V_{(2)}\Gamma^C_{(2)}\Gamma^{V^{-1}}_{(2)}](: k_1\underline{\mathbf{q}}_{n-1} - \underline{\mathbf{q}}_n). \tag{4.62}$$

The system (4.62) has two possibilities of dimension scaling, that we consider further.

Dimension scaling for initially coupled systems

The coupling mechanism (4.62) for the system (4.49) can be generalized for a high-dimensional case in the following form

$$q^i_{n+1} = aq^i_n - (q^i_n)^2 + \varepsilon(q^{i-1}_n + q^{i+1}_n) + k\sum_{j=1}^{m} r^i_j(k_1 q^j_{n-1} - q^j_n), \tag{4.63}$$

where $r^j_i = (\Gamma^V_{(2)}\Gamma^C_{(2)}\Gamma^{V^{-1}}_{(2)})_{ij}$. The main problem, encountered here, consists in a determination of the tensor $\Gamma^V_{(2)}$ for high-dimensional systems. Although there are different numerical methods for a determination of eigenvectors we would like to simplify this method from the procedural viewpoint. This simplification implies a replacement of global couplings on local ones.

In order to construct the local coupling mechanism, the function (4.56) has to be introduced into all unstable as well as stable mode equations (see [161]). In this case the tensor $(\Gamma^C_{(2)})$ has the following components

$$(\Gamma^C_{(2)})_{ii} = 1, \quad (\Gamma^C_{(2)})_{ij} = 0, \quad i \neq j. \tag{4.64}$$

Calculating once more the eigenvalues of the Jacobian of the system (4.58), we get

$$\lambda_{1,2} = 0.5a + \varepsilon - 0.5k \pm 0.5\sqrt{(a + 2\varepsilon - k)^2 + 4kk_1}, \tag{4.65a}$$

$$\lambda_{3,4} = \lambda_{5,6} = 0.5a - 0.5\varepsilon - 0.5k \pm 0.5\sqrt{(a - \varepsilon - k)^2 + 4kk_1}. \tag{4.65b}$$

Choosing k and k_1 as in (4.60), the eigenvalues acquire then

$$\lambda_{1,2} = -(a + 2\varepsilon), \quad \lambda_{3,5} = -(a + 2\varepsilon) + \psi_1, \quad \lambda_{4,6} = -(a + 2\varepsilon) - \psi_2, \tag{4.66}$$

where

$$\psi_1 = \frac{-3\varepsilon + \sqrt{12a\varepsilon + 33\varepsilon^2}}{2}, \quad \psi_2 = \frac{3\varepsilon + \sqrt{12a\varepsilon + 33\varepsilon^2}}{2}. \tag{4.67}$$

As we can see in (4.66) the unstable eigenvalues $\lambda_{4,6}$ have changed own sign, but moreover they are shifted backward on the value of coefficient ψ_2. This coefficient arises because the control function is introduced into the stable and unstable modes. Choosing this function for stabilization of the unstable mode, we change hereby the stable modes too. This forced modification of stable modes leads to an appearance of the coefficient ψ_2. Since ψ_2 is small and moreover always calculable, we suppose that this method fulfil our initial task.

Thus the second scheme of a dimension scaling has the following form

$$q^i_{n+1} = aq^i_n - (q^i_n)^2 + \varepsilon(q^{i-1}_n + q^{i+1}_n) + k(k_1 q^i_{n-1} - q^i_n). \tag{4.68}$$

As we can see, now the tensor of eigenvectors $\Gamma^V_{(2)}$ is not included in this scheme.

Dimension scaling for initially uncoupled systems

Here we formalize the scaling scheme (4.48) presented in Sec. 4.3.1.1.The main idea is that the coupling mechanism and the control itself can be combined together in the global control scheme. Obtained hereby coupling differs from the one- or two-way ring coupling (4.50). However the delayed function (4.39), being globally applied, can be also viewed as a specific coupling method suitable for a propagation of interactions in CML.

Let us rewrite (4.33), coupled by means of delayed function (4.39), in the following generalized form

$$\underline{\mathbf{q}}_{n+1} = \Gamma_{(1)} + \Gamma^I_{(2)}(:\underline{\mathbf{q}}_n) + \sum_{r=2}^{p} \Gamma^I_{(r+1)}(:\underline{\mathbf{q}}_n)^r + k\Gamma^V_{(2)}\Gamma^C_{(2)}\Gamma^{V^{-1}}_{(2)}(:\underline{\mathbf{q}}_{n-1} - \underline{\mathbf{q}}_n). \tag{4.69}$$

Here $(\Gamma^I_{(2)})_{i\neq j} = 0$, $(\Gamma^I_{(2)})_{ii} = a$, $(\Gamma^I_{(3)})_{i\neq j\neq l} = 0$, $(\Gamma^I_{(3)})_{iii} = a$ and so on, where a is some function and p maximal order of nonlinear terms. Tensor $\Gamma^C_{(2)}$ is determined like in the case of Eq. (4.57). Remark that eigenvectors $\Gamma^V_{(2)}$ in Eq. (4.69) are not yet determined.

Now repeating (4.51)-(4.55), we transform (4.69) into the form of mode equations, writing separately basic system and the coupling part (see [161])

$$\underline{\boldsymbol{\xi}}_{n+1} = \Gamma^{V^{-1}}_{(2)}\Gamma^{IL}_{(2)}\Gamma^V_{(2)}(:\underline{\boldsymbol{\xi}}_n) + \Gamma^{V^{-1}}_{(2)} \sum_{r=2}^{p} \Gamma^{IN}_{(r+1)}(\Gamma^V_{(2)}(:\underline{\boldsymbol{\xi}}_n))^r + k\Gamma^C_{(2)}(:\underline{\boldsymbol{\xi}}_{n-1} - \underline{\boldsymbol{\xi}}_n), \tag{4.70}$$

where the tensor $\Gamma^{IL}_{(2)}$ has a diagonal form. Simplifying the linear part, Eq. (4.70) can be rewritten as

$$\underline{\boldsymbol{\xi}}_{n+1} = \Gamma^{IL}_{(2)}(:\underline{\boldsymbol{\xi}}_n) + \sum_{r=2}^{p} \Gamma^{\tilde{N}}_{(r+1)}(:\underline{\boldsymbol{\xi}}_n)^r + k\Gamma^C_{(2)}(:\underline{\boldsymbol{\xi}}_{n-1} - \underline{\boldsymbol{\xi}}_n). \tag{4.71}$$

We can see from (4.70), the tensor $\Gamma^V_{(2)}$ influences only the nonlinear part. We can conclude that $\Gamma^V_{(2)}$ does not exert any influence on the linear properties of coupled system (4.69) and consequently can be chosen arbitrary. Strictly speaking, we have additionally to take into account the change of subcritical-supercritical conditions of shifted bifurcation in $\Gamma^{\tilde{N}}$ that could be caused by a global control. The Jacoblan of Eq. (4.71) has a block-diagonal form, consequently the coefficient k can be calculated by a linear stability analysis of only one equation, that the delayed function was introduced into.

Returning to (4.69), we assert that the globally coupled CML has the following form

$$q^i_{n+1} = N^i(q^i_n) + k\sum_{j=1}^{m} r^i_j(q^j_{n-1} - q^j_n), \quad i = 1, ..., m, \tag{4.72}$$

where

$$r^i_j = \left(\Gamma^V_{(2)}\Gamma^C_{(2)}\Gamma^{V^{-1}}_{(2)}\right)_{ij}, \tag{4.73}$$

and $\Gamma^V_{(2)}$ is chosen arbitrary (for that $\Gamma^{V^{-1}}_{(2)}$ has to exist). In particular we get the following well-known scheme

$$r^i_j = \frac{1}{m}, \tag{4.74}$$

where m is a dimension of CML.

Considering the case of global control scheme (4.72), we assumed that in CML there are a few fixed unstable modes, which have to be stabilized. It is not always true, forasmuch as any mode in CML can be unstable. This is occurred because all linear parts are equal and decoupled from each other. This essential remark restricts an application of a global control. Shifting the instability point forwards, we stabilize hereby only one equation. However instability points of other equations still remain at the same place, causing an unshifted bifurcation. Therefore the scheme (4.72) is suitable only for the shift of the instability point backwards.

Modifying this scheme to be suitable to the shift forwards, we rewrite the tensor $\Gamma^C_{(2)}$ in the following form

$$(\Gamma^C_{(2)})_{11} = 1, \quad (\Gamma^C_{(2)})_{ii} = (1+p), \quad i = 2, ..., m, \tag{4.75}$$

whose other components are equal to zero. Then the coefficients r^i_j of (4.73), using $\Gamma^V_{(2)}$ chosen as in the case of (4.74), can be presented as the following matrix

$$\underline{\mathbf{r}} = \frac{1}{\tilde{m}} \begin{pmatrix} \tilde{m}(1+\tilde{p}) & -\tilde{p} & \vdots & -\tilde{p} \\ -\tilde{p} & \tilde{m}(1+\tilde{p}) & \vdots & -\tilde{p} \\ ... & ... & \ddots & ... \\ -\tilde{p} & -\tilde{p} & \vdots & \tilde{m}(1+\tilde{p}) \end{pmatrix}, \tag{4.76}$$

where p and $\tilde{p}$ are some coefficients, $0 < p < 1$, $0 < \tilde{p} < 1$ and $\tilde{m} = m - 1$. Meaning of scheme (4.76) is that the other equations of CML are shifted a bit further than the "initially" unstable equation. Therefore this equation, even being shifted, still remains a single unstable equation of the coupled system. Let us point out that the bifurcation point of the whole CML can be calculated on the base of only one equation.

4.3.1.3 Complex-conjugate eigenvalues: The Neimark-Sacker bifurcation

Explained above approach, shown on two examples, enable us to consider the second possibility of the eigenvalues modification. In this case we intend to create the complex-conjugate eigenvalues that cause the Naimark-Sacker bifurcation. Moreover, it is expected to derive the delayed coupling that is able to modify additional conditions determining the subcritical or supercritical type of bifurcation. This will be achieved, introducing a correction function into the expression, which describes the sub- or supercritical type of instability. Further, the coupled system has to be so transformed, that such a function would be inserted only by means of couplings between initial systems.

Let us consider the mode equations (3.2a) that are already decoupled from the stable modes and are written in the following form up to the fourth order

$$\xi_{n+1}^{\pm} = \lambda^{\pm}\xi_n^{\pm} + k_{20}^{\pm}(\xi_n^{\pm})^2 + k_{11}^{\pm}\xi_n^{\pm}\xi_n^{\mp} + k_{02}^{\pm}(\xi_n^{\mp})^2 + k_{30}^{\pm}(\xi_n^{\pm})^3 + k_{21}^{\pm}(\xi_n^{\pm})^2\xi_n^{\mp} + \\ +k_{12}^{\pm}\xi_n^{\pm}(\xi_n^{\mp})^2 + k_{03}^{\pm}(\xi_n^{\mp})^3 + O(4), \quad (4.77)$$

where k are coefficients and

$$|\lambda^{\pm}| = 1, \quad (\lambda^{\pm})^n \neq 1, \quad n = 1, ..., 4. \quad (4.78)$$

Carrying out the nonlinear coordinate transformation (3.9) (see [87]) the system (4.77) can be transformed into the following system

$$\varphi_{n+1}^{\pm} = \lambda^{\pm}\varphi_n^{\pm} + \mu_{21}^{\pm}(\varphi_n^{\pm})^2\varphi_n^{\mp} + O(4), \quad (4.79)$$

where coefficient $\mu_{21}^{\pm}$ is given by

$$\mu_{21}^{\pm} = k_{21}^{\pm} + \frac{k_{20}^{\pm}k_{11}^{\pm}}{\lambda^{\pm}}\left(\frac{2\lambda^{\pm} + \lambda^{\mp} - 3}{(\lambda^{\mp} - 1)(\lambda^{\pm} - 1)}\right) + \frac{|k_{11}^{\pm}|^2}{1 - \lambda^{\mp}} + 2\frac{|k_{02}^{\pm}|^2}{(\lambda^{\pm})^2 - \lambda^{\mp}}. \quad (4.80)$$

Transforming (4.79) into polar coordinates by letting

$$\varphi_n^{\pm} = p_n e^{\pm I\phi_n}, \quad (4.81)$$

we get

$$\begin{cases} p_{n+1} = |\lambda^{\pm}|p_n + Re(\mu_{21}^{\pm})p_n^3 + O(p_n^5) \\ \phi_{n+1} = \phi_n + \theta + Im(\mu_{21}^{\pm})/|\lambda^{\pm}|p_n^2 + O(p_n^4) \end{cases}. \quad (4.82)$$

The first equation in (4.82) is the normal form of the pitchfork bifurcation, whose supercritical condition $Re(\mu_{21}^{\pm})$ has to be negative

$$Re(\mu_{21}^{\pm}) < 0. \quad (4.83)$$

The second equation of (4.82) is the circle map, where the variable ϕ_n is defined modulo 2π. Expression

$$\omega = \theta + Im(\mu_{21}^{\pm})/|\lambda^{\pm}|p_n^2 + O(p_n^4) \quad (4.84)$$

defines number and rotation of the points that will be mapped in the bifurcation diagram of the Neimark-Sacker bifurcation.

Considering (4.80), we arrive at the conclusion that a sign of $Re(\mu_{21}^{\pm})$ can be affected only by means of the coefficients $k_{20}^{\pm}$, $k_{11}^{\pm}$ corresponding to the second order variables $(\xi_n^{\pm})^2$, $\xi_n^{\pm}\xi_n^{\mp}$ and by coefficient $k_{21}^{\pm}$ at the third order term $(\xi_n^{\pm})^2\xi_n^{\mp}$.

In order to influence the coefficients at these nonlinear terms, the coupling methods have also to possess nonlinear terms. These nonlinear terms have to be included into the function $\underline{\mathbf{F}}$ of (4.33) so that they further appear in the mode equation (4.77) at the terms $(\xi_n^{\pm})^2$, $\xi_n^{\pm}\xi_n^{\mp}$, $(\xi_n^{\pm})^2\xi_n^{\mp}$.

As we can see from (4.68), the coupling mechanism is split up into two parts (both parts are dimension-scalable): the connection part $\varepsilon(q_n^{i-1} + q_n^{i+1})$ itself and the local time delayed control $k(k_1 q_{n-1}^i - q_n^i)$. In this case the connection part provides the interactions between basic systems and the control part transforms bifurcations in a desired manner. Following this scheme further, we intend to modify the control part of coupling mechanism (4.68) so that the system (4.33) undergoes the Neimark-Sacker bifurcation, taking into account the conditions (4.37) and (4.83).

Forasmuch as $\underline{\mathbf{q}}_{st} = 0$ we simplify the control part (4.56), by setting $k_1\{\beta\} = 1$ and $k_2\{\beta\} = 0$ so that

$$\underline{\mathbf{F}} = k\underline{\mathbf{q}}_{n-1}. \quad (4.85)$$

Substituting (4.85) into the system (4.42) and carrying out the linear stability analysis, we yield two eigenvalues

$$\lambda_{1,2} = \frac{a \pm \sqrt{a^2 + 4\,k}}{2}. \tag{4.86}$$

Being interested to create instability at $a_{cr} = 1$, we require $|\lambda^{\pm}_{1,2}| = 1$ at this critical value of bifurcation parameter a. The desired eigenvalues $\lambda^{\pm}$ can be found in the form $a\vartheta^{\pm}$ where coefficient $|\vartheta^{\pm}| = 1$. Performing the calculation, we get

$$k = -a^2. \tag{4.87}$$

Then the Jacobian of (4.42) with included (4.85) has the following form

$$\underline{\underline{\mathbf{J}}} = a \begin{pmatrix} \cos(-\frac{\pi}{3}) & -\sin(-\frac{\pi}{3}) \\ \sin(-\frac{\pi}{3}) & \cos(-\frac{\pi}{3}) \end{pmatrix} \tag{4.88}$$

with the eigenvalues

$$\lambda^{\pm} = \frac{a}{2}(1 \pm I\sqrt{3}), \tag{4.89}$$

that satisfies the conditions (4.78). Introducing the complex-conjugate variable $\xi^{\pm}_n$ and performing the linear coordinate transformation we get the corresponding mode equation

$$\xi^{\pm}_{n+1} = ae^{\mp\frac{\pi}{3}I}\xi^{\pm}_n - \frac{1}{4}(\xi^{\pm}_n + \xi^{\mp}_n)^2. \tag{4.90}$$

Applying the near identity transformation (3.9), we derive the normal form (4.79) for mode equations (4.90), where the eigenvalues are determined by (4.89) and the coefficient $\mu^{\pm}_{21}$ has the following form

$$\mu^{\pm}_{21} = \frac{1}{8((\lambda^{\pm})^2 - \lambda^{\mp})} + \frac{4\lambda^{\pm}\lambda^{\mp} - 3\lambda^{\mp} + (\lambda^{\mp})^2 - 2\lambda^{\pm}}{8\lambda^{\pm}(\lambda^{\mp} - 1)(\lambda^{\pm} - 1)\lambda^{\mp}}. \tag{4.91}$$

At $a_{cr} = 1$ and $k = -a^2$ the coefficient $\mu^{\pm}_{21}$ receives the numerical value $\mu^{\pm}_{21} = 0.09375 \pm I0.05412$. As can see $Re(\mu^{\pm}_{21})$ is positive, meaning the subcritical Neimark-Sacker bifurcation. In order to influence on the sign of $Re(\mu^{\pm}_{21})$ we have to introduce into the initial system (4.33) some functions (or coefficients) that will appear at the second order terms $(\xi^{\pm}_n)^2$, $\xi^{\pm}_n\xi^{\mp}_n$ of the mode equations. To this effect let us rewrite the delayed function (4.85) as follows

$$\underline{\mathbf{F}} = k_1\underline{\mathbf{q}}_n\underline{\mathbf{q}}_{n-1} + k\underline{\mathbf{q}}_{n-1}, \tag{4.92}$$

where k_1 is a coefficient. The nonlinear term $k_1\underline{\mathbf{q}}_n\underline{\mathbf{q}}_{n-1}$ does not affect on the eigenvalues $\lambda^{\pm}$ of (4.89). Now calculating once more the mode equation (4.90) with function (4.92), we yield

$$\xi^{\pm}_{n+1} = ae^{\mp\frac{\pi}{3}I}\xi^{\pm}_n - \frac{1}{4}(\xi^{\pm}_n + \xi^{\mp}_n)^2 \mp \frac{1}{4}k_1 I((\xi^{\pm}_n)^2 + (\xi^{\mp}_n)^2) \tag{4.93}$$

The NF, derived from (4.93), has the coefficient $\mu^{\pm}_{21}$ given by

$$\mu^{\pm}_{21} = \frac{((\lambda^{\mp})^2 - 2\lambda^{\pm} + 4\lambda^{\pm}\lambda^{\mp} - 3\lambda^{\mp})}{(8\lambda^{\pm}\lambda^{\mp}(\lambda^{\pm} - 1)(\lambda^{\mp} - 1))} - \frac{(k_1^2 + 1)}{8(\lambda^{\mp} - (\lambda^{\pm})^2)} + I\frac{k_1(\lambda^{\mp} + 2\lambda^{\pm} - 3)}{8h(\lambda^{\pm} - 1)(\lambda^{\mp} - 1)}. \tag{4.94}$$

The condition (4.83) of supercritical bifurcation is determined by the following inequality

$$\frac{((\lambda^{\mp})^2 - 2\lambda^{\pm} + 4\lambda^{\pm}\lambda^{\mp} - 3\lambda^{\mp})}{(8\lambda^{\pm}\lambda^{\mp}(\lambda^{\pm}-1)(\lambda^{\mp}-1))} < \frac{(k_1^2+1)}{8(\lambda^{\mp} - (\lambda^{\pm})^2)}. \tag{4.95}$$

Setting for example $k_1 = 10$ and $a_{cr} = 1$, we get for the coefficient $\mu_{21}^{\pm}$ the numerical value $-.86618 \pm I5.46678$.

Here we make one remark concerning a rotations number (4.84) determined by $Im(\mu_{21}^{\pm})$ (see [87]). The condition (4.83) requires that $\mu_{21}^{\pm}$ has to be negative, but does not impose any conditions on its numerical value. Consequently, varying numerical value of $Im(\mu_{21}^{\pm})$, we are also able to modify both the number of points, that are created by Neimark-Sacker bifurcation, and their rotation frequency, as shown in Fig. 4.11.

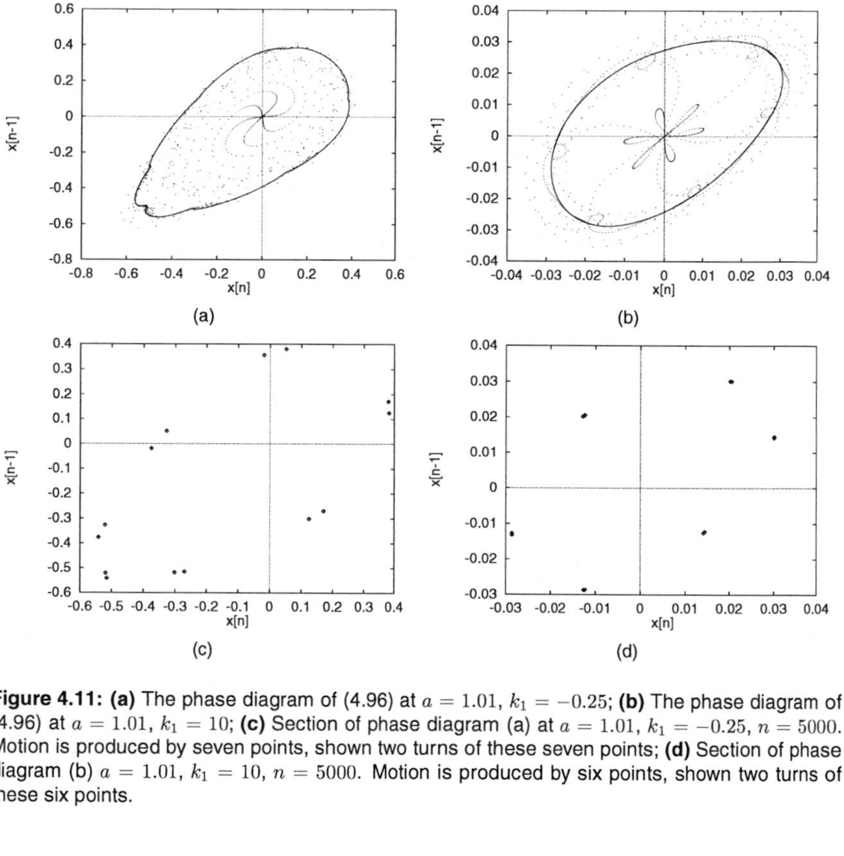

Figure 4.11: (a) The phase diagram of (4.96) at $a = 1.01$, $k_1 = -0.25$; **(b)** The phase diagram of (4.96) at $a = 1.01$, $k_1 = 10$; **(c)** Section of phase diagram (a) at $a = 1.01$, $k_1 = -0.25$, $n = 5000$. Motion is produced by seven points, shown two turns of these seven points; **(d)** Section of phase diagram (b) $a = 1.01$, $k_1 = 10$, $n = 5000$. Motion is produced by six points, shown two turns of these six points.

Dimension scaling

The derived control (4.92) is calculated for one-dimensional system with only one unstable eigenvalue. If the couplings in CML possess the connection part, that does not modify this unstable eigenvalue or this modification is small, the local control (4.92) can be used at the high-dimensional CML. In this case the control is introduced into all equations of CML, but the coefficients k, k_1 are calculated only for one unstable eigenvalue, as already shown in Sec. 4.3.1.2. Thus, choosing the basic systems in the form of Eq. (4.42) and coupling them by the coupling methods preserving unstable eigenvalue (see [131], [128], [129]), we get, for example, the following system

$$q^i_{n+1} = aq^i_n - (q^i_n)^2 + \varepsilon(q^{i-1}_n + q^{i+1}_n) + k_1 q^i_n q^i_{n-1} + kq^i_{n-1}, \;\; i = 1, ..., m. \tag{4.96}$$

Here, taking into account the boundary condition for q^{i-1}_n, q^{i+1}_n, m is the dimension of (4.96).

Generalizing this consideration, we show in Fig. 4.12 a spatio-temporal behavior of the system (4.49) with different coupling methods: firstly, the two-way ring coupling (4.49) allowing a non-periodical behavior, then the coupling method (4.96) with limit cycle, and finally, the coupling (4.68) allowing a period-two motion.

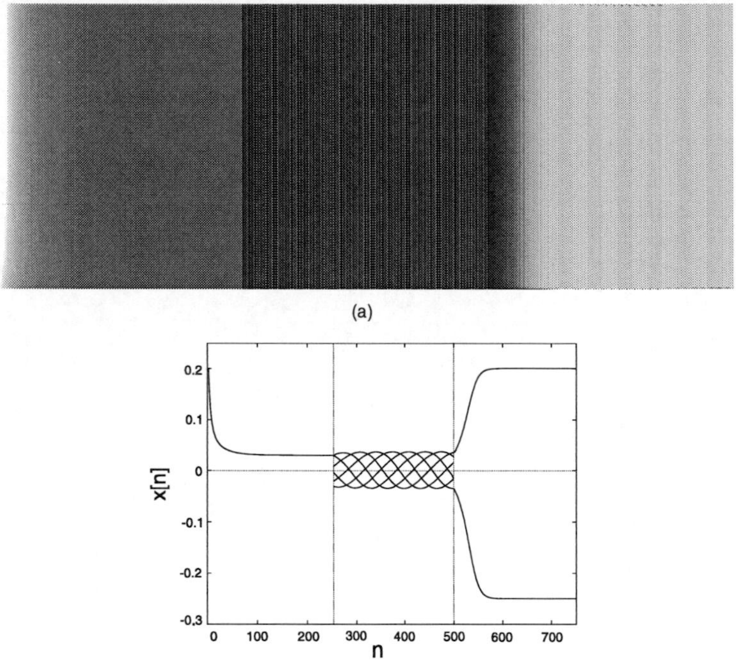

Figure 4.12: (a) Spatio-temporal behavior of the system (4.49) with $m = 512$, $a = 1.01$, $\varepsilon = 0.01$. At every 250th time step the coupling method is changed: at first the two-way ring coupling (4.49) (nonperiodical behavior), then the coupling method (4.96) (limit cycle), and finally the coupling (4.68) (period-two motion); **(b)** The temporal behavior of the 400th maps in the CML shown in Fig. 4.12 (a)

4.3.2 Modification of the nonlinear terms of normal forms

Three following sections deal with the cases where coupling methods influence nonlinear terms of NF. The first and the most simple case in Sec. 4.3.2.1 arises when the coefficient at a term of the order r is equal to zero. This is illustrated by an example with a dimension scaling. In the next two sections 4.3.2.1 and 4.3.2.2 we discuss more generally the determinancy order and unfolding of NF. Forasmuch as these cases are mainly caused by specific coupling methods, they will be illustrated by only one common example, shown in Sec. 4.3.2.3.

4.3.2.1 Change of determinancy order

Introducing into the main points of this section, we would like briefly to consider one example given by Golubitsky and Schaeffer in [114]. Considering the bifurcation problem in the form

$$\varepsilon\varphi^2 - \lambda_u\varphi + p(\varphi) = 0, \tag{4.97}$$

where $\varepsilon = 0.001$ and $p(\varphi)$ includes cubical terms, containing $\mu\varphi^3$ and $\mu \gg 0.001$, authors argue that it makes more sense to view (4.97) as a perturbation of pitchfork $\varphi^3 - \lambda_u\varphi$. Motivation can be clarified by two reasons. Firstly, the neighbourhood, on which the performed local analysis is valid, will be extremely small. The second argument in favor of a pitchfork is that this bifurcation problem may have the additional solution namely a limit point close to the origin. Accordingly, in order to correctly analyze this solution, we have to take into account the cubical terms, i.e. to increase the determinancy order of (4.97). The change of the determinancy order of NF is also mentioned by some other authors (see [162], [171]).

According to [42], [114], the reduced bifurcation problem $g(\varphi, \lambda_u)$ derived from the mode equations (3.2) is k-determined by the condition

$$\frac{d^k g}{d\varphi^k} \neq 0, \frac{d^{k-1} g}{d\varphi^{k-1}} = 0, \frac{d^{k-2} g}{d\varphi^{k-2}} = 0, \dots . \tag{4.98}$$

In other words, the bifurcation problem $g(\varphi, \lambda_u)$ is determined by the nonlinear resonant term in (4.35) of the lowest order with respect to φ. Generally speaking, the determinancy order is essential point in the local analysis because it finally defines the maximal order of Taylor expansions, of near identity transformation and of NF itself.

In this section we are interested in the question when the determinancy order of the bifurcation problem has to be changed. Before we begin to treat this matter, it needs to remark that the nonlinear terms of the time discrete NF are determined by resonant conditions imposed by unstable eigenvalues (e.g [42]). In case the unstable eigenvalue $\lambda_u = 1$, every term of one-dimensional NF is resonant. When the unstable eigenvalue is getting $\lambda_u = -1$, some terms of NF are becoming non-resonant, i.e. they do not influence the dynamics in the vicinity of bifurcations. It can be quite simply illustrated by considering the following NF

$$\varphi_{n+1} = \lambda_u\varphi_n + \mu_2\varphi_n^2 + \mu_3\varphi_n^3, \tag{4.99}$$

where μ are some functions. At $\lambda_u = -1$ this system undergoes the period-doubling bifurcation. Now we determine the orbits of the period-two motion, considering the stationary states of the following second iterated system up to the fourth order

$$\varphi_{n+1} = \lambda_u^2\varphi_n - (1 + \lambda_u)\lambda_u\mu_2\varphi_n^2 - \lambda_u(\mu_3\lambda_u^2 + \mu_3 + 2\mu_2^2)\varphi_n^3. \tag{4.100}$$

As we can see, the second order term at $\lambda_u = -1$ is always equal to zero. Moreover, from the viewpoint of bifurcation theory, any modifications of the coefficient at φ_n^3 does not perturb this normal form in the sense of NF equivalency [114]. It means, that changes of the coefficient μ_2 in (4.99), caused by the modification of couplings, does not affect this type of instability. Accordingly, in this case the possibilities of perturbation of the period-doubling bifurcation by the coupling mechanism are essentially restricted.

In the case of complex-conjugate eigenvalues $\lambda_u^{\pm}$ there are more restrictions on the possible perturbations. Therefore, being motivated by the most representative case for our approach, we will further consider only the NF with the unstable eigenvalue $\lambda_u = 1$.

Now we return to the initial question about a determinancy order of NF with $\lambda_u = 1$. In this case we assume that the main reason causing an increase of determinancy order, consists in the qualitative topological equivalency between the initial system (4.33) and reduced one (3.8) in the neighbourhood of the origin. Consequently, the correct reduction procedure has to take into account all stationary states of initial systems that are close to the origin. This is especially important when the system (4.33) immediately after bifurcation jumps to such additional stationary states. Therefore, further examples are intended to show, that it is sometimes reasonably to modify the determinancy order of initial bifurcation problem and then to consider an unfolding of this bifurcation.

1. The coefficients at the terms of order r are equal to zero

In this section we focus on the coefficients of NF at nonlinear terms of the order r. If these coefficients are equal to zero, the determinancy order r of NF has to be increased at least up to the order $r+1$. In this case the NF of basic systems has the determinancy order r, but after they are coupled, we obtain the NF of the coupled system with the determinancy order $r+1$.

Now let us consider the following low-dimensional system that is linearly coupled as the one-way ring $F_{coupl} = \varepsilon q_n^{i+1}$

$$x_{n+1} = ax_n - x_n^2 - \varepsilon y_n, \tag{4.101a}$$

$$y_{n+1} = ay_n - y_n^2 - \varepsilon z_n, \tag{4.101b}$$

$$z_{n+1} = az_n - z_n^2 - \varepsilon w_n, \tag{4.101c}$$

$$w_{n+1} = aw_n - w_n^2 - \varepsilon x_n, \tag{4.101d}$$

where $\varepsilon > 0$. Performing the linear stability analysis for this system, we get four eigenvalues. They, being evaluated on the zero stationary states, have the following form

$$\lambda_{u,s_1} = a \pm \varepsilon, \quad \lambda^{\pm}_{s_2,s_3} = a \pm I\varepsilon, \tag{4.102}$$

with instability $\lambda_u = 1$. Transforming the system (4.101) into the form (3.2) (see (4.51)-(4.55)), we use the tensor $\Gamma^V_{(2)}$ given by the eigenvectors

$$\underline{\mathbf{v}}_1 = \begin{pmatrix} -1 \\ 1 \\ -1 \\ 1 \end{pmatrix}, \underline{\mathbf{v}}_2 = \begin{pmatrix} 1 \\ 1 \\ 1 \\ 1 \end{pmatrix}, \underline{\mathbf{v}}_3 = \begin{pmatrix} -I \\ -1 \\ I \\ 1 \end{pmatrix}, \underline{\mathbf{v}}_4 = \begin{pmatrix} I \\ -1 \\ -I \\ 1 \end{pmatrix}, \tag{4.103}$$

and finally get the following mode amplitude equations

$$\xi_{n+1}^{u} = \lambda_u \xi_n^u - 2\xi_n^u \xi_n^{s_1} - (\xi_n^{s_2})^2 - (\xi_n^{s_3})^2, \tag{4.104a}$$
$$\xi_{n+1}^{s_1} = \lambda_{s_1} \xi_n^{s_1} - 2\xi_n^{s_2} \xi_n^{s_3} - (\xi_n^{s_1})^2 - (\xi_n^{u})^2, \tag{4.104b}$$
$$\xi_{n+1}^{s_2} = \lambda_{s_2} \xi_n^{s_2} - 2\xi_n^{u} \xi_n^{s_3} - 2\xi_n^{s_1} \xi_n^{s_2}, \tag{4.104c}$$
$$\xi_{n+1}^{s_3} = \lambda_{s_3} \xi_n^{s_3} - 2\xi_n^{u} \xi_n^{s_2} - 2\xi_n^{s_1} \xi_n^{s_3}. \tag{4.104d}$$

As shown in Sec.3.2, carrying out the normal form reduction, we are able to remove some nonlinear terms of (4.104) so that the unstable mode (4.104a) will be decoupled from other mode equations

$$\varphi_{n+1}^{u} = \lambda_u \varphi_n^u + \mu_{130} (\varphi_n^u)^3 + O(5), \tag{4.105a}$$
$$\varphi_{n+1}^{s_1} = \lambda_{s_1} \varphi_n^{s_1} + \mu_{221} (\varphi_n^u)^2 \varphi_n^{s_1} + O(4), \tag{4.105b}$$
$$\varphi_{n+1}^{s_2} = \lambda_{s_2} \varphi_n^{s_2} + \mu_{322} (\varphi_n^u)^2 \varphi_n^{s_2} + O(4), \tag{4.105c}$$
$$\varphi_{n+1}^{s_3} = \lambda_{s_3} \varphi_n^{s_3} + \mu_{423} (\varphi_n^u)^2 \varphi_n^{s_3} + O(4), \tag{4.105d}$$

where the coefficients μ are given by

$$\mu_{130} = \frac{2}{\lambda_u^2 - \lambda_{s_1}}, \quad \mu_{221} = \mu_{130} + \frac{4}{\lambda_u(\lambda_{s_1} - 1)},$$
$$\mu_{322} = \mu_{130} + \frac{4}{(\lambda_u \lambda_{s_2} - \lambda_{s_3})}, \quad \mu_{423} = \mu_{130} + \frac{4}{(\lambda_u \lambda_{s_3} - \lambda_{s_2})}. \tag{4.106}$$

The Eq. (4.105a) represents the NF of the pitchfork bifurcation. Thus, the NF of basic equations (4.42) has the determinancy order 2, but the linear couplings increase it up to the order 3. Now setting $\varepsilon > 0$, e.g., $\varepsilon = 0.1$, we get the numerical value $\mu_{130} > 0$ ($\mu_{130} = 10$) meaning a subcritical bifurcation. In order to obtain the supercritical pitchfork bifurcation, we apply the approach described in Sec. 4.3.1.3, introducing the following nonlinear coupling

$$q_{n+1}^{i} = a q_n^i - (q_n^i)^2 - \varepsilon q_n^{i+1} + \beta q_n^i q_n^{i+1}, \quad i = 1, ..., m. \tag{4.107}$$

so that

$$x_{n+1} = a x_n - x_n^2 + \beta x_n y_n - \varepsilon y_n, \tag{4.108a}$$
$$y_{n+1} = a y_n - y_n^2 + \beta y_n z_n - \varepsilon z_n, \tag{4.108b}$$
$$z_{n+1} = a z_n - z_n^2 + \beta z_n w_n - \varepsilon w_n, \tag{4.108c}$$
$$w_{n+1} = a w_n - w_n^2 + \beta w_n x_n - \varepsilon x_n. \tag{4.108d}$$

Applying the normal form reduction again to the system (4.108), we get the following values for the coefficient μ_{130}

$$\mu_{130} = 2 \frac{\beta + 1}{\lambda_u^2 - \lambda_{s_1}}. \tag{4.109}$$

The supercritical condition (4.83) for (4.109) is satisfied when $\beta < -1$. Carrying out this idea further, we could set $\beta = -1$ and so to achieve the further increase of the determinacy order of NF (4.105a), e.g., $r = 4$ and so on.

Dimension scaling

The previous schemes of dimension scaling were founded on the preservation of unstable eigenvalues. But in case the coupling methods transform nonlinear terms of NF, there are no general principles, that such a scheme of dimension scaling could be based on.

Considering one-way coupled ring with the negative ε (4.101), we remark that a quadratic term in NF (4.105a) vanishes because of an absence of term $(\xi_n^u)^2$ in the mode equation (4.104a). This is caused by the coordinate transformation (4.54)-(4.55) (below only the $(\xi_n^u)^2$,..., $(\xi_n^{s_3})^2$ components of these tensors are shown)

$$\Gamma_{(2)}^{V^{-1}}\Gamma_{(3)}\left(\Gamma_{(2)}^{V}(:\underline{\boldsymbol{\xi}}_n)\right)^2 = -\Gamma_{(2)}^{V^{-1}}\left(\Gamma_{(2)}^{V}(:\underline{\boldsymbol{\xi}}_n)\right)^2 =$$
$$= -\frac{1}{4}\begin{pmatrix} -1 & 1 & -1 & 1 \\ 1 & 1 & 1 & 1 \\ I & -1 & -I & 1 \\ -I & -1 & I & 1 \end{pmatrix}\begin{pmatrix} (\xi_n^{s_1} - \xi_n^u - I\xi_n^{s_2} + I\xi_n^{s_3})^2 \\ (\xi_n^u + \xi_n^{s_1} - \xi_n^{s_2} - \xi_n^{s_1})^2 \\ (\xi_n^{s_1} - \xi_n^u + I\xi_n^{s_2} - I\xi_n^{s_3})^2 \\ (\xi_n^u + \xi_n^{s_1} + \xi_n^{s_2} + \xi_n^{s_3})^2 \end{pmatrix}. \tag{4.110}$$

In order to prevent an appearance of $(\xi_n^u)^2$ in (4.105a), the eigenvector $\underline{\mathbf{v}}_1$ has to possess equal components in the absolute value. Moreover, the first row in the inverse matrix of eigenvectors has also to possess the equal number of positive and negative components. Considering this linear problem, we ascertain the interchange of -1 and $+1$ at $\underline{\mathbf{v}}_1 = (-1, +1, -1, +1, ...)^T$ independently of dimension of (4.101). However, the difficulty here is that we are not always able to determine analytically the inverse matrix of eigenvalues. In order to get round this problem, we have numerically calculated this matrix. It was hereby confirmed that for the dimension $2^m = \{8, 16, 32, 64, ...\}$ we will always get the equal number of positive and negative components in the first row. Thus, the linear one-way ring coupling (4.101) can be scaled at least to the dimension 2^m.

Now considering the nonlinear coupling terms in (4.108), we would like to know, whether their modification also leads to the change of NF (4.105a). For that, let us first scale down the dimension of the nonlinear coupling ring in (4.108), rewriting the nonlinear coupling $\underline{\mathbf{q}}_n^i\underline{\mathbf{q}}_n^{i+1}$ in the following form

$$x_{n+1} = ax_n - x_n^2 + \beta x_n y_n - \varepsilon y_n, \tag{4.111a}$$
$$y_{n+1} = ay_n - y_n^2 + \beta y_n x_n - \varepsilon z_n, \tag{4.111b}$$
$$z_{n+1} = az_n - z_n^2 + \beta z_n w_n - \varepsilon w_n, \tag{4.111c}$$
$$w_{n+1} = aw_n - w_n^2 + \beta w_n z_n - \varepsilon x_n. \tag{4.111d}$$

Calculating once more the NF of the system (4.111), we ascertain that this change of nonlinear coupling does not influence neither on the nonlinear terms of one-dimensional NF (4.105a) nor on the coefficients at these terms. Consequently, we can arrive at the conclusion that increasing a dimension of nonlinear coupling ring (4.108) also does not lead to changes of NF.

Thus, the scaling rule for the linear and nonlinear terms of the system (4.108) is

$$q_{n+1}^i = aq_n^i - (q_n^i)^2 + \beta q_n^i q_n^{i+1} - \varepsilon q_n^{i+1}, \quad i = 1, ..., 2^m\,, \tag{4.112}$$

where $\beta < -1$, ε is small and $\varepsilon > 0$, taking into account the boundary condition for q_n^{i+1}.

The numerical investigations are shown that every 2^m-scaled system demonstrates the same qualitative type of behavior in the vicinity of $a_{cr} = 0.9$ at $\varepsilon = 0.1$ and $\beta = -1.1$. Remark, that the obtained systems are dependent on the initial conditions, that in our experiment were selected equal for every even and uneven equations of system (4.112).

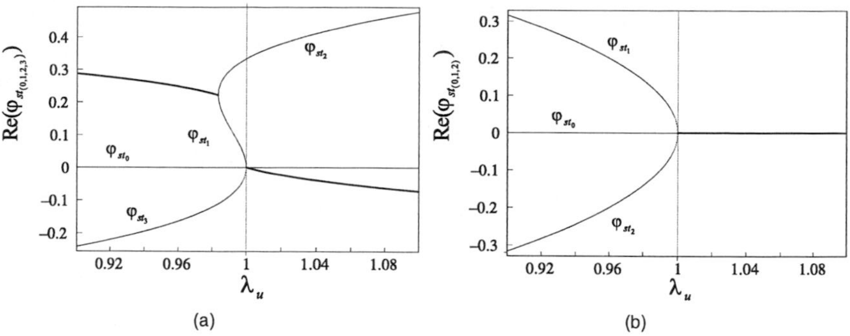

Figure 4.13: (a) The real parts of four stationary states $\varphi_{st_{0,1,2,3}}$ of the system (4.113). After the instability point at $\lambda_u = 1$ the system has the local, stable, complex-conjugate state (bold) and the real state φ_{st_2}; **(b)** The real parts of three stationary states $\varphi_{st_{0,1,2}}$ of the system (4.114). After the instability point at $\lambda_u = 1$ the system has only one stable complex-conjugate state (bold).

2. Complex-conjugate stationary states connected to the origin

Simplifying the original system and deriving the NF, we take always into account the topological adequacy of the simplified bifurcation problem with the initial one. We assume, that the normal form which does not coincide topologically with the initial problem, is false derived. In this section we intend to show that in some particular situations, caused in fact by the coupling method, the adequacy of NF with the initial problem can be achieved increasing a determinancy order of NF.

Let us first consider some hypothetical one-dimensional bifurcation problem with the unstable eigenvalue $\lambda_u = 1$, that possesses the local and nonlocal stationary states and can be described by the following NF up to the fifth order

$$\varphi_{n+1} = \lambda_u \varphi_n + \varphi_n^3 - 3\varphi_n^4 + O(5), \tag{4.113}$$

where $\varphi_n \in \mathbb{R}$ (that makes sense in practical applications). The stationary states of (4.113) are shown in Fig. 4.13(a).

We can see that the system (4.113) has in a neighbourhood of the critical point $\lambda_u = 1$ a pair of complex-conjugate stationary states and an additional branch of real states.

Applying the normal form reduction to the bifurcation problem,which initiates the NF (4.113), the transformation should be terminated on the first nonzero order term with respect to φ_n (see [42], [87], [114]). Therefore we obtain the following normal form

$$\varphi_{n+1} = \lambda_u \varphi_n + \varphi_n^3 + O(4). \tag{4.114}$$

The stationary states of (4.114) are shown in Fig. 4.13(b), where the additional branch of a real stationary state is lost and we get finally the subcritical pitchfork bifurcation. Here one can protest, being motivated that (4.113) and (4.114) coincide in the neighbourhood of instability. However, we point to very essential difference between them, being important for control purposes, namely in contrast to (4.113) the system (4.114) collapses after $\lambda_u = 1$.

These cases often arise at non-homogeneously coupled maps. We consider such an example in Sec. 4.3.2.3, where the stationary states connected to the origin are also complex. In this case only the change of determinancy order allows keeping the topological adequacy of the reduced bifurcation problem to the original one.

Generalizing, we remark that at the unstable eigenvalue $\lambda_u = 1$ the discrete system undergoes the stability change of stationary states. If the original system has local and nonlocal stationary states we can change the determinancy order of NF in order to cover nonlocal solutions of original systems. Correspondingly, in such cases the maximal determinancy order is equal to the number of distinct stationary states of initial system.

4.3.2.2 Unfolding of normal forms and an increase of complexity

The determinancy order, being once increased, introduces new high-order terms into consideration. The changing of the coupling coefficients in CML modifies the coefficients at these high-order terms of NF, causing the perturbation of the given bifurcation problem [114]. Such a perturbed bifurcation looks completely different compared to the non-perturbed original bifurcation. Correspondingly, we are interested in estimating how coupling methods can perturb the local dynamics of CML. In other words, we would like to offer a systematic characterization of the coupling methods from the viewpoint of their causation by the perturbation effect.

In considering this problem, we can apply the concept of qualitative equivalence provided by the unfolding theory [114], [115]. In this case, every perturbed bifurcation has a "maximal" universal perturbation, denoted as a universal unfolding. All other perturbations can be absorbed by a nonlinear change of coordinates. Universal unfolding points to perturbed terms that can generally be seen in the given bifurcation problem.

Moreover, the number of such terms is denoted as a codimension of NF [42], [114].

To characterize the perturbations of nonlinear terms we can consistently modify all coefficients of an initial system and then try to systematize the bifurcation diagrams arising. However, if there are more than three such coefficients, such an approach seems very awkward. Another approach to a systematic characterization may be to construct an organizing center [114] that is associated with the noted values of the coupling coefficients. Modification of these values in a very local neighborhood leads to different behaviors of the CML. However, such an organizing center is very particular to every system and therefore cannot be generalized.

Our suggestion is that the couplings can be characterized by the number of perturbed NF terms[3] (caused by this coupling) that build a non-equivalent NF in the sense of singularity theory, that is, by a codimension of instability. In this case the codimension of NF can be viewed as the "degree of freedom" of the local bifurcations that is affected by the coupling methods. Evidently, the greater the codimension is, the more possibilities to perturb the local dynamics of CML by means of couplings exist.

The codimension of local bifurcation contrasts with such essential characterization of systems as complexity. Different definitions and values of complexity exist (see [173]). As shown in Sec. 4.3.1 the CMLs, coupled homogeneously, possess a regular structure and their local dynamics is governed by a low-dimension NF contained in the initial system. Following Ebeling and Klimontovich [174], [175], it is assumed that the complexity of a CML's local dynamics is still qualitatively the same as in the initial map. However, if the coupling perturbs the NF, leading thereby to the increase of a dimension or codimension, the local dynamics is no longer governed by the NF of the initial systems. In this case, the regularity of the CML structure is disturbed, indicating that the local dynamics of CML becomes qualitatively more complex than in the initial maps. That is distinctly seen on maps coupled non-homogeneously. Thus, an increased codimension can be viewed as

[3]The terms that build non-equivalent NF in sense of singularities theory.

an increase in the complexity of the local dynamics (concerning initial maps). Forasmuch as the complexity of local dynamics is connected with the complexity of a system in the whole, it is expected that a modification of codimension also changes the complexity of the overall system.

We would like to illustrate the ideas of Secs. 4.3.2.2, and 4.3.2.1 by one example for which the notion of complex-conjugate stable states and codimensions are defined.

4.3.2.3 Nonhomogeneous couplings

The two previous sections have dealt with the essential points, arising mainly at nonhomogeneous couplings, such as the change of determinancy order of NF and the following perturbation of nonlinear terms. The mentioned homogeneous (regular) and nonhomogeneous (nonregular) couplings are different in principle, both from the viewpoint of system dynamics and their physical interpretation.

Dynamics of homogeneous systems is mainly defined by dynamics of their basic maps and by specific properties of coupling, as shown in Sec. 4.3.1 and 4.3.2. However, at nonhomogeneous coupling we obtain a completely new system, so-called kernel, whose properties can be defined only by an analysis of the whole system. The nonhomogeneous couplings usually lead to more complicated dynamics and moreover such a system can not be simplified by means of dimension scaling.

Both types of coupling can be viewed in a sense as "strong" and "weak" interactions. A few basic maps can be "strongly" connected into a kernel, further the kernels build by means of "weak" coupling a high-dimensional hybrid CML. In this way we get a hierarchy of structures, that in accordance with [174], [175] points to the increased complexity of the overall system. Remark, that in this case the dynamics of the kernel mainly determines the dynamics of the whole hybrid CML.

In order to exemplify Sec. 4.3.2.1, Sec. 4.3.2.2 and to show the specific properties of system connected by nonhomogeneous coupling, we continue analyzing the OLL map (4.1) started in Sec. (4.2). As it will be shown, the dynamics of (4.1) has little in common with an initial logistic map. Therefore the (4.1) can be viewed as a kernel with the internal "strong" coupling. The system (4.1), connected further by means of homogeneous coupling with other equal systems, build a high-dimensional hybrid CML. Moreover, an analysis of such CML can be reduced to the analysis of the system (4.1).

In the following section we investigate the cases III and IV shown by a linear analysis of (4.1) in Sec. 4.2.1. In these cases the coupling has already exerted the influence on the OP so that the behavior of the coupled OLL map is completely different from the initial logistic map. As it will be shown, these changes of OP are caused by mechanisms described in previous sections, where simultaneously a few of them affect the dynamics of two interacting systems.

1. The NF reduction for one unstable eigenvalue $\lambda_u = 1$

The further analysis will be performed separately for cases when only one eigenvalue $\lambda_1 = \lambda_u$ is unstable and when both eigenvalues $\lambda_1 = \lambda_{u_1}$, $\lambda_2 = \lambda_{u_2}$ are unstable. Accordingly the modes ξ_n^1, ξ_n^2 will be denoted in term of synergetics as the unstable ξ_n^u and the stable ξ_n^s modes for the first case (Sec. 4.3.2.3) and as $\xi_n^{u_1}$, $\xi_n^{u_2}$ for second one (Sec. 4.3.2.3).

Undergoing the bifurcation at unstable eigenvalue $\lambda_u = 1$ (case III), a system changes the stability of stationary states, so to say, it "jumps" from one to another. Considering

Fig. 4.14(a), we remark that system (4.1) possesses five stationary states. At $a = 0.64$

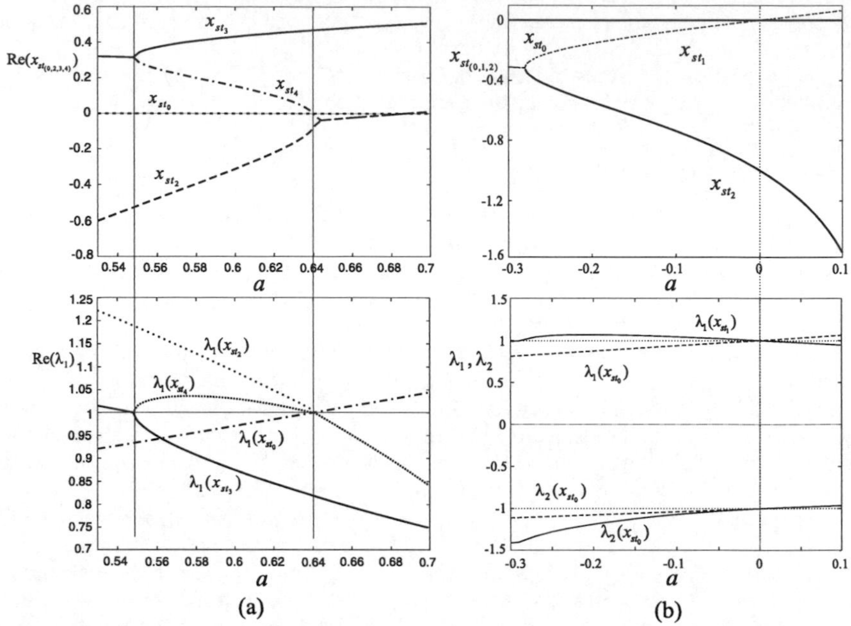

Figure 4.14: (a) The stationary states $x_{st_{(0,2,3,4)}}$ and the relevant eigenvalue λ_1 of system (4.1) at parameters value $b = 2$, $c = 0.6$. The stationary states x_{st_2} x_{st_4} after $a_{cr} = 0.64$ are complex-conjugate. The stationary state x_{st_1} lies far away in the area of negative numbers and is therefore not shown; **(b)** The stationary states $x_{st_{(0,1,2)}}$ and the relevant eigenvalues λ_1, λ_2 of system (4.1) at parameters value $b = -1.5$, $c = -1$. As can see, there are two unstable eigenvalues $\lambda_{1,2} = \lambda_{u_{1,2}} = \pm 1$ at x_{st_0} as well as at x_{st_1}.

the stationary state x_{st_0} loses stability and the system has to proceed to x_{st_4} that is connected to the origin. But in the vicinity of $a = 0.64$ the stationary states x_{st_4} and x_{st_2} are becoming complex-conjugate and the system jumps on the nonlocal stable state x_{st_3}. In order to reconstruct, at least qualitatively, the dynamics of OLL map, the reduced system at $\lambda_u = 1$ has also to possess five stationary states. It can be achieved by increasing the determinancy order of NF, in this case from two (transcritical bifurcation of logistic map) to five, that also changes a codimension of this local bifurcation.

Now we introduce the following near identity transformation

$$\underline{\xi}_n = \underline{\varphi}_n + \underline{\mu}_2 \underline{\varphi}_n^2 + \underline{\mu}_3 \underline{\varphi}_n^3 + \underline{\mu}_4 \underline{\varphi}_n^4 + \underline{\mu}_5 \underline{\varphi}_n^5 + O(6), \tag{4.115}$$

where $\underline{\mu}$ are the vector of coefficients, $\underline{\varphi}_n = (\varphi_n^u, \varphi_n^s)^T$ and then substitute (4.115) into the mode equations (4.15). Performing the simplification, we are able to remove the terms contained ξ_n^2 in (4.15a) so that to decouple unstable mode equation (4.15a) from the stable one (4.15b). Finally, we get the following reduced system calculated for the case $\lambda_1 = \lambda_u = 1$

$$\varphi_{n+1}^u = \lambda_1 \varphi_n^u + \mu_{12}(\varphi_n^u)^2 + \mu_{13}(\varphi_n^u)^3 + \mu_{14}(\varphi_n^u)^4 + \mu_{15}(\varphi_n^u)^5 + O(6), \tag{4.116a}$$
$$\varphi_{n+1}^s = \lambda_2 \varphi_n^s + \mu_{21}\varphi_n^u \varphi_n^s + \mu_{22}(\varphi_n^u)^2 \varphi_n^s + \mu_{23}(\varphi_n^u)^3 \varphi_n^s + \mu_{24}(\varphi_n^u)^4 \varphi_n^s + O(6). \tag{4.116b}$$

The equation (4.116a) represents a normal form of the system (4.1) up to sixth order, where the coefficients μ calculated for $x_{st} = 0$, $y_{st} = 0$ are summarized in Table 4.14. Numerically obtained stationary states of NF (4.116a) for the parameters values $b = 2$,

Terms	μ	Coefficients
$(\varphi_n^u)^2$	μ_{12}	$-(\lambda_1(a\lambda_1 + \lambda_2 b))\mathcal{S}$
$(\varphi_n^u)^3$	μ_{13}	$-(\lambda_2 b\lambda_1 A_{21} + 2a\lambda_1\lambda_2 A_{21} + \lambda_2^2 bA_{21} - \lambda_2 b\lambda_1)\mathcal{S}$
$(\varphi_n^u)^4$	μ_{14}	$-\lambda_2(\lambda_2 bA_{21}^2 + b\lambda_1 A_{24} - \lambda_2 bA_{21} + 2a\lambda_1 A_{24} - 2b\lambda_1 A_{21} + a\lambda_2 A_{21}^2 +$ $+\lambda_2 bA_{24})\mathcal{S}$
$(\varphi_n^u)^5$	μ_{15}	$-\lambda_2(-2\lambda_2 bA_{21}^2 + \lambda_2 bA_{28} - \lambda_2 bA_{24} + 2\lambda_2 bA_{21}A_{24} + 2\lambda_2 aA_{21}A_{24} -$ $-2b\lambda_1 A_{24} - b\lambda_1 A_{21}^2 + b\lambda_1 A_{28} + 2a\lambda_1 A_{28})\mathcal{S}$

$$\mathcal{S} = \frac{1}{(\lambda_1 - \lambda_2)c}$$

$$A_{21} = \frac{\lambda_1^2(a+b)}{c(\lambda_1 - \lambda_2)(-\lambda_2 + \lambda_1^2)}$$

$$A_{24} = -\frac{\lambda_1(-2a\lambda_2 A_{21} + 2A_{21}\mu_{12}c\lambda_1 - 2A_{21}\mu_{12}c\lambda_2 - \lambda_2 bA_{21} - \lambda_1 bA_{21} + b\lambda_1)}{c(\lambda_1 - \lambda_2)(-\lambda_2 + \lambda_1^3)}$$

$$A_{28} = -\frac{A_{21}(\mu_{12})^2 c\lambda_1 - A_{21}(\mu_{12})^2 c\lambda_2 + 3A_{24}\lambda_1^3\mu_{12}c - 3A_{24}\lambda_1^2\mu_{12}c\lambda_2 - \lambda_1^2 bA_{24} - \lambda_2 b\lambda_1 A_{24}}{c(\lambda_2^2 + \lambda_1^5 - \lambda_2\lambda_1 - \lambda_1^4\lambda_2)} -$$
$$-\frac{\lambda_2 b\lambda_1 A_{21} - 2a\lambda_1\lambda_2 A_{24} + 2\lambda_1^2 bA_{21} - a\lambda_2^2 A_{21}^2 + 2A_{21}\lambda_1^2\mu_{13}c - 2A_{21}\lambda_1\mu_{13}c\lambda_2 - \lambda_2 b\lambda_1 A_{21}^2}{c(\lambda_2^2 + \lambda_1^5 - \lambda_2\lambda_1 - \lambda_1^4\lambda_2)}$$

Table 4.14: The coefficients of the normal forms (4.116a) obtained from the mode equations (4.15) with $\lambda_1 = 1$, $x_{st} = 0$, $y_{st} = 0$.

$c = 0.6$ are shown in Fig. 4.15(a). As we can see, they qualitatively coincide with the stationary states of the original system (4.1) at least in the vicinity of $a = 0.64$.

The stationary states $\varphi_{st_{0,2,3,4}}$ of (4.116a) at $b = 0.5$, $c = -0.8$ are shown in Fig. 4.15(b). In the neighbourhood of $a = 0.36$ they have the similar to the case of Fig. 4.15(a) qualitative form, but here the state φ_{st_3} is unstable. That explains the loss of dissipation in the system (4.1) after the point $a = 0.36$ in Fig. 4.4(d).

Now we would like to explain the ideas of increasing a codimension caused by the coupling methods. The NF (4.116a) describes also the transcritical bifurcation in Fig. 4.4(a). In this case the cubic and other high-order terms does not exert an essential influence on this local bifurcation and therefore can be theoretically removed from this NF. But even at a small variation of coupling coefficients b, c the dynamics of (4.1) is noticeably perturbed by these nonlinear terms. Therefore our assumption is that the coupling generally increases the determinancy order of NF contained in the initial maps, but the high-order nonlinear terms do not influence on dynamics at some values of coupling coefficients. Modifying these coefficients, the dynamics will be changed, and hereby the coupling methods determine how much terms will perturb the dynamics. In a sense the coupling method determines how the "degree of freedom" of a local bifurcation will be changed by a coupling of initial maps into CML.

In order to obtain all possible perturbed nonlinear terms of NF with the determinancy order 5 ($\lambda_u = 1$), it needs to calculate the universal unfolding that is given e.g., by

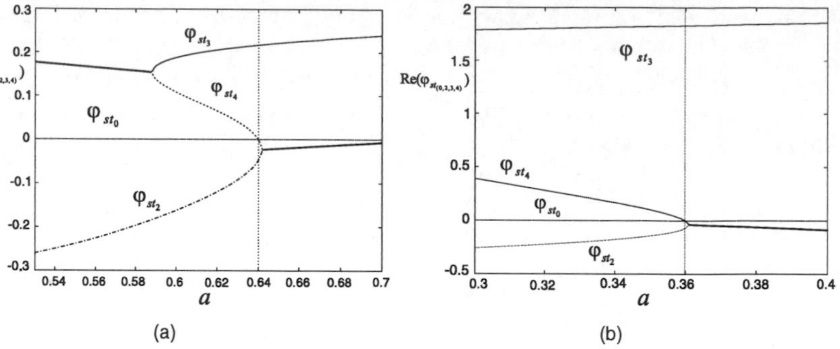

Figure 4.15: **(a)** The stationary states $\varphi_{st_{(0,2,3,4)}}$ at parameters value $b = 2$, $c = 0.6$ obtained from the normal form (4.116a). The stationary states φ_{st_2}, φ_{st_4} after $a_{cr} = 0.64$ are complex-conjugate; **(b)** The stationary states $\varphi_{st_{(0,2,3,4)}}$ at parameters value $b = 0.5$, $c = -0.8$. The nonlocal stationary states φ_{st_3} in unstable. The stationary state φ_{st_1} is not shown in both figures.

$$G(\varphi_n, \lambda_u) = \alpha_1 + \lambda_u \varphi_n + \alpha_2 \varphi_n^2 + \alpha_3 \varphi_n^3 + \alpha_4 \varphi_n^4 + \varphi_n^5 \tag{4.117}$$

with the codimension 4, where α are coefficients [4]. Therefore we can see that the nonhomogeneous coupling method of OLL map changes the codimension of a local bifurcation ($\lambda_u = 1$) from 1 (transcritical bifurcation contained in logistic map) to 4. Thus, from the viewpoint of local dynamics, the coupled system (4.1) is more complex than the initial logistic map.

2. The NF reduction for two unstable eigenvalues $\lambda_{u_1} = 1$, $\lambda_{u_2} = -1$

As shown by a linear stability analysis the system (4.1) possesses two unstable eigenvalues at $c = \pm 1$ and $a_{cr} = 0$ (case IV) as shown in Fig. 4.14(b). Such cases have been discussed for the systems continuous in time [171], [172], especially the cases when the original system cannot be represented in the form of mode equations (see [42], [176], [177]).

In order to derive NF with two unstable eigenvalues we apply the standard approach of normal form reduction [42], [45], [113]. The resonant conditions, given by the unstable eigenvalues, specify the nonlinear terms that can be removed from the mode equations (4.15), determining hereby the dimension of the derived NF (see Sec. 3.2).

Introducing the NIT (4.115), where $\underline{\varphi}_n = (\varphi_n^{u_1}, \varphi_n^{u_2})^T$, we have first to determine the maximal order of transformations. As usually, we take into account the qualitative equivalence between the original bifurcation problem and reduced one, and gradually increase the order of performed transformation until we achieve such an equivalence. As shown by a numerical investigation, the transformation up to the third order does not satisfy this requirement. By the transformation up to the fourth order we achieve a qualitative equivalence. The codimension of the obtained hereby NF, calculated by the "k-jet" technique [42], [97] is equal to 12.

However, in this case a chaotic window occurs in the vicinity of a bifurcation point that cannot be avoided by a change of initial conditions. Therefore, we will perform the trans-

[4] In cases the original system can be transformed into the form (3.2), the terms of zero order with respect to state variables (such as α_1) cannot be contained in the unfolding of the NF.

formation up to the fifth order to reach, firstly, the qualitative equivalence and, secondly, to avoid the chaotic window. The obtained two-dimensional NF has the following form

$$\begin{aligned}\varphi_{n+1}^{u_1} &= \lambda_1 \varphi_n^{u_1} + \mu_{32}(\varphi_n^{u_1})^2 + \mu_{33}(\varphi_n^{u_2})^2 + \mu_{34}\varphi_n^{u_1}(\varphi_n^{u_2})^2 + \mu_{35}(\varphi_n^{u_1})^3 + \\ &+ \mu_{36}(\varphi_n^{u_1})^2(\varphi_n^{u_2})^2 + \mu_{37}(\varphi_n^{u_1})^4 + \mu_{38}(\varphi_n^{u_2})^4 + O(5), \qquad (4.118a)\\ \varphi_{n+1}^{u_2} &= \lambda_2 \varphi_n^{u_2} + \mu_{42}\varphi_n^{u_1}\varphi_n^{u_2} + \mu_{43}(\varphi_n^{u_1})^2\varphi_n^{u_2} + \mu_{44}(\varphi_n^{u_2})^3 + \mu_{45}(\varphi_n^{u_1})^3\varphi_n^{u_2} + \\ &+ \mu_{46}\varphi_n^{u_1}(\varphi_n^{u_2})^3 + O(5), \qquad (4.118b)\end{aligned}$$

where μ are coefficients, summarized in Table 4.15.

The numerically obtained 3D bifurcation diagram of the system (4.118) is shown in Fig. 4.16. As we can see, the system has a period-two motion before the bifurcation

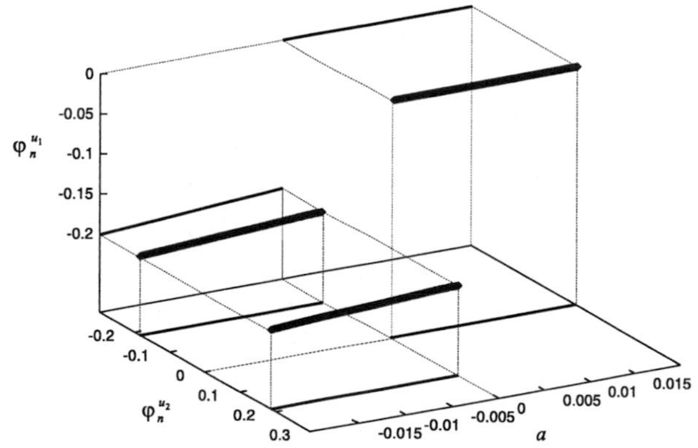

Figure 4.16: 3D bifurcation diagram of two-dimensional normal form (4.118) of system (4.1) at parameters value $b = -1.5$, $c = -1$ with two unstable eigenvalues $\lambda_{u_1} = 1$, $\lambda_{u_2} = -1$. One can observe period-two motion that at $a_{cr} = 0$ is abruptly changed to nonperiodical behavior.

point $a_{cr} = 0$. Immediately after this point the dynamics is abruptly changing to the non-periodical motion $x_{st} = 0$, $y_{st} = 0$. The same type of behavior can be observed in the original OLL map in Fig. 4.4 (e),(f) (allowing for other stationary states).

Generalizing our treatment with the nonhomogeneous coupling at OLL map, we would like to summarize the main results. Firstly, varying the coupling coefficients b and c, the NF of (4.1) undergoes all possible modifications. Different unstable eigenvalues cause different resonant terms. Determinancy order is increased from the second to the fifth order. Finally, two unstable eigenvalues cause a two-dimensional normal form, that demonstrates some non-generic bifurcation types. In this way we have obtained the system, having little in common with the initial logistic map. From this viewpoint, the nonhomogeneous couplings (and similar to them) have been yet not enough researched and such investigations have to be continued further.

Secondly, the required calculations even for the two-dimensional coupled map are pretty tedious and are mainly carried out by means of symbolic manipulation programs. The analytical treatment of really high-dimensional maps, coupled non-homogeneously, seems to be very hard or even impossible. From this viewpoint, the example of the OLL map was mainly intended to show increasing the system's complexity, occurring at nonhomogeneous couplings. Finally, characterizing the couplings in the OLL map,

Terms	μ	Coefficients
$(\varphi_n^{u_1})^2$	μ_{32}	$(aN_2^2 + N_1bN_2 - 2N_1by_{st_1}N_2 + N_1bx_{st_1}P)\mathcal{S}$
$(\varphi_n^{u_2})^2$	μ_{33}	$N_1(bN_1 + aN_1 - 2y_{st_1}bN_1 + bx_{st_1}P)\mathcal{S}$
$\varphi_n^{u_1}\varphi_n^{u_2}$	μ_{42}	$-N_2(bN_2 + bN_1 + 2aN_1 - 2by_{st_1}N_2 + 2bx_{st_1}P - 2y_{st_1}bN_1)\mathcal{S}$
$(\varphi_n^{u_1})^3$	μ_{35}	$N_1(A_{21}(2aN_2 + 2bx_{st_1}P + N_1b - 2N_1by_{st_1} +$ $+bN_2 - 2by_{st_1}N_2) - bN_2)\mathcal{S}$
$\varphi_n^{u_1}(\varphi_n^{u_2})^2$	μ_{34}	$N_1(-b(N_2 + 2N_1) + A_{23}(N_1b + 2bx_{st_1}P + 2aN_2 - 2N_1by_{st_1} +$ $bN_2 - 2by_{st_1}N_2) + A_{12}(2bx_{st_1}P - 2by_{st_1}N_2 - 2N_1by_{st_1} +$ $+N_1b + bN_2 + 2aN_2))\mathcal{S}$
$(\varphi_n^{u_2})^3$	μ_{44}	$(-2aN_1^2A_{23} + 4N_1by_{st_1}N_2A_{23} - 2N_1bN_2A_{23} - 2N_2bx_{st_1}A_{23}P +$ $+N_1bN_2)\mathcal{S}$
$(\varphi_n^{u_1})^2\varphi_n^{u_2}$	μ_{43}	$(-A_{12}(-4N_2^2by_{st_1} + 2N_2bx_{st_1}P + 2N_2^2b + 2aN_2^2) - A_{21}(2N_2bx_{st_1}P +$ $+2aN_1^2 - 4N_1by_{st_1}N_2 + 2N_1bN_2) + 2bN_2^2 + N_1bN_2)\mathcal{S}$
$(\varphi_n^{u_1})^2(\varphi_n^{u_2})^2$	μ_{36}	$(A_{12}(-2N_1^2b - 4N_1bN_2 + A_{12}(-2N_1by_{st_1}N_2 + N_1bx_{st_1}P +$ $+aN_2^2 + N_1bN_2)) + A_{15}(N_1^2b + 2N_1bx_{st_1}P - 2N_1^2by_{st_1} +$ $+N_1bN_2 - 2N_1by_{st_1}N_2 + 2aN_1N_2) - A_{21}3N_1^2b - A_{23}(N_1^2b +$ $+2N_1bN_2) + A_{21}A_{23}(2N_1^2b + 2N_1bx_{st_1}P + 2aN_1^2 - 4N_1^2by_{st_1}) +$ $+A_{26}(-2N_1by_{st_1}N_2 + 2N_1bx_{st_1}P + N_1^2b + 2aN_2N_1 - 2N_1^2by_{st_1} +$ $+N_1bN_2))\mathcal{S}$
$(\varphi_n^{u_1})^4$	μ_{37}	$N_1(A_{24}(N_1b + 2bx_{st_1}P - 2N_1by_{st_1} + 2aN_2 - 2by_{st_1}N_2 + bN_2) +$ $+A_{21}(A_{21}(bx_{st_1}P - 2N_1by_{st_1} + aN_1 + N_1b) - N_1b - 2bN_2))\mathcal{S}$
$(\varphi_n^{u_2})^4$	μ_{38}	$N_1(A_{17}(bN_2 - 2by_{st_1}N_2 + N_1b - 2N_1by_{st_1} + 2aN_2 + 2bx_{st_1}P) +$ $+A_{23}(A_{23}(aN_1 - 2N_1by_{st_1} + N_1b + bx_{st_1}P) - 3N_1b))\mathcal{S}$
$(\varphi_n^{u_1})^3\varphi_n^{u_2}$	μ_{45}	$(-A_{12}(A_{21}(N_2^2b + N_1bN_2 - 2N_1by_{st_1}N_2 + 2aN_2N_1 + 2N_2bx_{st_1}P -$ $-2N_2^2by_{st_1}) + 3N_2^2b) - A_{15}(2aN_2^2 - 4N_2^2by_{st_1} + 2N_2^2b + 2N_2bx_{st_1}P) -$ $-A_{21}(-2N_2^2b - 4N_1bN_2) - A_{24}(-4N_1by_{st_1}N_2 + 2N_1bN_2 + 2aN_1^2 +$ $+2N_2bx_{st_1}P))\mathcal{S}$
$\varphi_n^{u_1}(\varphi_n^{u_2})^3$	μ_{46}	$(-A_{12}(A_{23}(-2N_1by_{st_1}N_2 + 2aN_2N_1 + N_1bN_2 + 2N_2bx_{st_1}P -$ $-2N_2^2by_{st_1} + N_2^2b) - (2N_1bN_2 + N_2^2b)) - A_{17}(2aN_2^2 + 2N_2bx_{st_1}P +$ $+2N_2^2b - 4N_2^2by_{st_1}) + A_{23}(2N_2^2b + 4N_1bN_2) - A_{26}(-4N_1by_{st_1}N_2 +$ $+2aN_1^2 + 2N_1bN_2 + 2N_2bx_{st_1}P))\mathcal{S}$

Table 4.15: The coefficients of the two-dimensional normal forms (4.118) obtained from the mode equations (4.15) with $\lambda_1 = 1$, $\lambda_2 = -1$, $x_{st} = x_{st_1}$, $y_{st} = y_{st_1}$.

$$\mathcal{S} = \frac{1}{P(\lambda_1 - \lambda_2)}$$

$$A_{12} = \frac{N_1(-2by_{st_1}N_2 + 2bx_{st_1}P + 2aN_2 + bN_2 + bN_1 - 2y_{st_1}bN_1)}{P\lambda_1(\lambda_1 - \lambda_2)(\lambda_2 - 1)}$$

$$A_{21} = -\frac{N_2(aN_2 + bx_{st_1}P - 2by_{st_1}N_2 + bN_2)}{P(\lambda_1 - \lambda_2)(\lambda_1^2 - \lambda_2)}$$

$$A_{23} = -\frac{aN_1^2 + N_1bN_2 + N_2bx_{st_1}P - 2N_1by_{st_1}N_2}{P\lambda_2(\lambda_1 - \lambda_2)(\lambda_2 - 1)}$$

$$A_{24} = -\frac{A_{21}(N_2^2b - 2N_1by_{st_1}N_2 + 2N_2bx_{st_1}P - 2N_2^2by_{st_1} + 2\lambda_1^2\mu_{32}P - 2\lambda_1\mu_{32}P\lambda_2)}{P(\lambda_1 - \lambda_2)(\lambda_1^3 - \lambda_2)} - \frac{A_{21}(N_1bN_2 + 2aN_2N_1) - bN_2^2}{P(\lambda_1 - \lambda_2)(\lambda_1^3 - \lambda_2)}$$

$$A_{17} = -\frac{A_{23}(-4N_1^2by_{st_1} + 2N_1^2b + 2aN_1^2 + 2N_1bx_{st_1}P) - A_{12}(\mu_{33}\lambda_2P\lambda_1 + \mu_{33}\lambda_2^2P) - N_1^2b}{P(\lambda_1 - \lambda_2)(\lambda_1 - \lambda_2^3)}$$

$$A_{15} = -\frac{A_{12}(P(\lambda_1 - \lambda_2)(\mu_{32}\lambda_2 + \mu_{42}\lambda_1) + N_1(4by_{st_1}N_2 - 2bx_{st_1}P - 2bN_2) - 2aN_2^2)}{P\lambda_1(\lambda_1 - \lambda_2)(\lambda_1\lambda_2 - 1)} - \frac{A_{21}(4N_1^2by_{st_1} - 2N_1^2b - 2aN_1^2 - 2N_1bx_{st_1}P) + N_1^2b + 2bN_1N_2}{P\lambda_1(\lambda_1 - \lambda_2)(\lambda_1\lambda_2 - 1)}$$

$$A_{26} = -\frac{A_{23}(2\lambda_2\mu_{42}P\lambda_1 - 2\lambda_2^2\mu_{42}P + N_2^2b - 2N_1by_{st_1}N_2 + 2aN_2N_1 + 2N_2bx_{st_1}P)}{P\lambda_2(\lambda_1 - \lambda_2)(\lambda_1\lambda_2 - 1)} - \frac{A_{23}(-2N_2^2by_{st_1} + N_1bN_2)}{P\lambda_2(\lambda_1 - \lambda_2)(\lambda_1\lambda_2 - 1)} - \frac{A_{12}N_2(b(2x_{st_1}P - 2N_2y_{st_1} + N_2) - N_1(2by_{st_1} - b - 2a))}{P\lambda_2(\lambda_1 - \lambda_2)(\lambda_1\lambda_2 - 1)} - \frac{2A_{21}P\lambda_1\mu_{33}(\lambda_1 - \lambda_2) - bN_2(N_2 + 2N_1)}{P\lambda_2(\lambda_1 - \lambda_2)(\lambda_1\lambda_2 - 1)}$$

Table 4.16: Prolongation of Table 4.15.

we remark that they increase the codimension of a local bifurcation from 1 to 4 at one unstable eigenvalue and from 1 to 12 at two unstable eigenvalues.

4.4 Summary

In this chapter we have considered the influence of coupling methods on the local dynamics of interacting systems. The main focus of our consideration lies on normal forms, where couplings can modify four elements: eigenvalues, the determinancy order, the dimension and finally the perturbation of nonlinear terms. Moreover, couplings can change the number of degrees of freedom and complexity (at least of calculations) of the system. The last effect is distinctly seen in homogeneous and nonhomogeneous couplings. In the first case, the dimension of the coupled systems is not important because the structure of the couplings is regular and the local dynamics is governed by a low-dimension NF contained in the initial systems. Therefore, the whole coupled system can be scaled down to these equations. In the second case, the couplings fuse the initial systems into a completely new system with an irregular structure, whose local dynamics is no longer governed by a low-dimension NF of the initial systems. An analytical treatment of even a two-dimension coupled map is very difficult. This change of complexity is seen by changing a codimension of local bifurcations that can be viewed in a sense as a qualitative characterization of the coupling method.

In summarizing this chapter, we point to two primary results. First, we have improved and numerically compared different techniques of the OP derivation begun in Chap. 3. Second, we have described the origin of how couplings modify the collective activity of interacting autonomous systems. The information thus obtained, in particular, the macroscopic order parameters, may be applied to the control and design of the desired cooperative phenomena in collective systems. The development of external control using the order parameter is described in Chap. 5, whereas the modification of interactions is discussed in Chap. 6.

Chapter 5

External control mechanisms

Das Ziel der Gewandten: ohne Gewalt zu herrschen.
Luc de Clapiers Vauvenargues.

The order parameter equations, describing behavior in regions of instability, can be used not only for analysis but also for control of nonlinear time-discrete dynamical systems. Usually, the dimension of order parameters equations is less than the dimension of the original evolution equations. Therefore, it is reasonable to introduce control mechanisms primarily into order parameter equations, and then to use the results obtained for the control of the original systems.

The first control strategy is to avoid chaotic behavior. This is achieved by shifting the appropriate points of the first bifurcation in a period-doubling cascade. The second control strategy is focused on the control of periodical motion by using a feedback mechanism with a large time delay. In the method proposed here, the delayed feedback is introduced in the n-th iterated system to shift only the first bifurcation, as proposed by the first technique. Then, going back to the original system by changing the time delay in the feedback functions, the successive bifurcations of the period-doubling cascade can be also selectively shifted. Therefore, combining the results of both techniques, we are able to modify several first bifurcations of the period-doubling cascade. Using the synergetic concept, different local and global control schemes are derived which can be successfully applied to a control of collective activity in autonomous distributed systems even in the high-dimensional case.

5.1 Motivation

Considering different processes and phenomena in technical areas, in computer science as well as in physics, biology, and chemistry, we can ascertain they are often discrete in time (see [121], [178], [179], [180]). The mathematical models, in the form of discrete equations, are more suitable for describing these processes than models obtained by ordinary differential equations [154]. Thus, the analytical and numerical analyses of discrete dynamical systems, and the control of their behavior, have become an important point of modern applied research. This control may be primarily thought of as the control of chaotic behavior often occurring in such systems and the control of bifurcations.

Chaotic behavior in real dynamical systems may lead to failure, overload or even damage, and therefore has to be strictly avoided. The control of bifurcations points to the control of collective behavior as being the main focus of this thesis; both problems are connected, as noted in Chap. 2. Many general approaches (see [145], [147], [181], [182], [183]) as well as application-oriented control mechanisms (see [129], [146], [184], [185]) have recently been suggested as methods for solving these problems.

There are some differences between the work cited and the work presented here. First, the calculation of parameters for control is made analytically, and is based on the synergetic concept of order parameters. Second, we do not stabilize the unstable periodic orbit of a strange attractor, but extend the working area in the parameter space by shifting appropriate bifurcation points, as suggested in [184]. The concept of order parameters allows on the one hand an analysis of a system in the vicinity of instability, and on the other hand the development of a general control mechanism, which makes it possible to change the behavior of a system in an appropriate way.

The derived control scheme may be implemented in a collective system in two ways, as sketched in Fig. 5.1.

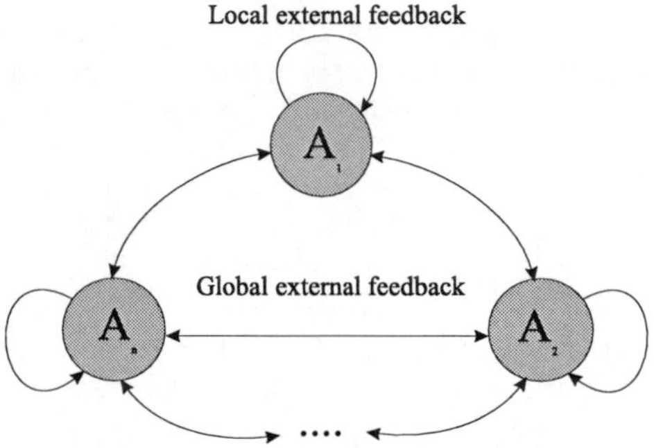

Figure 5.1: Local and global external feedbacks.

First, each subsystem might possess its own local feedback. The modification of global collective behavior is achieved through an aggregation of the modified behaviors of every subsystem. This control approach is very efficient, especially in the case where the interactions are inaccessible, or their modifications are very expensive. Second, the external control may be performed by means of global interaction among autonomous subsystems. An advantage of this scheme is that the global control feedback may perform specific interactions that are destined for information and interaction transfer, and simultaneously form the collective behavior in the desired way. Both schemes will be implemented for the examples introduced in chapter 3 and 4.

From the technical viewpoint, the investigations in this chapter are primarily focused on two important strategies for allowing the control of non-periodical and periodical motion. The stabilization of non-periodical motion is accomplished by shifting the first period-doubling bifurcation. This is achieved by applying a specific delayed feedback scheme to the original system. The advantage of this control mechanism lies in their ability to be applied without any a priori knowledge of the system, for example for the use of the well-known method suggested by [146]. The bifurcation control (see [145], [170],) may be also applied to the stabilization of periodical motion. The need for this arises for example in the discrete rhythmic control of robot motion dynamics (see [186]). Here, a dynamical behavior is changed from a high-periodical motion to a low-periodical one. Due to the dimension scaling, this method can be applied both to low- and to high-dimensional systems.

The underlying ideas can be generalized as follows. We begin with the synergetic order parameter equations [43], [104] that describe local behavior. Applying the control to these equations, for example changing the stability conditions by inserting the delayed feedback control (DFC), we can shift the bifurcation point towards increasing or decreasing the control parameter, and thus stabilize this motion in a dedicated area of the parameter space. As shown in [161], we achieve the same stabilization effect, controlling only the unstable mode amplitude equations. Use of this technique noticeably simplifies the calculation of control feedback in high-dimensional systems [161] in contrast to some other approaches (for example [129]). In addition, by stabilizing either only the unstable amplitudes or a few stable amplitudes, different global and local feedback mechanisms [187] can be analytically derived, as shown in Sec. 5.2.

The stabilization of periodical motion in the original discrete system is achieved by the delayed feedback introduced first into the n-th iterated system. After calculating the necessary parameters in the iterated system for the case of non-periodical motion, we return to the original system, changing the time delay in the feedback functions. Such an approach allows an exact definition of the period of stabilization and also a new critical value of the control parameter in the original system. In addition, in this case, completely different time delays may be obtained compared to those, for example, suggested in [185]. As shown below, all the mechanisms obtained are completely applicable to a pure period-doubling scenario as well as to scenarios that also contain other bifurcation types, for example the Naimark-Sacker bifurcation. An interesting feature of this method is the use of combined control mechanisms with different time delays, thereby stabilizing a few periodical motions with different periods.

5.2 A synergetic approach toward control

In the following sections we consider the theoretical framework which leads to the control of non-periodical and periodical motion: the synergetic approach towards control, an extension to the high-dimensional case and the change of the time delay in feedback functions. The essential remark at this step is that the control mechanisms, being once derived, can be further applied directly to a bifurcation control of time discrete systems, not requiring an additional transformation of these systems.

5.2.1 Delayed feedback control method

Here we intend to show an application of the synergetic concept to a control purpose. The underlying idea of the developed method mainly consists of the fact that low-dimensional equations (order parameter equations) of motion describe a dynamics of high-dimensional systems in the vicinity of bifurcations. Considering a problem of bifurcation control in some area of a parameter space, we can assert that the order parameters play the key role, namely *it is enough to control only the order parameters to stabilize the dynamics of the original system near the bifurcation* [161].

For a derivation of these reduced equations, the original system should be first transformed into the mode amplitude equations with the decoupled linear parts. As already demonstrated in the previous sections, the order parameter equations can be obtained from the equations with unstable eigenvalues by elimination of mode amplitudes with stable eigenvalues in linear parts.

As shown in [161], some control methods, such as the time-delay feedback [141], [184]

$$\underline{\mathbf{F}} = k(\underline{\mathbf{q}}_{n-1} - \underline{\mathbf{q}}_n), \tag{5.1}$$

where k is the control coefficient and $\underline{\mathbf{q}}$ is the state vector, have *the same effect of stabilization when the control functions are inserted into the order parameter equations or into the mode amplitude equations with the unstable eigenvalues.* In other words, for the stabilization of the system's behavior the time delay control scheme can be directly introduced into the mode amplitude equations.

Thus, being guided by the aforementioned idea, we propose to transform the original system into the form of the mode amplitude equations and then to implement the time delayed feedback in the unstable mode equations. After this, the changed system together with the introduced control has to be transformed back into the original form.

5.2.2 Implementation of the control method

The mode amplitude equations Eqs. 3.78 have the decoupled linear parts which are given by $\Gamma^{\Lambda}_{(2)}$. From the conditions $|\lambda| \geq 1$ and $|\lambda| < 1$ the eigenvalues are denoted as unstable λ_u and correspondingly as stable λ_s. For simplification, henceforth the mode amplitude equations will be denoted as the mode equations and the mode amplitudes $\underline{\boldsymbol{\xi}}_n$ as the amplitudes. In the terms of the synergetics the amplitude with the stable eigenvalue λ_s in the linear part is denoted as the stable amplitude ξ^s_n and accordingly with the unstable eigenvalue λ_u as the unstable amplitude ξ^u_n.

As already mentioned, we write the equations of the unstable and stable amplitudes separately and then use the unstable mode amplitude equations for a stabilization in the following form [141], [161]

$$\underline{\boldsymbol{\xi}}^u_{n+1} = \Gamma^{\Lambda_u}_{(2)}(:\underline{\boldsymbol{\xi}}^u_n) + \sum_{r=2}^{p} \Gamma^{\tilde{N}_u}_{(r+1)}(:\underline{\boldsymbol{\xi}}^u_n)^r + k\Gamma^{C_u}_{(2)}(:\underline{\boldsymbol{\xi}}^u_{n-1} - \underline{\boldsymbol{\xi}}^u_n), \tag{5.2a}$$

$$\underline{\boldsymbol{\xi}}^s_{n+1} = \Gamma^{\Lambda_s}_{(2)}(:\underline{\boldsymbol{\xi}}^s_n) + \sum_{r=2}^{p} \Gamma^{\tilde{N}_s}_{(r+1)}(:\underline{\boldsymbol{\xi}}^s_n)^r, \tag{5.2b}$$

where k is the coefficient and $\Gamma^{\Lambda_u}_{(2)}$, $\Gamma^{\Lambda_s}_{(2)}$ are the tensors of the unstable and correspondingly of the stable eigenvalues. Extending this scheme further, we can also insert DFC into several or all of the stable mode equations

$$\underline{\boldsymbol{\xi}}^u_{n+1} = \Gamma^{\Lambda_u}_{(2)}(:\underline{\boldsymbol{\xi}}^u_n) + \sum_{r=2}^{p} \Gamma^{\tilde{N}_u}_{(r+1)}(:\underline{\boldsymbol{\xi}}^u_n)^r + k\Gamma^{C_u}_{(2)}(:\underline{\boldsymbol{\xi}}^u_{n-1} - \underline{\boldsymbol{\xi}}^u_n), \tag{5.3a}$$

$$\underline{\boldsymbol{\xi}}^s_{n+1} = \Gamma^{\Lambda_s}_{(2)}(:\underline{\boldsymbol{\xi}}^s_n) + \sum_{r=2}^{p} \Gamma^{\tilde{N}_s}_{(r+1)}(:\underline{\boldsymbol{\xi}}^s_n)^r + k\Gamma^{C_s}_{(2)}(:\underline{\boldsymbol{\xi}}^s_{n-1} - \underline{\boldsymbol{\xi}}^s_n), \tag{5.3b}$$

where $\Gamma^{C_u}_{(2)}$, $\Gamma^{C_s}_{(2)}$ are tensors and

$$(\Gamma^{C_u}_{(2)})_{ij} = \delta_{ij}, \quad (\Gamma^{C_s}_{(2)})_{ij} = \delta_{ij}. \tag{5.4}$$

For the case of Eq. (5.2) we can also introduce $\Gamma^{C_s}_{(2)}$ and delayed feedback into the stable mode equations, but then all components of this tensor should be equal to zero. Such representation will be used again in the backward transformation into the original system.

Now we intend to show that the systems (5.2) and (5.3) lead to the local or correspondingly to the global control mechanisms [188] in the original system (3.1). As the first step, the control coefficient k should be suitably calculated. To this effect, we introduce new state vector $\underline{\eta}_n$ and then the delayed equations (5.2) are equivalent to the following system

$$\underline{\xi}^u_{n+1} = \Gamma^{\Lambda_u}_{(2)}(:\underline{\xi}^u_n) + \sum_{r=2}^{p} \Gamma^{\tilde{N}_u}_{(r+1)}(:\underline{\xi}^u_n)^r + k\Gamma^{C_u}_{(2)}(:\underline{\eta}_n - \underline{\xi}^u_n), \tag{5.5a}$$

$$\underline{\xi}^s_{n+1} = \Gamma^{\Lambda_s}_{(2)}(:\underline{\xi}^s_n) + \sum_{r=2}^{p} \Gamma^{\tilde{N}_s}_{(r+1)}(:\underline{\xi}^s_n)^r, \tag{5.5b}$$

$$\underline{\eta}_{n+1} = \underline{\xi}^u_n. \tag{5.5c}$$

From the conditions $|\lambda_u(\alpha_{cr})| = 1$ at $\alpha_{cr} = \alpha^{new}_{cr}$, where α^{new}_{cr} is the new critical value of the control parameter α, we find the corresponding values of the parameter k. At this value the system (5.5) and accordingly (5.2) lose stability i.e. the bifurcation point will be appropriately shifted. This leads to the avoidance of a high periodic and chaotic behaviour at least in a confined region of parameter's space, which we call "working area" of control parameters i.e. it leads to a stabilization of the system in this working area. Similar calculations can be also carried out for the system (5.3), but in this case two new state vectors $\underline{\eta}^1_n$, $\underline{\eta}^2_n$ have to be introduced into the system (5.5).

Now we briefly sketch a calculation from changed mode equations backward to the original system. Since the time delay feedback functions are linear, the tensor of the nonlinear term $\Gamma^{\tilde{N}}$ in Eq. (3.78) remains unchanged and the mode equations can be written in the following form

$$\underline{\xi}_{n+1} = \Gamma^{\Lambda}_{(2)}(:\underline{\xi}_n) + \sum_{r=2}^{p} \Gamma^{\tilde{N}}_{(r+1)}(:\underline{\xi}_n)^r + k\Gamma^{C}_{(2)}(:\underline{\xi}_{n-1} - \underline{\xi}_n), \tag{5.6}$$

where $\Gamma^C_{(2)}$ consists of the tensors $\Gamma^{C_u}_{(2)}$ and $\Gamma^{C_s}_{(2)}$, and where all components of $\Gamma^{C_s}_{(2)}$ are equal to zero in the case of Eq. (5.2). Multiplying (5.6) from the left side with the tensor $\Gamma^V_{(2)}$, by the analogy to the equation (3.77), we yield

$$\Gamma^V_{(2)}(:\underline{\xi}_{n+1}) = \Gamma^L_{(2)}\Gamma^V_{(2)}(:\underline{\xi}_n) + \sum_{r=2}^{p} \Gamma^N_{(r+1)}(\Gamma^V_{(2)}(:\underline{\xi}_n))^r +$$
$$+ [k\Gamma^V_{(2)}\Gamma^C_{(2)}\Gamma^{V^{-1}}_{(2)}]\Gamma^V_{(2)}(:\underline{\xi}_{n-1} - \underline{\xi}_n). \tag{5.7}$$

The emergence of the expression in the square brackets in the rhs of (5.7) is caused by the control mechanism (5.1). Now adding the stationary state $\underline{\mathbf{q}}_{st}$, we finally get

$$\underline{\mathbf{q}}_{n+1} = \underline{N}(\underline{\mathbf{q}}_n, \{\alpha\}) + [k\Gamma^V_{(2)}\Gamma^C_{(2)}\Gamma^{V^{-1}}_{(2)}](:\underline{\mathbf{q}}_{n-1} - \underline{\mathbf{q}}_n). \tag{5.8}$$

Calculating the terms in square brackets $k[\Gamma^V_{(2)}\Gamma^C_{(2)}\Gamma^{V^{-1}}_{(2)}] = k\Gamma^{cnt}_{(2)}$, using $\Gamma^C_{(2)}$ for the case of Eq. (5.2), we obtain the global control mechanism

$$\underline{\mathbf{q}}_{n+1} = \underline{\mathbf{N}}(\underline{\mathbf{q}}_n, \{\alpha\}) + k\Gamma^{cnt}_{(2)}(:\underline{\mathbf{q}}_{n-1} - \underline{\mathbf{q}}_n), \tag{5.9}$$

or assuming $\underline{\mathbf{q}}_n = (q^1_n, q^2_n, ..., q^m_n)^T$ in components we get

$$q^i_{n+1} = N^i(q^1_n, q^2_n, ..., q^m_n, \{\alpha\}) + k\sum_{j=1}^{m} r^i_j(q^i_{n-1} - q^i_n), \tag{5.10}$$

where $r^i_j = (\Gamma^{cnt}_{(2)})_{ij}$ are coefficients.

Otherwise, using $\Gamma^C_{(2)}$ for the Eq. (5.3), the system (5.8) will be transformed into the following local control scheme

$$\underline{\mathbf{q}}_{n+1} = \underline{\mathbf{N}}(\underline{\mathbf{q}}_n, \{\alpha\}) + k(\underline{\mathbf{q}}_{n-1} - \underline{\mathbf{q}}_n), \tag{5.11}$$

or in components

$$q^i_{n+1} = N^i(q^1_n, q^2_n, ..., q^m_n, \{\alpha\}) + k(q^i_{n-1} - q^i_n). \tag{5.12}$$

Here $\Gamma^C_{(2)}$ shows, that the feedback functions are inserted into each of the unstable as well as the stable mode equations. Let us remark that deriving the global control mechanism, we generally do not need to calculate the mode equations; it is enough to calculate only the term $k[\Gamma^V_{(2)}\Gamma^C_{(2)}\Gamma^{V^{-1}}_{(2)}]$ determined mainly by tensor of eigenvectors $\Gamma^V_{(2)}$. These calculations are shown in the example of the global control scheme in the case of the Hénon map.

5.2.3 Application to the high-dimensional case

Parameters of control schemes applied to low-dimensional maps can be calculated completely analytically. Advantage of this approach is obvious especially in such cases where the goal of control consists in a complex transformation of local dynamics, e.g., a change of the type of local bifurcation. However, the main question of interest is whether these control mechanisms can achieve the goal of control if they are applied to high-dimensional maps. Considering the special case of the delayed feedback, we would assert it affects similarly a local dynamics of both low-dimensional maps and high-dimensional CML, if these CML are coupled by using special coupling methods. These coupling methods have to preserve the unstable eigenvalues independently of the dimension of the coupled system. In particular this relates to many homogeneous couplings, such as one- and two-way ring coupling with open or closed boundaries (see [128], [129]).

This thought can be illustrated by two-way ring coupling. Let λ be the eigenvalue of one-dimensional basic systems (3.1) that generally depends on α and $\underline{\mathbf{q}}_{st}$, and ε is the coupling coefficient where $\varepsilon > 0$ and $\varepsilon \ll 1$. The eigenvalues $\tilde{\lambda}$ of this CML are given by the following expression (see [131])

$$\tilde{\lambda}_i = \lambda(\alpha, \underline{\mathbf{q}}_{st}) + 2\varepsilon \cos\left(\frac{2\pi i}{2m}\right), \quad 1 \leq i \leq 2m, \tag{5.13}$$

where $2m$ is the dimension of the coupled system. Evidently, that the maximal in magnitude eigenvalue $\tilde{\lambda}_i$ and hence the parameter k are independent of m and can be analytically calculated as long as $\lambda_u(\alpha_{cr}, \underline{\mathbf{q}}_{st})$ is independent of $\underline{\mathbf{q}}_{st}$, i.e., $\underline{\mathbf{q}}_{st} = 0$. In the cases where $\underline{\mathbf{q}}_{st} \neq 0$ the coefficient k can be determined at least numerically.

This approach can be directly applied to the local control scheme (5.12). In this case the control, being introduced into all mode equations, has finally the same form in the transformed system (5.12) and in the mode equations (5.6), i.e. it is not modified by the linear coordinate transformation. Therefore only the unstable eigenvalue needs to be determined in order to calculate the coefficient k, i.e. we are able to apply the local controlling directly to the two-way ring coupled map lattice (5.12)

$$q_{n+1}^{i}=N^{i}(q_{n}^{i})+\varepsilon(q_{n}^{i-1}+q_{n}^{i+1})+k(q_{n-1}^{i}-q_{n}^{i}),\ \ 1\leq i\leq 2m, \tag{5.14}$$

where $2m$ is the dimension of the CML. In this way the parameters of a local control scheme can be calculated first for a low-dimensional system and then the dimension of a controlled system can be scaled up without modifications of control.

In contrast to a local control, the scaling of a global control is not so simple. The main difficulty here consists in calculating the tensor of eigenvectors $\Gamma^{V}_{(2)}$ for a high-dimensional CML coupled by any of conventional couplings. Forasmuch as a modification of dimension changes stationary states, the $\Gamma^{V}_{(2)}$ calculated first for the dimension m may have another form for $m+1$, i.e. scaling of a control mechanism is in fact not possible. In order to get round this problem, we can use the fact that the delayed feedback does not modify stationary states. Therefore if the coupling and the control can be combined together in a global "control-coupling" scheme we get the system whose local properties are dimension-invariant.

Let us write a CML coupled by means of delayed functions in the following form (see Eq. (3.73)-(3.78))

$$\underline{\mathbf{q}}_{n+1}=\Gamma_{(1)}+\Gamma^{I}_{(2)}(:\underline{\mathbf{q}}_{n})+\sum_{r=2}^{p}\Gamma^{I}_{(r+1)}(:\underline{\mathbf{q}}_{n})^{r}+k\Gamma^{V}_{(2)}\Gamma^{C}_{(2)}\Gamma^{V^{-1}}_{(2)}(:\underline{\mathbf{q}}_{n-1}-\underline{\mathbf{q}}_{n}), \tag{5.15}$$

where $(\Gamma^{I})_{i\neq j}=0$, $(\Gamma^{I})_{ii}=a$, a is some function of control parameters and $(\Gamma^{C_s}_{(2)})_{ij}=0$. This notation means that every basic equation of this CML is dependent only on its own variable $q_{n+1}^{i}=N^{i}(q_{n}^{i})$ and these basic equations are then globally coupled by the delayed function. Remark that the tensor $\Gamma^{V}_{(2)}$ in Eq. (5.15) is not yet determined.

Now repeating the steps (3.75)-(3.78), we transform (5.15) into the form of mode equations, writing the basic system and the coupling part separately

$$\underline{\boldsymbol{\xi}}_{n+1}=\Gamma^{V^{-1}}_{(2)}\Gamma^{IL}_{(2)}\Gamma^{V}_{(2)}(:\underline{\boldsymbol{\xi}}_{n})+\Gamma^{V^{-1}}_{(2)}\sum_{r=2}^{p}\Gamma^{IN}_{(r+1)}(\Gamma^{V}_{(2)}(:\underline{\boldsymbol{\xi}}_{n}))^{r}+k\Gamma^{C}_{(2)}(:\underline{\boldsymbol{\xi}}_{n-1}-\underline{\boldsymbol{\xi}}_{n}), \tag{5.16}$$

where the tensors $\Gamma^{IL}_{(2)}$, $\Gamma^{IN}_{(2)}$ generally have a block-diagonal form. Simplifying the linear part, Eq. (5.16) can be rewritten as

$$\underline{\boldsymbol{\xi}}_{n+1}=\Gamma^{IL}_{(2)}(:\underline{\boldsymbol{\xi}}_{n})+\sum_{r=2}^{p}\Gamma^{\tilde{N}}_{(r+1)}(:\underline{\boldsymbol{\xi}}_{n})^{r}+k\Gamma^{C}_{(2)}(:\underline{\boldsymbol{\xi}}_{n-1}-\underline{\boldsymbol{\xi}}_{n}). \tag{5.17}$$

We can see the tensor $\Gamma^{V}_{(2)}$ influences only the nonlinear part of every basic equation. Hence we can conclude that $\Gamma^{V}_{(2)}$ does not exert any influence on linear properties of the coupled system (5.15) and consequently can be chosen arbitrary. Moreover, the Jacobian of Eq. (5.17) has a block-diagonal form, consequently the coefficient k can be calculated by a linear stability analysis of only one equation with the introduced delayed function.

Generally speaking, we have additionally to take into account a change of other characteristic properties of bifurcations, like sub- or super-criticality, caused by a global control. Since the most significant condition for a successful control is the modification of eigenvalues, we correspondingly restrict the focus of consideration on this topic.

Returning to the initial CML (5.15), we assert that the globally coupled CML has the following form

$$q_{n+1}^i = N^i(q_n^i) + k\sum_{j=1}^{m} r_j^i(q_{n-1}^j - q_n^j), \;\; 1 \le i \le m, \tag{5.18}$$

where

$$r_j^i = \left(\Gamma_{(2)}^V \Gamma_{(2)}^C \Gamma_{(2)}^{V^{-1}}\right)_{ij}, \tag{5.19}$$

and $\Gamma_{(2)}^V$ is chosen arbitrary (such that $\Gamma_{(2)}^{V^{-1}}$ exists). Using the following tensor, that the inverse tensor exists for

$$(\Gamma_{(2)}^V) = \begin{pmatrix} 1 & 0 & \cdots & 0 & -1 \\ 1 & 0 & \cdots & -1 & 0 \\ \vdots & \vdots & \ddots & \vdots & \vdots \\ 1 & -1 & \cdots & 0 & 0 \\ 1 & 1 & \cdots & 1 & 1 \end{pmatrix}, (\Gamma_{(2)}^{V^{-1}}) = \frac{1}{m}\begin{pmatrix} 1 & 1 & \cdots & 1 & 1 \\ 1 & 1 & \cdots & 1-m & 1 \\ \vdots & \vdots & \ddots & \vdots & \vdots \\ 1 & 1-m & \cdots & 1 & 1 \\ 1-m & 1 & \cdots & 1 & 1 \end{pmatrix}, \tag{5.20}$$

we get the following well-known scheme

$$r_j^i = \frac{1}{m}, \tag{5.21}$$

where m is the dimension of the CML.

Now let us consider the restriction of the global control scheme. As already mentioned, the delayed function is inserted into the equation which is supposed to be unstable. Shifting the instability point forwards, we stabilize hereby this equation. Bifurcation points of other equations still remain at the same place, causing the instability at the old unshifted position. Therefore the scheme (5.18) is suitable only for the shift of instability backwards.

Modifying this scheme to be suitable to the shift forwards, we rewrite the tensor $\Gamma_{(2)}^C$ in the following form

$$(\Gamma_{(2)}^{C_u})_{ij} = \delta_{ij}, \;\; (\Gamma_{(2)}^{C_s})_{ij} = (1+p)\delta_{ij}. \tag{5.22}$$

where p is a coefficient so that $0 < p < 1$. Then using (5.20), the coefficients r_j^i of (4.73) can be presented as the following matrix

$$\underline{\underline{\mathbf{r}}} = \frac{1}{m}\begin{pmatrix} \tilde{p}-p & -p & \cdots & -p \\ -p & \tilde{p}-p & \cdots & -p \\ \vdots & \vdots & \ddots & \vdots \\ -p & -p & \cdots & \tilde{p}-p \end{pmatrix}, \tag{5.23}$$

where $\tilde{p} = m(p+1)$.

The meaning of the scheme (5.23) is that other equations of the CML are shifted a bit further than the "initially" unstable equation. Therefore, even being shifted, this equation still remains the unstable equation in the coupled system. Let us point out, that the bifurcation point of the whole CML can be calculated on the basis of only one equation as it will be shown further in Sec. 5.4.2. Performing the transformation to a high-dimensional CML, it has been assumed basic systems are one-dimensional. In case they are multidimensional the common form of (5.20), (5.23) does not change, but every element of these matrices has to be replaced by the corresponding diagonal matrix.

The last point, needed to be mentioned here, concerns a transient time and a choice of control parameters reducing it. Generally, the transient time depends on three factors:

dimension of CML, type of coupling and, finally, the working area of bifurcation parameters. The underlying idea about a reduction of transient time is well-known: the closer the values of bifurcation parameters are to a super-stable point, the faster the dynamics relaxes to stationary states. Therefore, in order to reduce a transient time the control has to be so chosen that the working area of bifurcation parameters was shifted to the super-stable point as close as possible. Influence of dimension on a transient time depends on coupling methods: the global couplings conduct interactions much faster than the local (e.g., one or two neighbours) couplings. Correspondingly, the large the dimension is, the more "global" the coupling should be.

5.3 Stabilization of periodical motions

5.3.1 Controlling of the initial system

The underlying idea towards the stabilization of a periodical motion emanates from a bifurcation analysis of dynamical systems discrete in time; namely, for analysis of a periodical motion we use the iterated system, such that

$$\underline{\mathbf{q}}_{n+1} = \underbrace{\underline{\mathbf{N}}(...(\underline{\mathbf{N}}(\underline{\mathbf{q}}_n)))}_{r\ times} = \underline{\mathbf{N}}^{[r]}(\underline{\mathbf{q}}_n). \tag{5.24}$$

In this case, a periodical motion of the period-doubling cascade is first narrowed down to a non-periodical motion and then investigated by conventional methods of local analysis. Hereby, with the second iterated system, it is possible to analyze the second bifurcation point, with the third iterated - the third bifurcation point and so on. The impact of the iteration is that a system at the asymptotic dynamics depicts only several points on the bifurcation diagram. For instance, with the second iterated system, only every second point of the original system will be depicted and thus, instead of two bifurcation branches, only one branch will be shown on the bifurcation diagram. Accordingly, the period-doubling bifurcation is hereby transformed into the pitchfork bifurcation.

Now using the n-th iterated map as an initial system, we can construct a control mechanism, which shifts the bifurcation point along a working area of control parameters and thus stabilizes this nonperiodical behavior

$$\underline{\mathbf{q}}_{n+1} = \underline{\mathbf{N}}^{[r]}(\underline{\mathbf{q}}_n, \{\alpha\}) + k(\underline{\mathbf{q}}_{n-1} - \underline{\mathbf{q}}_n). \tag{5.25}$$

The coefficient k in the system (5.25) can be exactly calculated for the appropriate α_{cr}, thereby defining the desired local (5.11) or global (5.9) control scheme.

5.3.2 Alternative control scheme

The question here arises is, how the mechanism explicated above can be applied for a stabilization of periodical motion of the non-iterated original system. Evidently, we need to expand the control mechanism, which is first introduced into the n-th iterated map for a stabilization of a non-periodical motion, so that it can then stabilize an appropriate periodical motion of the original system.

To answer this question, we have to consider in detail the influence of delayed feedback on the dynamics of systems discrete in time. At the long-time dynamics, the time-delay term (5.1) constricts all neighbour points to a medial point defined by the coefficient

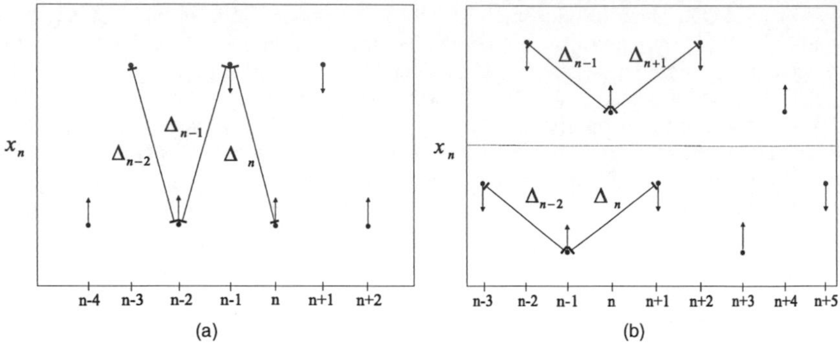

Figure 5.2: (a) Mechanism of the time delay feedback for a period-two motion; **(b)** Mechanism of the time delay feedback for a period-four motion.

k. This mechanism in the case of a period-two motion is shown in Fig. 5.2(a). Here for calculation of the point x_{n+1}, the distance between the points x_{n-1} and x_n should first be obtained that generally corresponding to the delayed term $(\underline{\mathbf{q}}_{n-1} - \underline{\mathbf{q}}_n)$ and then the point x_{n+1} has to be displaced down on the factor $k\Delta_n$. In the bifurcation diagram for $k < 0$ the instability point will be shifted forward in compliance with an increment of control parameters and for $k > 0$ correspondingly backwards.

At a long-time dynamics, all neighbour points eventually converge to one point, in other words, a periodical motion will be reduced to a non-periodical motion. Considering in a like manner the effect of the feedback (5.1), we arrive at the conclusion that this delayed function has a period two; namely the following functions

$$\underline{\mathbf{F}} = k(\underline{\mathbf{q}}_{n-1} - \underline{\mathbf{q}}_{n-2}), \tag{5.26a}$$

$$\underline{\mathbf{F}} = k(\underline{\mathbf{q}}_{n-3} - \underline{\mathbf{q}}_{n-2}), \tag{5.26b}$$

$$\underline{\mathbf{F}} = k(\underline{\mathbf{q}}_{n-3} - \underline{\mathbf{q}}_{n}) \tag{5.26c}$$

are equivalent to the original feedback $k(\underline{\mathbf{q}}_{n-1} - \underline{\mathbf{q}}_n)$. Validity of this statement can be verified in the following way. Let λ be the eigenvalue of one-dimensional system (3.1) whose value lies in some neighbourhood of -1. Inserting original feedback $k(q_{n-1} - q_n)$ into (3.1), we try to shift the bifurcation point so that the unstable eigenvalue of this system becomes $\tilde{\lambda}_u^{orig}(\{\alpha\}, k) = -1$. From this condition the corresponding control law $k = 0.5\lambda + 0.5$ for the inserted original feedback can be calculated. Now using this control law, the unstable eigenvalue $\tilde{\lambda}_u^{orig}$ can be shifted to the value -1 despite the value of λ. Inserting the functions (5.26) into (3.1) and utilizing the derived control law, the one-dimensional system (3.1) is expected to be also stabilized. Deriving the unstable eigenvalue for this system, we remark that the corresponding Jacobians are the companion matrices [189] whose eigenvalues can be calculated from

$$(\tilde{\lambda}^a)^3 - \lambda(\tilde{\lambda}^a)^2 - k\tilde{\lambda}^a + k = 0, \tag{5.27a}$$

$$(\tilde{\lambda}^b)^4 - \lambda(\tilde{\lambda}^b)^3 + k\tilde{\lambda}^b - k = 0, \tag{5.27b}$$

$$(\tilde{\lambda}^c)^4 - (\lambda - k)(\tilde{\lambda}^c)^3 - k = 0. \tag{5.27c}$$

Inserting $k = 0.5\lambda + 0.5$ and then solving (5.27) numerically, we get $\tilde{\lambda}_u^{orig} = \tilde{\lambda}_u^a = \tilde{\lambda}_u^b = \tilde{\lambda}_u^c = -1$ for $\lambda = [0.9(9), -1.3(9)]$. Moreover numerical investigations of the systems

(3.68) and (4.1) (see also Secs. 5.4.2 and 5.5) with the included functions (5.26) have become more sensitive to initial conditions.

Now we consider the motion of a system at period-doubling cascade with a period four. In this case there are two branches of the period-two motion, which are shown in the upper and lower parts of the Fig. 5.2(b). All points on this diagram are depicted in turn, at first in the upper part then in the lower one. Narrowing down the period-four motion to the period-two motion, we should separately control the upper and lower branches. In the case of Fig. 5.2(b), applying the control (5.1) in a similar way, we have first to obtain the distance between the points x_{n-3} and x_{n-1} and then correspondingly to displace the point x_{n+1}. Likewise for a calculation of the point x_{n+2} the distance between points x_{n-2} and x_n should be determined. Continuing such contemplations further, we find that this mechanism corresponds to the following time delay term

$$\underline{\mathbf{F}} = k(\underline{\mathbf{q}}_{n-3} - \underline{\mathbf{q}}_{n-1}). \tag{5.28}$$

After avoiding an unwanted period-four motion, the influence of function (5.28) is gradually becoming zero.

It is necessary to remark that the coefficients k calculated for the same α_{cr} in the upper and lower parts of Fig. 5.2(b) can be different. This reflects the fact that different stationary states are used by the linear stability analysis at the second iterated map. Another control scheme is based on this observation which uses the control of only one bifurcation's branch. In this case, the control mechanism should be in turn switched on in the one part and switched off in the other

$$\underline{\mathbf{F}} = \psi(n)k(\underline{\mathbf{q}}_{n-3} - \underline{\mathbf{q}}_{n-1}), \quad \psi(n) = n \; mod \; 2. \tag{5.29}$$

The function $\psi(n)$ must equal 1 at the stable state used by the linear stability analysis (in other words on the used branch) and accordingly equal 0 at the other. The numerical investigation shows that the delayed feedback (5.29) guarantees more stable controlling for various initial conditions than the function (5.28).

Generalizing this method for a stabilization of τ-periodical motion at the period-doubling scenario, the following time delay feedback can be introduced

$$\underline{\mathbf{F}} = k(\underline{\mathbf{q}}_{n-(2\tau-1)} - \underline{\mathbf{q}}_{n-(\tau-1)}). \tag{5.30}$$

Here τ is the period which has to be stabilized (for example, in Eq. (5.28) the motion, that has to be stabilized, is the period-two motion) and the coefficient k is calculated by the linear stability analysis of the τ-iterated map, as shown in Sec. 5.2. In other case the control is activated only on one of bifurcation branches

$$\underline{\mathbf{F}} = \psi(n)k(\underline{\mathbf{q}}_{n-(2\tau-1)} - \underline{\mathbf{q}}_{n-(\tau-1)}), \quad \psi(n) = \begin{cases} 1: & \textit{selected branch} \\ 0: & \textit{other} \end{cases}, \tag{5.31}$$

where $\psi(n)$ determines on which branch the control should be activated.

Now one last question remains: Can this mechanism be applied to other bifurcation scenarios? We have investigated some systems and arrived at the conclusion, that a non-periodical motion and a periodical motion of the period-doubling scenario can be stabilized despite the type of next bifurcations. This case is shown by the coupled map demonstrating the period-doubling bifurcation and next the Naimark-Sacker bifurcation.

Other types of periodical motion seemingly cannot be stabilized by the same method (some arguments for an application of the similar methods to non-period-doubling bifurcations are given at the end of Sec. 5.5). This fact can be explained by a characteristic behavior shown by a system undergoing e.g., the Naimark-Sacker bifurcation. In this case there is a lot of points (their amount depends on a distance from bifurcation point) that are moving along a circle in contrast to the period-doubling bifurcation where a defined amount of points (depends from the bifurcation) is moving between two (several) positions. Moreover, a conventional delayed control manipulates these points always pair wise. The delayed control scheme designed to control the Naimark-Sacker bifurcation has to manipulate all points separately, adapting to their amount.

5.4 Control mechanism applied to single agent

In this section we apply the developed control mechanism to the first example of a single agent introduced in Chap. 3, whose mathematical model is represented by the Hénon map. This example is intended to show the features of bifurcation control using the order parameter as well as the cases of control of periodical motion and control of spatio-temporal chaos.

5.4.1 Bifurcation control

Controlling the behavior of a dynamical system is of great practical interest, because it is often the case that the behavior is not suitable for a given purpose. So the question arises of how we can implement in a given dynamical system such control mechanisms which guarantee a suitable behavior. Our idea in this context is to exploit the knowledge obtained by the investigation of order parameters. Although they are only defined in the neighbourhood of instability, this is not a crucial restriction, because the control is required only in order to avoid the instability which leads to the unsuitable behavior. Thus it is expected, that we can use the fact that the number of degrees of freedom in the order parameter equation is reduced, compared to that in the original system. Our intention here is to proceed in two steps. In the first step we introduce suitable control mechanisms only in the order parameter equation and investigate its stability in order to validate the control mechanism. Whenever the control mechanism is found to be suitable enough, our second step is then to implement exactly the same control mechanism into the mode equations. This is due to the fact that it is not possible to derive the mode equations only from the knowledge of the order parameter equation by itself. After inserting the control mechanism into the mode equations, it is possible to calculate the corresponding original equations. To demonstrate the usefulness of this approach, we have selected a control mechanism which was suggested first by Pyragas [141], [184], and which uses feedback functions with delay of the (5.1). With these functions, we have attempted to avoid a high periodic and even chaotic behavior at least in a confined region in a parameter space which we call "working area" by shifting the bifurcation point. Hereby all other bifurcations in the scenario will be shifted also although not by the same value. This is due to the scaling properties of the bifurcation diagram.

5.4.1.1 Control of a order parameter equation

The simplest solution regarding the shift of a bifurcation point is the insertion of a parameter k only in the linear part of the order parameter equation (3.98) with the quadratic center manifold

$$\xi_{n+1}^u = (\lambda_u - k)\xi_n^u + \frac{a}{b(\lambda_s - \lambda_u)}(\lambda_s A_2(\xi_n^u)^2 + \lambda_u \xi_n^u)^2. \tag{5.32}$$

where A_2 is defined by (3.96). This solution however is trivial and it is usually not possible to considerably shift the bifurcation point due to the nonlinear part of (5.32).

Introducing the feedback functions (5.1) and by a variation of the parameter k, we are able to change the location of the critical point. This critical point is shifted in such a way, so that the stationary states remain stable in the whole "working area". The order parameter equation (3.98) has then the following form:

$$\xi^u_{n+1} = (\lambda_u - k)\xi^u_n + \frac{a}{b(\lambda_s - \lambda_u)}(\lambda_s A_2(\xi^u_n)^2 + \lambda_u \xi^u_n)^2 + k\xi^u_{n-1}. \tag{5.33}$$

Obviously, this one-dimensional equation is equivalent to the two-dimensional system

$$\xi^u_{n+1} = (\lambda_u - k)\xi^u_n + \frac{a}{b(\lambda_s - \lambda_u)}(\lambda_s A_2(\xi^u_n)^2 + \lambda_u \xi^u_n)^2 + k\eta_n \ , \tag{5.34a}$$

$$\eta_{n+1} = \xi^u_n \tag{5.34b}$$

with a new state variable η_n. In order to investigate the stability of the Eqs. (5.34) we perform a linear stability analysis and yield the following eigenvalues

$$\hat{\lambda}_{1,2} = \frac{\lambda_u - k}{2} \pm \frac{1}{2}\sqrt{(\lambda_u - k)^2 + 4k} \tag{5.35}$$

and the matrix of the eigenvectors

$$\underline{\underline{\hat{\mathbf{V}}}} = \begin{pmatrix} \hat{\lambda}_s & \hat{\lambda}_u \\ 1 & 1 \end{pmatrix}. \tag{5.36}$$

From the condition $|\ \hat{\lambda}_{1,2}\ | > 1$ at $a_{cr} = a^n_{cr}$ where a^n_{cr} is the new critical value of the control parameter a, we find the corresponding values of the parameter k. At this value the order parameter equation loses its stability. The dependence of the unstable eigenvalue $\hat{\lambda}_2$ from the parameters a and k is shown in Fig. 5.3.

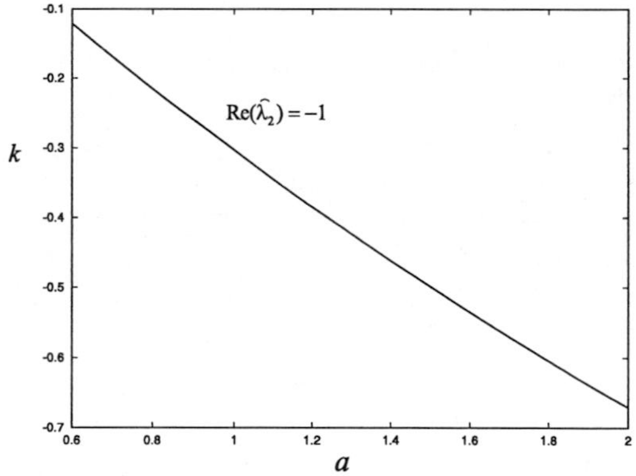

Figure 5.3: Real part of the unstable eigenvalue $\hat{\lambda}_2$ as a function of the parameters a, k.

Numerical investigations show, that with this method it is possible to shift the bifurcation point over a wide range, even further than the second bifurcation of period-doubling section which occurs at $a_{cr} = 0.9125$. To derive the order parameter equation for the system (5.34) with a quadratic center manifold we get the following mode equations:

$$\varphi^s_{n+1} = \hat{\lambda}_s \varphi^s_n + \hat{K}\hat{\Phi} \,, \tag{5.37a}$$
$$\varphi^u_{n+1} = \hat{\lambda}_u \varphi^u_n - \hat{K}\hat{\Phi} \,, \tag{5.37b}$$

with

$$\hat{K} = \frac{a}{b(\lambda_s - \lambda_u)(\hat{\lambda}_s - \hat{\lambda}_u)}, \quad \hat{\Phi} = (\lambda_u + \hat{\lambda}_s \lambda_s A_2 \varphi^s_n + \hat{\lambda}_s \lambda_u A_2 \varphi^u_n)^2 (\hat{\lambda}_s \varphi^s_n + \hat{\lambda}_u \varphi^u_n)^2 \,. \tag{5.38}$$

Here φ^s_n, φ^u_n are accordingly new stable and unstable modes. Now we apply again the center manifold theorem and restrict ourselves up to the second order with respect to the smallness parameter ε, which means:

$$\varphi^s_n = h(\varphi^u_n) = \hat{A}_2 (\varphi^u_n)^2 + O(3). \tag{5.39}$$

To proceed further we insert (5.39) in (5.37), and expand all ε-depended terms into a Taylor series with respect to ε. After that we neglect all term of the order $O(\varepsilon^{\frac{3}{2}})$ keeping in mind that for the period-doubling bifurcation $\varphi^u_n \sim \varepsilon^{\frac{1}{2}}$ holds. As result we get

$$\varphi^s_{n+1} = \hat{\lambda}_s \varphi^s_n + \hat{K}\hat{\Phi} = \hat{A}_2 \hat{\lambda}^0_s \varepsilon + (\hat{\lambda}^0_u)^2 \hat{K}^0 (\lambda^0_u)^2 \varepsilon + O(\varepsilon^{\frac{3}{2}}) \,, \tag{5.40a}$$
$$\varphi^u_{n+1} = \hat{\lambda}_u \varphi^u_n - \hat{K}\hat{\Phi} = \hat{\lambda}^0_u \varepsilon^{\frac{1}{2}} - (\hat{\lambda}^0_u)^2 \hat{K}^0 (\lambda^0_u)^2 \varepsilon + O(\varepsilon^{\frac{3}{2}}), \tag{5.40b}$$

where $\hat{K}^0$ is the first term of expressions (5.38) in the Taylor series with respect to the smallness parameter ε. From the analogy to Eq. (3.7) it follows then

$$\hat{A}_2 \varepsilon (\hat{\lambda}^0_u)^2 = ((\hat{\lambda}^0_u)^2 \hat{K}^0 (\lambda^0_u)^2 + \hat{A}_2 \hat{\lambda}^0_s) \varepsilon \,, \tag{5.41}$$

and from that for the coefficient of the center manifold

$$\hat{A}_2 = \hat{K}^0 \frac{(\hat{\lambda}^0_u \lambda^0_u)^2}{(\hat{\lambda}^0_u)^2 - \hat{\lambda}^0_s} \,. \tag{5.42}$$

Finally the order parameter equation has then the following form:

$$\varphi^u_{n+1} = \hat{\lambda}_u \varphi^u_n - \hat{K}(\lambda_u + \hat{\lambda}_s \lambda_s A_2 \hat{A}_2 (\varphi^u_n)^2 + \hat{\lambda}_s \lambda_u A_2 \varphi^u_n)^2 (\hat{\lambda}_s \hat{A}_2 (\varphi^u_n)^2 + \hat{\lambda}_u \varphi^u_n)^2. \tag{5.43}$$

Numerical investigations of Eq. (5.43) show, that with this method it is indeed possible to shift the bifurcation point remarkably, whereas the type of the bifurcation, i.e., the period-doubling bifurcation remains unchanged.

5.4.1.2 Control of mode equations

As already mentioned we turn now to the treatment of mode equations. In general, the number of mode equations is larger than the number of order parameter equations and so the question arises in which of these equations the control mechanism should be implemented. There exist at least two possibilities depending on the number of state variables: In the first alternative the control mechanism is implemented only in the unstable mode equation, whereas in the second one it is implemented in one or more of the stable mode equations as well as in the unstable mode equations. Controlling only the stable mode equations is not suitable because in this case the unstable mode equation remains unstable. This is a consequence of the fact, that the unstable eigenvalues are not affected by the control mechanism.

Concerning our two-dimensional example of the Hénon map, the first alternative leads to

$$\xi^s_{n+1} = \lambda_s \xi^s_n + N_s(\xi^u_n, \xi^s_n, \{\alpha\}) \,, \tag{5.44a}$$

$$\xi^u_{n+1} = \lambda_u \xi^u_n + N_u(\xi^u_n, \xi^s_n, \{\alpha\}) + F(k, \xi^u_n) \,, \tag{5.44b}$$

whereas in the second case we end up with

$$\xi^s_{n+1} = \lambda_s \xi^s_n + N_s(\xi^u_n, \xi^s_n, \{\alpha\}) + F_1(k, \xi^s_n) \,, \tag{5.45a}$$

$$\xi^u_{n+1} = \lambda_u \xi^u_n + N_u(\xi^u_n, \xi^s_n, \{\alpha\}) + F_2(k, \xi^u_n) \,. \tag{5.45b}$$

Different influence of the Eqs. (5.44) and (5.45) on the original system is shown at the end of this section, but first we have to investigate whether these two alternatives lead to a suitable behavior. In order to do this, we apply again the order parameter concept. First, we start with the system (5.44) with the inserted control function (5.1)

$$\xi^s_{n+1} = \lambda_s \xi^s_n + N_s(\xi^u_n, \xi^s_n, \{\alpha\}) \,, \tag{5.46a}$$

$$\xi^u_{n+1} = \lambda_u \xi^u_n + N_u(\xi^u_n, \xi^s_n, \{\alpha\}) + k(\xi^u_{n-1} - \xi^u_n). \tag{5.46b}$$

Introducing a new state variable η_n we are able to get rid of the delay. Hence we yield the following three-dimensional system:

$$\xi^s_{n+1} = \lambda_s \xi^s_n - K\Phi \,, \tag{5.47a}$$

$$\xi^u_{n+1} = \lambda_u \xi^u_n + K\Phi + k(\eta_n - \xi^u_n) \,, \tag{5.47b}$$

$$\eta_{n+1} = \xi^u_n, \tag{5.47c}$$

where K and Φ are defined by (3.92). The mode equations for the (5.47) read

$$\varphi^{s_1}_{n+1} = \tilde{\lambda}_{s_1} \varphi^{s_1}_n - K\tilde{\Phi} \,, \tag{5.48a}$$

$$\varphi^{s_2}_{n+1} = \tilde{\lambda}_{s_2} \varphi^{s_2}_n + \tilde{K}\tilde{\Phi} \,, \tag{5.48b}$$

$$\varphi^{u}_{n+1} = \tilde{\lambda}_{u} \varphi^{u}_n - \tilde{K}\tilde{\Phi} \,, \tag{5.48c}$$

with the abbreviations

$$\tilde{K} = \frac{a}{b(\lambda_s - \lambda_u)(\tilde{\lambda}_{s_2} - \tilde{\lambda}_u)} \,, \quad \tilde{\Phi} = \left(\lambda_s \varphi^{s_1}_n + \lambda_u(\tilde{\lambda}_{s_2}\varphi^{s_2}_n + \tilde{\lambda}_u \varphi^u_n)\right)^2 \,. \tag{5.49}$$

Hereby the $\varphi^{s_1}_n$,$\varphi^{s_2}_n$,φ^u_n are the amplitudes and the $\tilde{\lambda}_{s_1}$,$\tilde{\lambda}_{s_2}$,$\tilde{\lambda}_u$ the corresponding eigenvalues of the modes of the system (5.48). From the linear stability analysis we get for the eigenvalues

$$\tilde{\lambda}_{s_1} = \lambda_s, \;\; \tilde{\lambda}_{s_2} = \frac{\lambda_u - k + \sqrt{(\lambda_u - k)^2 + 4k}}{2}, \;\; \tilde{\lambda}_u = \frac{\lambda_u - k - \sqrt{(\lambda_u - k)^2 + 4k}}{2}, \tag{5.50}$$

where λ_s, λ_u are the eigenvalues (3.71) of the system (3.82). The matrix of the eigenvectors has the following form:

$$\underline{\underline{\tilde{V}}} = \begin{pmatrix} 1 & 0 & 0 \\ 0 & \tilde{\lambda}_{s_2} & \tilde{\lambda}_u \\ 0 & 1 & 1 \end{pmatrix}. \tag{5.51}$$

Our intention was to control the behavior of the system, in this case to shift the first period-doubling bifurcation, which occurs at the value $a_{cr} = \frac{147}{400} = 0.3675$ via the insertion of the function (5.1). As an example, we shift the bifurcation point to the value $a_{cr}^1 = 1$. From the condition $|\tilde{\lambda}_u| = 1$ we yield $k = k_{cr}^1 = -0.302893182$. The eigenvalues $\tilde{\lambda}_{s_1}$,$\tilde{\lambda}_{s_2}$ remain stable, i.e. $|\tilde{\lambda}_{s_1}| < 1$, $|\tilde{\lambda}_{s_2}| < 1$.

For two center manifolds in this case we use again

$$\varphi_n^{s_1} = h_1(\varphi_n^u) = \tilde{A}_2(\varphi_n^u)^2 + O(3), \tag{5.52a}$$

$$\varphi_n^{s_2} = h_2(\varphi_n^u) = \tilde{B}_2(\varphi_n^u)^2 + O(3), \tag{5.52b}$$

where we kept only terms up to the first order with respect to ε taking into account that $\varphi_n^u \sim \varepsilon^{\frac{1}{2}}$. A numerical experiment has confirmed as expected, that indeed $\alpha = \frac{1}{2}$ holds (see Table 5.1). With the expressions $\varphi_{n+1}^{s_1} = h_1(\varphi_{n+1}^u)$ and $\varphi_{n+1}^{s_2} = h_2(\varphi_{n+1}^u)$ one yields

Control	Order param. Eq. (5.33)	Mode Eqs. (5.46)	Mode Eqs. (5.56)	Eqs. (5.73)
$a_{cr} = 0.3675, k = 0$ $(a = 0.36751 - 0.36752)$	0.500684	0.500666	0.500666	0.498603
	0.499234	0.499215	0.499215	0.501297
$a_{cr} = 1, k = -0.302893182$	0.500703	0.500624	0.500625	0.499094
$(a = 1.00001 - 1.00002)$	0.499247	0.499330	0.499330	0.500871
$a_{cr} = 1.4, k = -0.461869429$	0.500637	0.500572	0.500573	0.499216
$(a = 1.40001 - 1.40002)$	0.499312	0.499379	0.499380	0.500745

Table 5.1: Coefficient α for Eqs. (5.33), (5.46), (5.56), (5.73)

from Eq. (3.7)

$$\tilde{A}_2(\tilde{\lambda}_u^0)^2\varepsilon^{2\alpha} = \tilde{\lambda}_{s_1}^0 \tilde{A}_2\varepsilon^{2\alpha} - \frac{a_{cr}(\lambda_u^0\tilde{\lambda}_u^0)^2\varepsilon^{2\alpha}}{b(\lambda_s^0 - \lambda_u^0)}, \tag{5.53a}$$

$$\tilde{B}_2(\tilde{\lambda}_u^0)^2\varepsilon^{2\alpha} = \tilde{\lambda}_{s_2}^0 \tilde{B}_2\varepsilon^{2\alpha} + \frac{a_{cr}(\lambda_u^0\tilde{\lambda}_u^0)^2\varepsilon^{2\alpha}}{b(\lambda_s^0 - \lambda_u^0)(\tilde{\lambda}_{s_2}^0 - \tilde{\lambda}_u^0)} \tag{5.53b}$$

and therefore for the coefficients of the center manifolds:

$$\tilde{A}_2 = K^0 \frac{(\lambda_u^0\tilde{\lambda}_u^0)^2}{\tilde{\lambda}_{s_1}^0 - (\tilde{\lambda}_u^0)^2}, \quad \tilde{B}_2 = -\tilde{K}^0 \frac{(\lambda_u^0\tilde{\lambda}_u^0)^2}{\tilde{\lambda}_{s_2}^0 - (\tilde{\lambda}_u^0)^2}. \tag{5.54}$$

Hereby K^0, $\tilde{K}^0$, $\tilde{\lambda}^0$ are the terms of the lowest order in the Taylor series with respect to the smallness parameter ε of expressions (3.92) and correspondingly (5.49), and λ_u^0 is exactly equal to $\lambda_u(a_{cr})$, where $a_{cr}^1 = 1$. In Fig. 5.4(a) one can see both the dependence of the center manifolds $\varphi_n^{s_1} = h_1(\varphi_n^u)$ and $\varphi_n^{s_2} = h_2(\varphi_n^u)$ on the unstable mode φ_n^u as well as the corresponding projections of the numerical solution of the mode Eqs. (5.48). Now we can derive the order parameter equation which has the following form:

$$\varphi_{n+1}^u = \tilde{\lambda}_u\varphi_n^u - \frac{a}{b(\lambda_s - \lambda_u)(\tilde{\lambda}_{s_2} - \tilde{\lambda}_u)}\left(\lambda_s(\tilde{A}_2(\varphi_n^u)^2 + \lambda_u(\tilde{\lambda}_{s_2}\tilde{B}_2(\varphi_n^u)^2 + \tilde{\lambda}_u\varphi_n^u)\right)^2, \tag{5.55}$$

where $\tilde{A}_2$, $\tilde{B}_2$ are the coefficients (5.54). In order to verify our analytical result, Fig. 5.4(b) shows the bifurcation diagram obtained numerically of both the order parameter equation and the mode equations. As we can see there is at least in the vicinity of the bifurcation at $a_{cr}^1 = 1$ a good coincidence of both solutions.

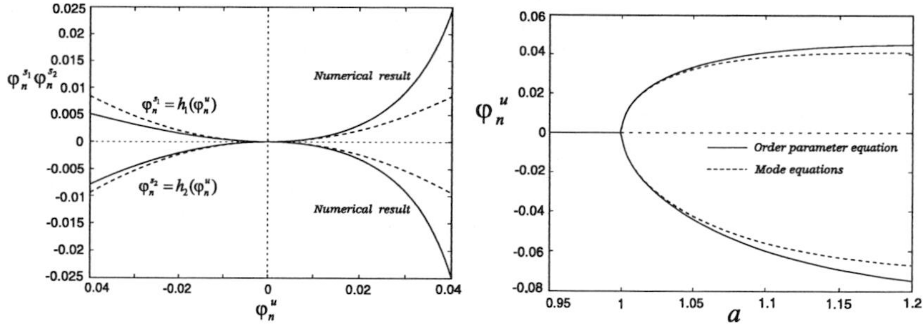

(a) (b)

Figure 5.4: (a) Dependencies $\varphi_n^{s_1} = h_1(\varphi_n^u)$ and $\varphi_n^{s_1} = h_1(\varphi_n^u)$ obtained analytically (5.52) and by simulation from the mode Eqs. (5.48); **(b)** Numerically obtained bifurcation diagram of the order parameter equation (5.55) and the mode equations (5.48).

Now we consider the system (5.45) and pursue the same approach. In the first step we get then

$$\xi_{n+1}^s = \lambda_s \xi_n^s + N_s(\xi_n^u, \xi_n^s, \{\alpha\}) + k(\xi_{n-1}^s - \xi_n^s), \tag{5.56a}$$
$$\xi_{n+1}^u = \lambda_u \xi_n^u + N_u(\xi_n^u, \xi_n^s, \{\alpha\}) + k(\xi_{n-1}^u - \xi_n^u)\,. \tag{5.56b}$$

The corresponding mode equation reads in detail

$$\varphi_{n+1}^{s1} = \check{\lambda}_{s1}\varphi_n^{s1} - \check{K}_1\check{\Phi}, \tag{5.57a}$$
$$\varphi_{n+1}^{s2} = \check{\lambda}_{s2}\varphi_n^{s2} + \check{K}_1\check{\Phi}, \tag{5.57b}$$
$$\varphi_{n+1}^{s3} = \check{\lambda}_{s3}\varphi_n^{s3} + \check{K}_2\check{\Phi}, \tag{5.57c}$$
$$\varphi_{n+1}^{u} = \check{\lambda}_{u}\varphi_n^{u} - \check{K}_2\check{\Phi}, \tag{5.57d}$$

with

$$\check{K}_1 = \frac{a}{b(\lambda_s - \lambda_u)(\check{\lambda}_{s_1} - \check{\lambda}_{s_2})},\ \check{K}_2 = \frac{a}{b(\lambda_s - \lambda_u)(\check{\lambda}_{s_3} - \check{\lambda}_{u})},$$
$$\check{\Phi} = \left(\lambda_s(\check{\lambda}_{s_1}\varphi_n^{s1} + \check{\lambda}_{s_2}\varphi_n^{s2}) + \lambda_u(\check{\lambda}_{s_3}\varphi_n^{s3} + \check{\lambda}_u\varphi_n^u)\right)^2. \tag{5.58}$$

The eigenvalues and the matrix of the eigenvectors of the Eqs. (5.57) have the following form:

$$\check{\lambda}_{s_{1,2}} = \frac{\lambda_s - k \pm \sqrt{(\lambda_s - k)^2 + 4k}}{2},\ \check{\lambda}_{s3,u} = \frac{\lambda_u - k \pm \sqrt{(\lambda_u - k)^2 + 4k}}{2} \tag{5.59}$$

and

$$\underline{\underline{\check{V}}} = \begin{pmatrix} \check{\lambda}_{s_1} & \check{\lambda}_{s_2} & 0 & 0 \\ 0 & 0 & \check{\lambda}_{s3} & \check{\lambda}_u \\ 1 & 1 & 0 & 0 \\ 0 & 0 & 1 & 1 \end{pmatrix}. \tag{5.60}$$

From (5.59) and the corresponding values of λ_s and $k = k_{cr}^{\perp}$ it can be shown, that the equations (5.57a), (5.57b) are complex conjugate. For the quadratic center manifolds

$$\varphi_n^{s_1} = h_1(\varphi_n^u) = \breve{A}_2(\varphi_n^u)^2 + O(3), \tag{5.61a}$$
$$\varphi_n^{s_2} = h_2(\varphi_n^u) = \breve{B}_2(\varphi_n^u)^2 + O(3), \tag{5.61b}$$
$$\varphi_n^{s_3} = h_3(\varphi_n^u) = \breve{C}_2(\varphi_n^u)^2 + O(3) \tag{5.61c}$$

we yield in this case from the analogy to Eq. (3.7)

$$\breve{A}_2(\breve{\lambda}_u^0)^2\varepsilon^{2\alpha} = \breve{\lambda}_{s_1}^0\breve{A}_2\varepsilon^{2\alpha} - \varepsilon^{2\alpha}\frac{a_{cr}(\lambda_u^0\breve{\lambda}_u^0)^2}{b(\lambda_s^0-\lambda_u^0)(\breve{\lambda}_{s_1}^0-\breve{\lambda}_{s_2}^0)}, \tag{5.62a}$$

$$\breve{B}_2(\breve{\lambda}_u^0)^2\varepsilon^{2\alpha} = \breve{\lambda}_{s_2}^0\breve{B}_2\varepsilon^{2\alpha} + \varepsilon^{2\alpha}\frac{a_{cr}(\lambda_u^0\breve{\lambda}_u^0)^2}{b(\lambda_s^0-\lambda_u^0)(\breve{\lambda}_{s_1}^0-\breve{\lambda}_{s_2}^0)}, \tag{5.62b}$$

$$\breve{C}_2(\breve{\lambda}_u^0)^2\varepsilon^{2\alpha} = \breve{\lambda}_{s_3}^0\breve{C}_2\varepsilon^{2\alpha} + \varepsilon^{2\alpha}\frac{a_{cr}(\lambda_u^0\breve{\lambda}_u^0)^2}{b(\lambda_s^0-\lambda_u^0)(\breve{\lambda}_{s_3}^0-\breve{\lambda}_u^0)} \tag{5.62c}$$

with the following coefficients $\breve{A}_2$, $\breve{B}_2$, $\breve{C}_2$:

$$\breve{A}_2 = \breve{K}_1^0\frac{(\lambda_u^0\breve{\lambda}_u^0)^2}{\breve{\lambda}_{s_1}^0-(\breve{\lambda}_u^0)^2}, \quad \breve{B}_2 = \breve{K}_1^0\frac{(\lambda_u^0\breve{\lambda}_u^0)^2}{(\breve{\lambda}_u^0)^2-\breve{\lambda}_{s_2}^0}, \quad \breve{C}_2 = \breve{K}_2^0\frac{(\lambda_u^0\breve{\lambda}_u^0)^2}{(\breve{\lambda}_u^0)^2-\breve{\lambda}_{s_3}^0}, \tag{5.63}$$

where $\breve{K}_1^0$, $\breve{K}_2^0$ are the terms of lowest order in the Taylor series with respect to the smallness parameter ε of expression (5.58).

Below are shown the numerical values of these coefficients for the values of the parameters $k_{cr}^1 = -0.302893182$, $a_{cr}^1 = 1$, $b = 0.3$

$$\begin{aligned}\breve{A}_2 &= -2.948179547 + I4.516842572\ ,\\ \breve{B}_2 &= -2.948179547 - I4.516842572\ ,\\ \breve{C}_2 &= 5.279103128.\end{aligned} \tag{5.64}$$

Substituting the obtained results in system (5.57), we obtain the order parameter equation

$$\begin{aligned}\varphi_{n+1}^u = \breve{\lambda}_u\varphi_n^u - &\frac{a}{b(\lambda_s-\lambda_u)(\breve{\lambda}_{s3}-\breve{\lambda}_u)}(\lambda_s(\breve{\lambda}_{s_1}\breve{A}_2(\varphi_n^u)^2 + \\ &+\breve{\lambda}_{s_2}\breve{B}_2(\varphi_n^u)^2) + \lambda_u(\breve{\lambda}_{s_3}\breve{C}_2(\varphi_n^u)^2 + \breve{\lambda}_u\varphi_n^u))^2\end{aligned} \tag{5.65}$$

In Figs. 5.5(a) and 5.5(b) the real and imaginary parts of the center manifolds are shown together with the projections of the numerically simulated mode equations (5.57). Fig.5.5(c) finally shows the bifurcation diagram of the order parameter equation and the numerically simulated mode equations. For the Eqs. (5.33), (5.46), (5.56) and (5.73) we calculate numerically in the Table 5.1 the coefficient α (see (3.90)) at the bifurcation points for various values of the parameter k in order to verify the assumption (3.90). The calculations were made with 500 000 iterations. In the intervals of the parameter a one hundred logarithmically equidistant points are chosen and the values of α were obtained by a least squares fit assuming the following functional dependence $\log(\xi_n^u) = \alpha * \log(\varepsilon) + \beta$ (respectively $\log(x_n - x_{st}) = \alpha * \log(\varepsilon) + \beta$ for the original Hénon map). In Table 5.1 the upper values correspond to the upper branch of the bifurcation and the lower values to the lower branch of the bifurcation.

To summarize our results, we can conclude that the stabilization of the unstable mode is important for this control mechanism in order to obtain the suitable behavior of the system. In Table 5.2 we have collected all coefficients of the order parameter equations (5.43), (5.55) and (5.65). As we can see they coincide in the first too lowest orders but differ in the higher orders which is a consequence of the dimensionality of the corresponding original system.

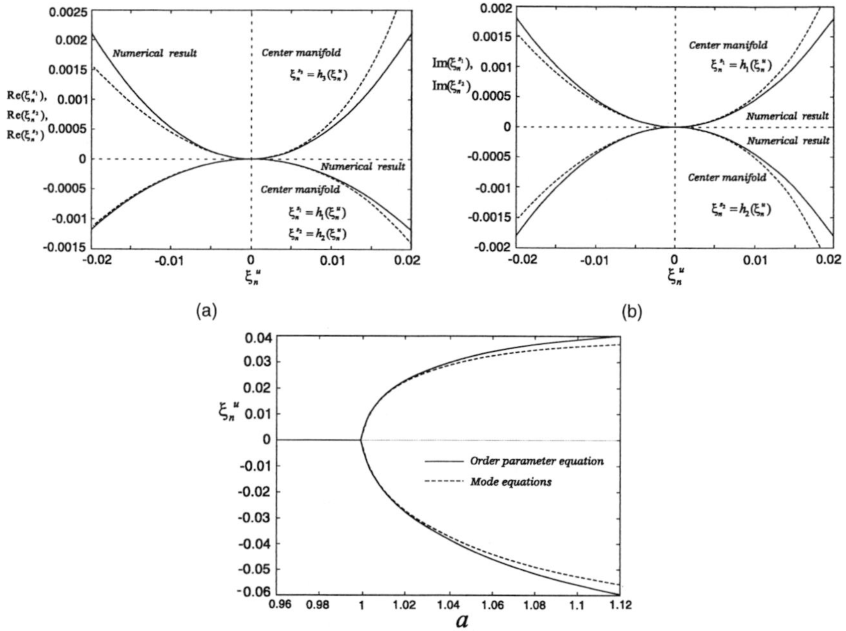

(c)

Figure 5.5: (a) Real part of the quadratic center manifold (5.63) together with the result obtained by a numerical simulation from the mode Eqs. (5.57); **(b)** Imaginary part of the quadratic center manifold (5.63) together with the result obtained by a numerical simulation from the mode Eqs. (5.57); **(c)** Numerically obtained bifurcation diagram of the order parameter equation (5.65) (dimension 1) and the mode equations (5.57) (dimension 4).

5.4.1.3 Transition from mode equations to initial equations

Although we have applied the control mechanism to the mode equations, we are mainly interested in the effect of the control mechanism in the original system. Therefore we consider now a transition from the modes equations (5.46), (5.56) back to the original equations. Starting from Eq. (3.78) and inserting the control mechanism, we get

$$\underline{\xi}_{n+1} = \Gamma^{\Lambda}_{(2)}(:\underline{\xi}_n) + \Gamma^{\tilde{N}}_{(3)}(:\underline{\xi}_n)^2 + k\Gamma^{S}_{(2)}(:\underline{\xi}_{n-1} - \underline{\xi}_n), \tag{5.66}$$

where we have introduced the tensor $\Gamma^{S}_{(2)}$. Components of this tensor are shown for Eq. (5.46) in the Table 5.3 and for Eq. (5.56) in the Table 5.4. Multiplying (5.66) from the left side with the tensor $\Gamma^{V}_{(2)}$ and obeying (3.79) and (3.80) from Chap. 3, we derive

$$\begin{aligned} \Gamma^{V}_{(2)}(\underline{\xi}_{n+1}) &= \Gamma^{L}_{(2)}\Gamma^{V}_{(2)}(:\underline{\xi}_n) + \Gamma^{N}_{(3)}\left(\Gamma^{V}_{(2)}(:\underline{\xi}_n)\right)^2 + \\ &+ \left[k\Gamma^{V}_{(2)}\Gamma^{S}_{(2)}\Gamma^{V^{-1}}_{(2)}\left(\Gamma^{V}_{(2)}(:\underline{\xi}_{n-1} - \underline{\xi}_n)\right)\right]. \end{aligned} \tag{5.67}$$

The Eq. (5.67) is analogue to the Eq. (3.77) of Chap. 3 but contains in addition the expression in square brackets which is caused by the control mechanism. If we add the

Terms	OP Eq. (5.43)	OP Eq. (5.55)	OP Eq. (5.65)
φ_n^u	$\overline{\lambda}$	$\overline{\lambda}$	$\overline{\lambda}$
$(\varphi_n^u)^2$	$K_e\lambda_u\overline{\lambda}$	$K_e\lambda_u\overline{\lambda}$	$K_e\lambda_u\overline{\lambda}$
$(\varphi_n^u)^3$	$2K_e(\lambda_u\overline{\overline{\lambda}}\hat{A}_2 + \lambda_s\overline{\lambda}^2 A_2)$	$2K_e(\lambda_u\overline{\overline{\lambda}}\tilde{B}_2 + \lambda_s\tilde{A}_2)$	$2K_e(\lambda_u\overline{\overline{\lambda}}\check{C}_2 + \lambda_s M)$
$(\varphi_n^u)^4$	$\dfrac{K_e((\lambda_u\overline{\overline{\lambda}}\hat{A}_2 + \lambda_s\overline{\lambda}^2 A_2)^2 + K_d)}{\lambda_u\overline{\lambda}}$	$\dfrac{K_e(\lambda_u\overline{\overline{\lambda}}\tilde{B}_2 + \lambda_s A_2)^2}{\lambda_u\overline{\lambda}}$	$\dfrac{K_e(\lambda_u\overline{\overline{\lambda}}\check{C}_2 + \lambda_s M^2)}{\lambda_u\overline{\lambda}}$

$$\overline{\lambda} = \tilde{\lambda}_u = \hat{\lambda}_u = \check{\lambda}_u = \frac{\lambda_u - k - \sqrt{(\lambda_u - k)^2 + 4k}}{2},$$

$$\overline{\overline{\lambda}} = \tilde{\lambda}_{s_2} = \hat{\lambda}_s = \check{\lambda}_{s_3} = \frac{\lambda_u - k + \sqrt{(\lambda_u - k)^2 + 4k}}{2}$$

$$K_e = -\frac{a\lambda_u\overline{\lambda}}{b(\lambda_s - \lambda_u)\sqrt{(\lambda_u - k)^2 + 4k}}, \quad K_d = 4\lambda_u\overline{\overline{\lambda}}\hat{A}_2\lambda_s A_2\overline{\lambda}^2$$

$$M = (\check{\lambda}_{s_1}\check{A}_2 + \check{\lambda}_{s_2}\check{B}_2)$$

Table 5.2: The coefficients of the order parameter equations (5.43), (5.55), (5.65).

$\Gamma^S_{(2)}$
$(\Gamma^S_{(2)})_{11} = 1$
The remaining components are equal to zero

Table 5.3: Components of tensor $\Gamma^S_{(2)}$ for Eq. (5.46).

stable stationary state $\underline{\mathbf{q}}_{st}$ to Eq. (5.67) we finally yield

$$\underline{\mathbf{x}}_{n+1} = \underline{\mathbf{N}}(\underline{\mathbf{x}}_n, \{\alpha\}) + \left[k\Gamma^V_{(2)}\Gamma^S_{(2)}\Gamma^{V^{-1}}_{(2)}(:\underline{\mathbf{x}}_{n-1} - \underline{\mathbf{x}}_n)\right] \tag{5.68}$$

with

$$\underline{\mathbf{N}}(\underline{\mathbf{x}}_n, \{\alpha\}) = \begin{pmatrix} 1 + y_n - ax_n^2 \\ bx_n \end{pmatrix}. \tag{5.69}$$

From Eq. (5.68) and Table 5.3, i.e., the control scheme of Eq. (5.46), we get the following matrix of control coefficients for the product of the tensors:

$$k\Gamma^V_{(2)}\Gamma^S_{(2)}\Gamma^{V^{-1}}_{(2)} = k\begin{pmatrix} A & B \\ C & D \end{pmatrix}, \tag{5.70}$$

where the coefficients A, B, C, D in general are different and depend on the control parameter k and the stationary state $\underline{q}_{st}$. With this the components of Eq. (5.68) will have the following form

$$x_{n+1} = 1 + y_n - ax_n^2 + k\left(A(x_{n-1} - x_n) + B(y_{n-1} - y_n)\right), \tag{5.71a}$$

$$y_{n+1} = bx_n + k\left(C(x_{n-1} - x_n) + D(y_{n-1} - y_n)\right). \tag{5.71b}$$

For the second control scheme, i.e. Eq. (5.56) we obtain from Eq. (5.68) and Table 5.4 the following matrix

$$k\Gamma^V_{(2)}\Gamma^S_{(2)}\Gamma^{V^{-1}}_{(2)} = k\begin{pmatrix} 1 & 0 \\ 0 & 1 \end{pmatrix} \tag{5.72}$$

$\Gamma^S_{(2)}$
$(\Gamma^S_{(2)})_{11} = 1$
$(\Gamma^S_{(2)})_{22} = 1$
The remaining components are equal to zero

Table 5.4: Components of tensor $\Gamma^S_{(2)}$ for Eq. (5.56).

and for the components of Eq. (5.68) the following form

$$x_{n+1} = 1 + y_n - ax_n^2 + k\,(x_{n-1} - x_n)\,, \tag{5.73a}$$

$$y_{n+1} = bx_n + k\,(y_{n-1} - y_n)\,. \tag{5.73b}$$

Looking at the Eqs. (5.71) and (5.73) it is interesting to remark, that more complex feedback mechanism corresponding to Eq. (5.56) leads to the simpler system (5.73). Numerical investigations of both systems show however, that this control mechanism leads to a behavior which is generally more stable. We assume that the reason for that lies in the stabilization of both modes. In Table 5.5 we have collected the coefficients for both control mechanisms. Here a_{cr} represents the old value of the bifurcation point,

Critical points		Parameter of the control (5.73)	Parameter of the extended control (5.71)				
a_{cr}	a^n_{cr}	k	k	A	B	C	D
	1	-0.30289	-0.30289	0.89578	-0.55784	-0.16735	0.10421
0.3675	1.4	-0.46186	-0.46186	0.92501	-0.48084	-0.14425	0.07498
	1.6	-0.53491	-0.53491	0.93455	-0.45151	-0.13545	0.06544

Table 5.5: Coefficients of the control (5.71), (5.73).

whereas a^n_{cr} represents the new one.

In Fig. 5.6(a) the original bifurcation diagram of the Hénon-map is shown, whereas in the Figs. 5.6(b) and 5.6(c) the bifurcation diagrams for the cases $a^n_{cr} = 1$ and $a^n_{cr} = 1.6$ for both control mechanisms are presented. As we can see it is possible to shift the first bifurcation point of the period-doubling cascade with the applied control mechanism.

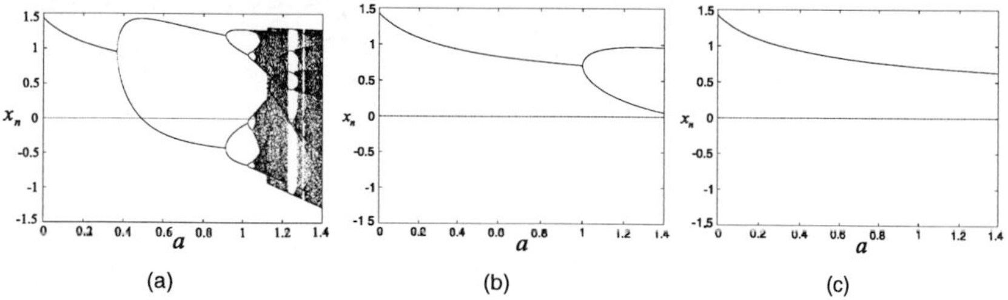

Figure 5.6: **(a)** The bifurcation diagram of the original Hénon map $x_0 = 0.5$, Eq. (3.68); **(b)** Shifted value of the first bifurcation $a^n_{cr} = 1$, $x_0 = 0.5$, Eqs. (5.71), (5.73); **(c)** The bifurcation diagram of completely stabilized Hénon map $a^n_{cr} = 1.6$, $x_0 = 0.5$, Eqs. (5.71), (5.73).

5.4.2 Control of periodical motion

The goal here is to stabilize the period-two motion of the period-doubling scenario by a shift of the second bifurcation point forward so far as possible, using the local, global, and combined control schemes.

The second-iterated Hénon map has the following form

$$x_{n+1} = 1 + bx_n - a(1 + y_n - ax_n^2)^2, \tag{5.74a}$$

$$y_{n+1} = b(1 + y_n - ax_n^2). \tag{5.74b}$$

To calculate the control mechanism the system (5.74) should be first transformed into the form of the mode equations[1]

$$\xi_{n+1}^s = \lambda_s \xi_n^s + \tilde{N}_s(\xi_n^u, \xi_n^s, a), \tag{5.75a}$$

$$\xi_{n+1}^u = \lambda_u \xi_n^u + \tilde{N}_u(\xi_n^u, \xi_n^s, a), \tag{5.75b}$$

where $\tilde{N}_s, \tilde{N}_u$ are the nonlinear functions of the corresponding amplitudes ξ_n^s, ξ_n^u and λ_s, λ_u are the eigenvalues of the linearized system (5.74). Performing this transformation in accordance with (3.73)-(3.78) of Sec. 3.7.3, we use the following tensor of the eigenvectors $\Gamma_{(2)}^V$, which has the following components

$$\begin{aligned} (\Gamma_{(2)}^V)_{11} &= \tfrac{b-\lambda_s}{2bax_{st}}, \quad & (\Gamma_{(2)}^V)_{12} &= \tfrac{b-\lambda_u}{2bax_{st}}, \\ (\Gamma_{(2)}^V)_{21} &= 1, & (\Gamma_{(2)}^V)_{22} &= 1. \end{aligned} \tag{5.76}$$

Here x_{st} is the stable stationary state of (5.74). Following the above described concept, the control mechanism (5.1) can be inserted into the unstable mode equation only

$$\xi_{n+1}^s = \lambda_s \xi_n^s + \tilde{N}_s(\xi_n^u, \xi_n^s, a), \tag{5.77a}$$

$$\xi_{n+1}^u = \lambda_u \xi_n^u + \tilde{N}_u(\xi_n^u, \xi_n^s, a) + k(\xi_{n-1}^u - \xi_n^u), \tag{5.77b}$$

or into both mode equations

$$\xi_{n+1}^s = \lambda_s \xi_n^s + \tilde{N}_s(\xi_n^u, \xi_n^s, a) + k(\xi_{n-1}^s - \xi_n^s), \tag{5.78a}$$

$$\xi_{n+1}^u = \lambda_u \xi_n^u + \tilde{N}_u(\xi_n^u, \xi_n^s, a) + k(\xi_{n-1}^u - \xi_n^u). \tag{5.78b}$$

Considering first the system (5.78) and introducing new state variables η_n^1, η_n^2, we get the following system

$$\xi_{n+1}^s = \lambda_s \xi_n^s + \tilde{N}_s(\xi_n^u, \xi_n^s, a) + k(\eta_n^1 - \xi_n^s), \tag{5.79a}$$

$$\xi_{n+1}^u = \lambda_u \xi_n^u + \tilde{N}_u(\xi_n^u, \xi_n^s, a) + k(\eta_n^2 - \xi_n^u), \tag{5.79b}$$

$$\eta_{n+1}^1 = \xi_n^s, \tag{5.79c}$$

$$\eta_{n+1}^2 = \xi_n^u. \tag{5.79d}$$

Forasmuch as the stationary states of the amplitudes ξ_{st}^s, ξ_{st}^u are equal to zero, the nonlinear functions $\tilde{N}_s, \tilde{N}_u$ do not exert an influence on the calculation of the eigenvalues of the system (5.79). The Jacobian of this system can be then written as

$$\begin{bmatrix} \lambda_s - k & 0 & k & 0 \\ 0 & \lambda_u - k & 0 & k \\ 1 & 0 & 0 & 0 \\ 0 & 1 & 0 & 0 \end{bmatrix}, \tag{5.80}$$

[1] In this two-dimensional case in place of the tensor $\Gamma_{(r)}^{\tilde{N}}$ we will write the nonlinear function $\tilde{N}$.

and has four eigenvalues $\tilde{\lambda}_{1-4}$

$$\tilde{\lambda}_{1,2} = 0.5\Big(\lambda_s - k \pm \sqrt{(\lambda_s - k)^2 + 4k}\Big), \ \ \tilde{\lambda}_{3,4} = 0.5\Big(\lambda_u - k \pm \sqrt{(\lambda_u - k)^2 + 4k}\Big). \quad (5.81)$$

The unstable eigenvalue $\tilde{\lambda}_u$ is determined by the condition $|\tilde{\lambda}_{1-4}| \geq 1$. The function $\tilde{\lambda}_u(a, k) = -1$ in the dependence of the coefficients a and k is shown in Fig. 5.7. Returning

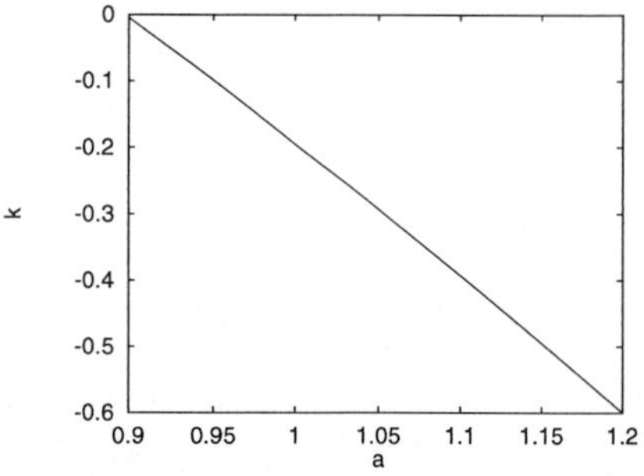

Figure 5.7: The function $\tilde{\lambda}_u(a, k) = -1$ in dependence of the coefficients k, a for the second iterated Hénon map with the inserted feedback.

from the mode equations to the two-iterated map (5.74), we use the transformation of the delayed terms $k\Gamma^{V}_{(2)}\Gamma^{C}_{(2)}\Gamma^{V^{-1}}_{(2)}$ where the tensor $\Gamma^{C}_{(2)}$ is defined for (5.77) by

$$(\Gamma^{C}_{(2)})_{11} = 0, \ \ (\Gamma^{C}_{(2)})_{12} = 0, \ \ (\Gamma^{C}_{(2)})_{21} = 0, \ \ (\Gamma^{C}_{(2)})_{22} = 1, \quad (5.82)$$

and correspondingly for (5.78)

$$(\Gamma^{C}_{(2)})_{11} = 1, \ \ (\Gamma^{C}_{(2)})_{12} = 0, \ \ (\Gamma^{C}_{(2)})_{21} = 0, \ \ (\Gamma^{C}_{(2)})_{22} = 1. \quad (5.83)$$

Considering (5.82) and using the tensor (5.76), we are able to calculate the tensor $\Gamma^{cnt}_{(2)}$, which represents the control terms

$$\begin{aligned} (\Gamma^{cnt}_{(2)})_{11} &= A = \tfrac{b-\lambda_u}{\lambda_s-\lambda_u}, & (\Gamma^{cnt}_{(2)})_{12} &= B = -\tfrac{(b-\lambda_u)(b-\lambda_s)}{2bax_{st}(\lambda_s-\lambda_u)}, \\ (\Gamma^{cnt}_{(2)})_{21} &= C = \tfrac{2bax_{st}}{\lambda_s-\lambda_u}, & (\Gamma^{cnt}_{(2)})_{22} &= D = -\tfrac{b-\lambda_s}{\lambda_s-\lambda_u}. \end{aligned} \quad (5.84)$$

The global control mechanism (5.77) introduced into the second-iterated map then has the following form

$$x_{n+1} = 1 + bx_n - a(1 + y_n - ax_n^2)^2 + k(A(x_{n-1} - x_n) + B(y_{n-1} - y_n)), \quad (5.85a)$$
$$y_{n+1} = b(1 + y_n - ax_n^2) + k(C(x_{n-1} - x_n) + D(y_{n-1} - y_n)), \quad (5.85b)$$

where the coefficients A, B, C, D are determined by (5.84) and the coefficient k is determined by the linear stability analysis of the equations (5.77). Now, with the calculated control coefficients, this mechanism can finally be expressed in the form of the original system (3.68)

$$x_{n+1} = 1 + y_n - ax_n^2 + k(A(x_{n-3} - x_{n-1}) + B(y_{n-3} - y_{n-1})), \quad (5.86a)$$
$$y_{n+1} = bx_n + k(C(x_{n-3} - x_{n-1}) + D(y_{n-3} - y_{n-1})). \quad (5.86b)$$

Considering the second case (5.78), we write the control functions in the following form

$$x_{n+1} = 1 + bx_n - a(1 + y_n - ax_n^2)^2 + k(x_{n-1} - x_n), \tag{5.87a}$$
$$y_{n+1} = b(1 + y_n - ax_n^2) + k(y_{n-1} - y_n). \tag{5.87b}$$

Accordingly for the original non-iterated system we get the local control scheme

$$x_{n+1} = 1 + y_n - ax_n^2 + k(x_{n-3} - x_{n-1}), \tag{5.88a}$$
$$y_{n+1} = bx_n + k(y_{n-3} - y_{n-1}). \tag{5.88b}$$

The numerical investigations have shown that there are no significant differences between the control methods (5.86) and (5.88). In our estimation the difference between both methods lies only in a practical usage of these mechanisms.

The stabilization of motion with another period can be found in the same way: an iteration of the original system, the linear stability analysis, and then the corresponding control scheme implemented in the original system. For example, for the local stabilization scheme of a period-four motion with the inserted function $\psi(n)$ we get

$$x_{n+1} = 1 + y_n - ax_n^2 + \psi(n)k(x_{n-7} - x_{n-3}), \tag{5.89a}$$
$$y_{n+1} = bx_n + \psi(n)k(y_{n-7} - y_{n-3}), \tag{5.89b}$$
$$\psi(n) = n \; mod \; 2. \tag{5.89c}$$

Generally speaking, a shift of the bifurcation point can be carried out in the direction of increasing of the control parameter a and also backwards by means of an appropriately chosen value of the coefficient k. Thus, we can shift a few motions with different periods in different directions. To this effect we construct a combined mechanism, which consists of the feedback functions with different time delay schemes

$$x_{n+1} = 1 + y_n - ax_n^2 + k_1(x_{n-1} - x_n) + k_2(x_{n-3} - x_{n-1}), \tag{5.90a}$$
$$y_{n+1} = bx_n + k_1(y_{n-1} - y_n) + k_2(y_{n-3} - y_{n-1}). \tag{5.90b}$$

In this case, the effect of the complete control function consists in a combined effect of two functions: a shift back of a non-periodical motion using $k_1(\underline{q}_{n-1} - \underline{q}_n)$ and then the stabilization of period two-motion by using $k_2(\underline{q}_{n-3} - \underline{q}_{n-1})$. Here the coefficient k_1 has to be calculated on the basis of the original system and the coefficient k_2 on the basis of the second-iterated system. In Fig. 5.8 the specified types of stabilizations (5.86), (5.88), (5.89), (5.90) in the case of the Hénon map are shown and the corresponding coefficients of these controls are summarized in Table 5.6.

Control	a_{cr}	k	A	B	C	D
Eqs.(5.88)	1.170	-0.540	-	-	-	-
Eqs.(5.86)	1.170	-0.540	1.168	1.225	-0.161	-0.168
Eqs.(5.89)	1.088	-0.230	-	-	-	-
Eqs.(5.90)	0.037	$k_1 = 0.200$	-	-	-	-
	0.796	$k_2 = -0.163$	-	-	-	-

Table 5.6: Summarized coefficients of the control schemes (5.86), (5.88), (5.89), (5.90).

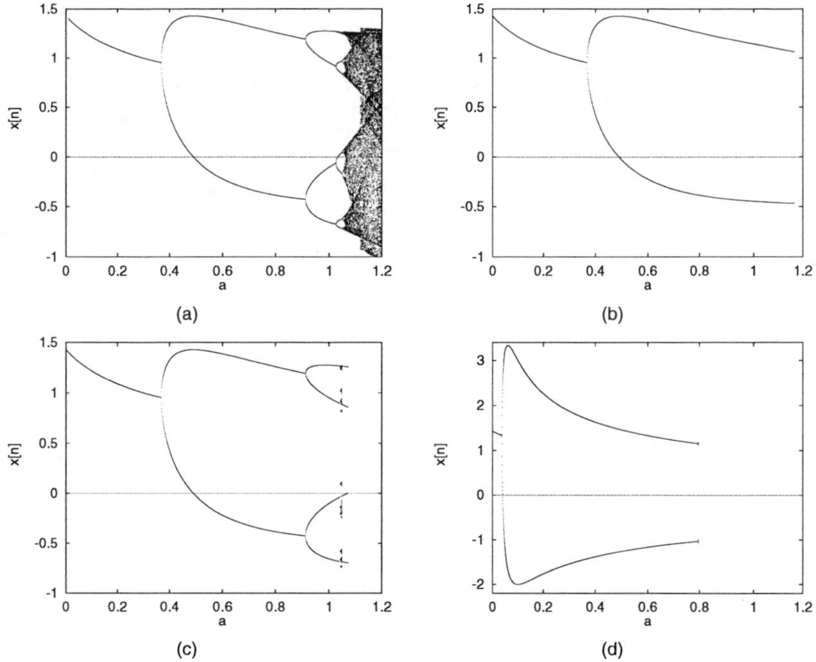

Figure 5.8: **(a)** Bifurcation diagram of the original Hénon map at $b = 0.3$; **(b)** Stabilized period-two motion at $k = -0.54, x(0) = y(0) = 0.05$, Eqs. (5.86), (5.88) using the scheme (5.29) with $\psi(n) = n \, mod \, 2$; **(c)** Stabilized period-four motion at $k = -0.23, x(0) = y(0) = 0.4, \psi(n) = n \, mod \, 2$; **(d)** Combined control mechanism - shifted nonperiodical motion at $k_1 = 0.2$ and stabilized period-two motion at $k_2 = -0.163$, $x(0) = y(0) = 0.1$ using the scheme (5.90).

5.4.3 Control of spatio-temporal chaos in high-dimensional case

Extending this example to a high-dimensional case, we use the global control scheme (5.18) and the two-dimensional Hénon map as a basic system

$$x^i_{n+1} = 1 - a(x^i_n)^2 + y^i_n + k \sum_{j=1}^{m} r^i_j (x^j_{n-3} - x^j_{n-1}), \tag{5.91a}$$

$$y^i_{n+1} = b x^i_n + k \sum_{j=1}^{m} r^i_j (y^j_{n-3} - y^j_{n-1}), \;\; 1 \leq i \leq m, \tag{5.91b}$$

where r^i_j is determined as (5.23) for $p = 0.2$, $m = 512$. The coefficient $k = -0.4$, calculated on the base of (5.81), guarantees a stabilized period-two motion in the area $a = [0.3675, 1.1025]$. The spatiotemporal behavior of coupled two-dimensional Hénon systems at $a = 1.1$ is shown in Fig. 5.9(a). In order to apply the local control scheme to a high-dimensional case, we first rewrite the system (3.68) as a one-dimensional delayed map and then couple $2m$ of such one-dimensional Hénon maps by means of a two-way ring with closed boundaries. Now introducing the local control scheme (5.14), we get

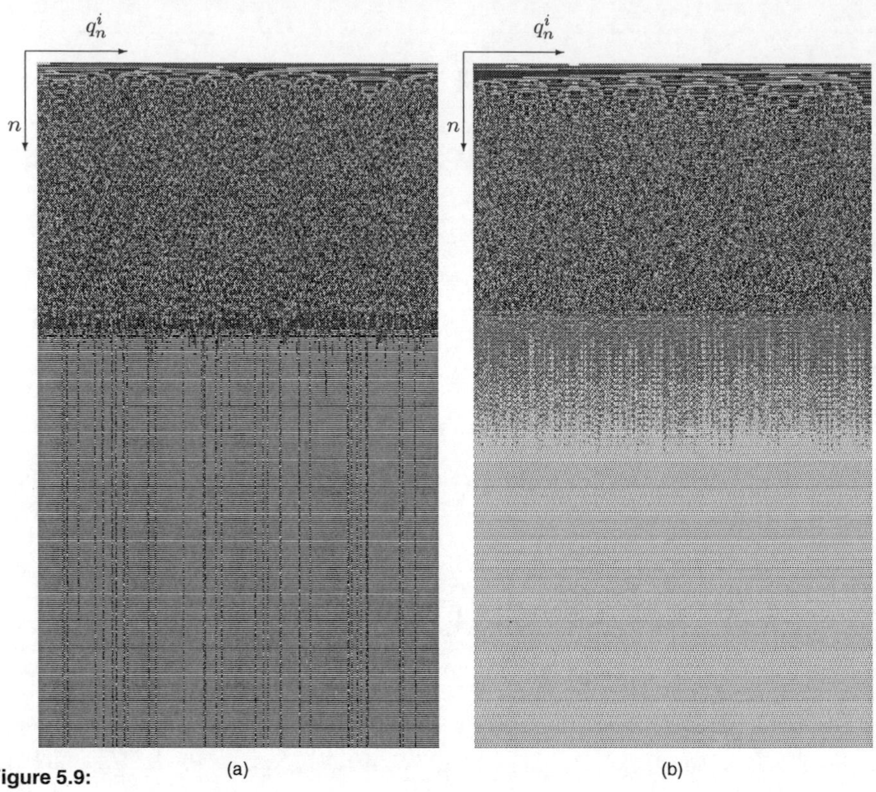

Figure 5.9:

(a) The spatiotemporal behavior of coupled system (5.91) at $a = 1.1$ for $m = 512$, $k = -0.4$. 700 time steps are shown (from top to bottom), after the 250-th time step the control is turned on. After the control is switched on, one can observe a stabilized period-two motion with a spatial clustering (small stripes); **(b)** The spatiotemporal behavior of coupled system (5.98) at $a = 1.2$ for $m = 512$, $k = 0.33$. 700 time steps are shown (from top to bottom), after the 250-th time step the control is turned on. In this case there exists only a spatial homogeneous period-two motion.

$$\begin{aligned} q^i_{n+1} &= 1 - a(q^i_n)^2 + bq^i_{n-1} + \varepsilon(q^{i+1}_n + q^{i-1}_n) + k\psi(n)(q^i_{n-3} + q^i_{n-1}), \\ \psi(n) &= n \, mod \, 2, \quad 1 \le i \le 2m. \end{aligned} \tag{5.92}$$

Because of $\underline{\mathbf{q}}_{st} \neq 0$ we are not able to calculate the coefficient k analytically, in contrast to the case of the global scheme (5.91). However, k can be calculated by numerical simulations, where for $k = -0.12$, $\varepsilon = 0.07$ and $m = 256$ the region of stabilization of the period-two motion is given by the interval $a = [0.32, 0.91]$.

5.5 Control mechanism applied to two interacting agents

The further features of the control mechanism will be shown on examples of two interacting agents whose mathematical model is given by the OLL map (4.1). Here we intend to show that the described control mechanism developed for the period-doubling scenario can be successfully applied to a shift of the Naimark-Sacker bifurcation as well as to the control of spatio-temporal chaos.

5.5.1 Bifurcation control

The behavior of the system (4.1), destined for the investigation, is achieved at the values $b = 2, c = 0.7$. Here, as shown in Fig. 5.10(a)-(b), the perturbed pitchfork bifurcation, the period-doubling bifurcation, and then the Naimark-Sacker bifurcations occur. Following

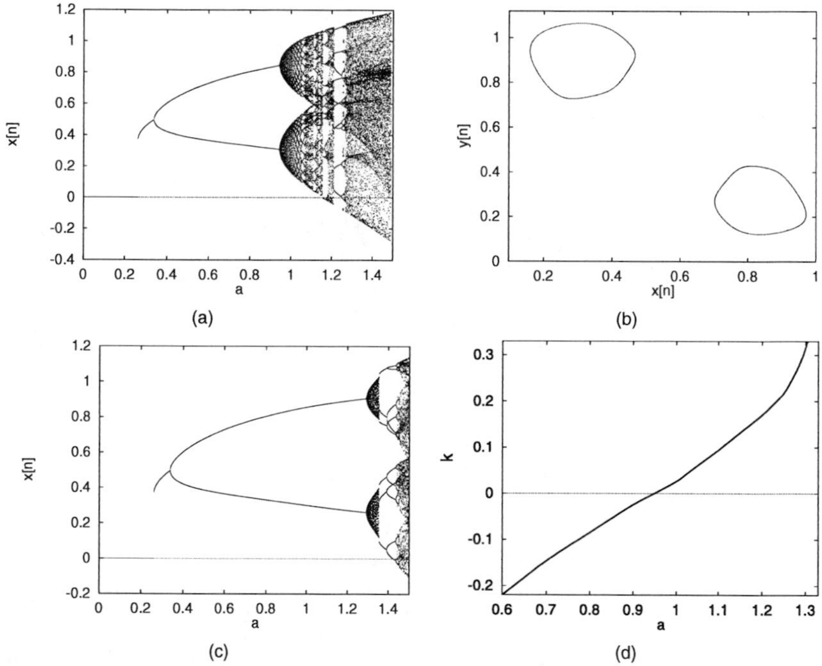

Figure 5.10: (a) Bifurcation diagram of the OLL map at $b = 2, c = 0.7, x(0) = 0.4$; **(b)** Phase diagram of the OLL map at the same $b, c, x(0)$ and $a = 1$ to demonstrate the limit cycle; **(c)** Shifted Naimark-Sacker bifurcation at $k = 0.33$ and at the same $b, c, x(0)$; **(d)** The function $|\lambda_u^{\pm}| = 1$ in dependence of the coefficients k, a for the second-iterated OLL map.

the method described in Sec. 5.2, we build the second-iterated system, include the control terms and then introduce a new state variables η_n^1, η_n^2 into these delayed equations, which can then be written in the following form

$$\begin{bmatrix} x_{n+1} \\ y_{n+1} \\ \eta_{n+1}^1 \\ \eta_{n+1}^2 \end{bmatrix} = \underbrace{\begin{bmatrix} N_x(N_x(x_n, y_n, a)) \\ N_y(N_y(x_n, y_n, a)) \\ 0 \\ 0 \end{bmatrix}}_{\underline{\mathbf{N}}^{[2]}(\underline{\mathbf{q}})} + \underbrace{\begin{bmatrix} k(\eta_n^1 - x_n) \\ k(\eta_n^2 - y_n) \\ x_n \\ y_n \end{bmatrix}}_{\underline{\mathbf{F}}(\underline{\mathbf{q}})}. \tag{5.93}$$

To simplify the required calculations, the chain rule of differentiation is used for the evaluation of the Jacobian $\underline{\underline{\mathcal{J}}}$ of the extended system (5.93), where $\underline{\mathbf{q}} = (x, y, \eta^1, \eta^2)^T$

$$\underline{\underline{\mathcal{J}}} = \frac{d(\underline{\mathbf{N}}^{[2]}(\underline{\mathbf{q}}) + \underline{\mathbf{F}}(\underline{\mathbf{q}}))}{d\underline{\mathbf{q}}} = \frac{d(\underline{\mathbf{N}}(\underline{\mathbf{N}}(\underline{\mathbf{q}})))}{d\underline{\mathbf{q}}} + \frac{d(\underline{\mathbf{F}}(\underline{\mathbf{q}}))}{d\underline{\mathbf{q}}}$$
$$= \left.\frac{d\underline{\mathbf{N}}}{d\underline{\mathbf{q}}}\right|_{\underline{\mathbf{N}}(\underline{\mathbf{q}}_{st})} \left.\frac{d\underline{\mathbf{N}}}{d\underline{\mathbf{q}}}\right|_{\underline{\mathbf{q}}_{st}} + \left.\frac{d\underline{\mathbf{F}}}{d\underline{\mathbf{q}}}\right|_{\underline{\mathbf{q}}_{st}} = \underline{\underline{\mathbf{J}}}|_{\underline{\mathbf{N}}(\underline{\mathbf{q}}_{st})} \underline{\underline{\mathbf{J}}}|_{\underline{\mathbf{q}}_{st}} + \underline{\underline{\mathbf{C}}}|_{\underline{\mathbf{q}}_{st}}. \tag{5.94}$$

Here $\underline{\underline{\mathbf{J}}}$ is the Jacobian obtained from (4.1)

$$\underline{\underline{\mathbf{J}}} = \begin{bmatrix} a\,(1-x) - ax & c & 0 & 0 \\ c + by\,(1-y) & bx(1-2y) & 0 & 0 \\ 0 & 0 & 0 & 0 \\ 0 & 0 & 0 & 0 \end{bmatrix}, \tag{5.95}$$

where x, y are either the functions N_x, N_y or the stationary states x_{st}, y_{st} of the system (4.1) and the matrix $\underline{\underline{\mathbf{C}}}$ is given by

$$\underline{\underline{\mathbf{C}}} = \begin{bmatrix} -k & 0 & k & 0 \\ 0 & -k & 0 & k \\ 1 & 0 & 0 & 0 \\ 0 & 1 & 0 & 0 \end{bmatrix}. \tag{5.96}$$

Calculating (5.94) and using (5.95),(5.96), we get for $\underline{\underline{\mathcal{J}}}$ four eigenvalues: $\lambda_{1,2}$ and two complex conjugate ones which are denoted $\lambda_u^{\pm}$. Solving $|\lambda_u^{\pm}| = 1$ (see [42], [154]) in dependence of k and a at fixed b, c, we are able to determine the value of the coefficient k for corresponding a_{cr}, as shown in Fig. 5.10(d).

Returning again to the original system (4.1) and applying the time delay scheme for the period-two stabilization (5.28), we finally yield

$$x_{n+1} = cy_n + ax_n(1-x_n) + k(x_{n-3} - x_{n-1}), \tag{5.97a}$$
$$y_{n+1} = cx_n + bx_ny_n(1-y_n) + k(y_{n-3} - y_{n-1}). \tag{5.97b}$$

behavior of the system (5.97) is shown in Fig. 5.10(c). Summarizing this section, we arrive at the conclusion that the functions with a large time delay can be applied to controlling such systems, which possess not only period-doubling scenario but also, for example, the Naimark-Sacker bifurcation (see [154]). However, an application of the functions with derived time delays (5.30) to the Naimark-Sacker bifurcation itself leads to unexpected effects, for example, to a change of a periodical motion after this bifurcation. Evidently, in this case the development of other time delays schemes in the feedback functions [190] is required.

5.5.2 Control of spatio-temporal chaos in a high-dimensional case

For a high-dimensional OLL map we use the time delayed coupling (5.18)

$$x^i_{n+1} = cy^i_n + ax^i_n(1 - x^i_n) + k \sum_{j=1}^{m} r^i_j (x^j_{n-3} - x^j_{n-1}), \tag{5.98a}$$

$$y^i_{n+1} = cx^i_n + bx^i_n y^i_n(1 - y^i_n) + k \sum_{j=1}^{m} r^i_j (y^j_{n-3} - y^j_{n-1}), \tag{5.98b}$$

where r^j_i is determined as (5.23) for $\tilde{p} = 0.2$, $m = 512$. The coefficient $k = 0.33$, calculated on the base of (5.95), guarantees a stabilized period-two motion in the area $a = [0.34, 1.3]$. The spatiotemporal behavior of the coupled system (5.98) for $a = 1.2$ is shown in Fig. 5.9(b).

5.6 Summary

In this chapter, we have presented some methods for control of non-periodical and periodical motion of time-discrete dynamical systems using time delay feedback functions. First, as shown, such control functions are periodical (in the sense of (5.26)) having a period τ, and second, it is possible to couple a few of these functions with different time delays to construct combined control mechanisms. In addition, corresponding to the synergetic concept, the control functions can be inserted either into the unstable mode equations or into all mode equations and thus different control mechanisms can arise in the original system. We refer to these mechanisms as local and global controls [185], [187]. In this case, the calculation of mode equations is not required; the calculation of the matrix of eigenvectors is sufficient. Moreover, the advantage of these methods lies in the analytical calculations of the required control coefficients, allowing an exact determination of a new value of a_{cr} and the period τ of stabilization even in a high-dimensional case. It should be noted that systems with inserted feedback become first, more sensitive to initial conditions and, second, chaotic windows are more frequently exhibited in the system dynamics, as shown in Fig. 5.8(c) in the case of the Hénon map.

From the technical viewpoint, we have investigated the influence of well-known control mechanisms [184] on the order parameter equation, the mode equations and, finally, the original equations. The calculations in this chapter demonstrate the analytic derivation of the order parameter equations for nonlinear maps of dimensions 2, 3, and 4. From the viewpoint of control theory, we have shown the efficiency of stabilization by shifting the bifurcation point using the method of delayed feedback and have derived the modified control scheme (see [170]). Furthermore,this can be formalized by using a symbolic manipulation program such as *Maple*, which makes it possible to derive order parameter equations for systems of four, or more, dimensions.

Finally, we would like to point out the method of implementation of time delayed control. In the first case, each system needs to be equipped with memory, and able to retain n numerical values corresponding to n time delays in feedback. The delayed output of other systems may be stored locally, that is, each system memorizes its own delay values and then exchanges these delayed values with the other systems. In another case, each system memorizes the values of all its communicating systems, that is, only actual data are exchanged within the group. Being simple enough to be implemented in real systems, this mechanism allows the realization of different forms of behavior control that may be categorized as:

- shift of local bifurcations allowing stabilization of non-periodical and periodical motion even with different period;

- control of chaos by using the scaling properties of a bifurcation diagram, as mentioned in Sec. 5.4.1;
- transforming different types of bifurcations, as described in Chap. 4.

As previously mentioned, local bifurcations can be viewed as generators of collective behavior; through controlling them, we control the collective activity. Though the control techniques described have demonstrated efficiency and reliability, some authors suppose they cause the forced behavior of the controlled systems [191]; their proposals are to use such regimes as would be most natural to the systems controlled. We expect that the most natural way toward achieving a desired behavior of distributed systems is to modify the interactions among autonomous systems (that is, agents) and so guarantee the collective behavior desired. This technique will be discussed in the next chapter.

Chapter 6

Collective decision making: Swarm agents

Dumb parts, properly connected into
a swarm, yield smart results.
Kevin Kelly

This chapter is devoted to the design of negotiation strategies, achieved by modification of local interactions in autonomous collective systems. From the viewpoint of the synergetic methodology, the taking of collective decisions changes the qualitative behavior of the agent coalition. Such a qualitative change is described by the order parameter. This characteristic value of collective behavior demonstrates whether the whole system is able to reach a mutual agreement in a finite time. Modifying the order parameter so as to guarantee such a common agreement, we in fact create a group social law that thereby determines the local interactions needed. Providing privacy and security for agents during negotiation is an additional advantage of the suggested techniques. This approach is demonstrated first theoretically, by deriving corresponding equations, then simulated with different evaluation criteria, and finally implemented and tested in a swarm of real micro-robots.

6.1 Introduction

As previously mentioned, interacting autonomous agents compose networks known as multi-agent systems (MAS). In robotics, such a network is denoted a multi-robot system. Since the process of interactions between agents is of a nonlinear nature, the complexity of a multi-agent system is much higher than the complexity of a separate agent [174], [192]. As a result, the overall multi-agent system shows new properties, in particular, demonstrating qualitatively new collective behavior [193].

The issue of multi-agent systems involves two essential concepts: the autonomy and rationality of elementary agents, and the interactions between them. The functionality of the whole MAS depends on both, where the basic agents determine the possible actions performed by a system, and the interactions form the activity of the overall system on the macroscopic level [192]. Such activity appears in different forms, in particular, as collective behavior. Consequently, modification of collective properties of such systems can be achieved largely by changing the interactions among agents.

An example of such interactions is the process of collective decision-making, see for example [26], [192], [194]. In considering this problem, it is important to note that taking a decision is independent of the execution of the decision. Therefore, the architecture of an agent has modules that provide the decisions, and modules that deliver the information needed, and modules that execute the decisions. The main requirement of this split is that the complexity of the negotiation module remains unchanged, whereas an increase of complexity in the whole system occurs mainly because of complexity growth in the execution modules.

Achievement of common agreement in multi-agent and multi-robot systems is a challenging task, because autonomous agents may possess different goals. Generally, it requires the development of specific negotiation strategies [5]. Moreover, it must guarantee that a collective system reaches agreement in a finite time. To solve this problem, we propose the application of analytical methods. In this case, the decision module is replaced by a dynamical system that determines the required negotiation strategy and guarantees the desired common decision. As a result, every elementary agent contains analytical and algorithmic components, consisting of a dynamical system and a computer program respectively.

The different decisions made by agents, that is, by their analytical component, correspond to qualitative changes in their collective behavior. From the viewpoint of synergetics, a modification of qualitative behavior, that is, the process of negotiation, is governed by a low-dimensional dynamical quantity denoted as the order parameter (OP) (for example [43], [104]). In this case, the behavior of the order parameter shows whether the process of negotiation is able to lead to mutual agreement. Since the OP possesses a reduced dimension, we can influence it to modify the agents's local interactions and to obtain the desired negotiation strategies. An additional advantage is that this approach provides privacy and security for agents in the communication network. Moreover, the process of negotiation is scalable in a wide range of parameters, see Sec. 6.5.2.2.

In order to handle real systems with synergetic approaches, first, the complexity of such systems must be reduced, and second, they should by expressed in the analytical form of dynamical equations. This means the building of a sequence of simplified models, see Fig. 6.1, until sufficient complexity is achieved in the model to allow the performing of an analysis.

The requirement to preserve a similar structure is imposed on this approach, that is they should have independent interacting components. We introduce the notion of three levels of complexity, denoted as the level of real world (RW), the level of multi-agent systems (MAS) and the level of mathematical models (MM).

Systems on the RW level are very complex; this is primarily related to the complexity of their environment, and different hardware and software components. As an example of such a system, we consider the "Jasmine" swarm of robots, focusing on the problem of collective energetic homeostasis, see more in Sec. 6.2.

This first algorithmic simplification is represented by an MAS, where an agent is a simplified model of a robot on the RW level. The second mathematical level is given by a mathematical object, such as the coupled map lattices (CML) [119]. In this case, the basic maps on the MM level represent a simplified (idealized) model of an autonomous agent. Accordingly, the interaction between agents corresponds to the coupling between maps.

From the viewpoint of synergetics, the collective behavior of coupled maps is a result of self-organization between the internal components of the initial maps. Such internal components are often denoted as "modes." It has been shown that only a few modes are needed in the generation of collective behavior. Therefore, we can reduce the dimension and degrees of freedom of coupled systems to these modes, and derive a compact analytical characterization of the collective phenomena – the order parameter. The pro-

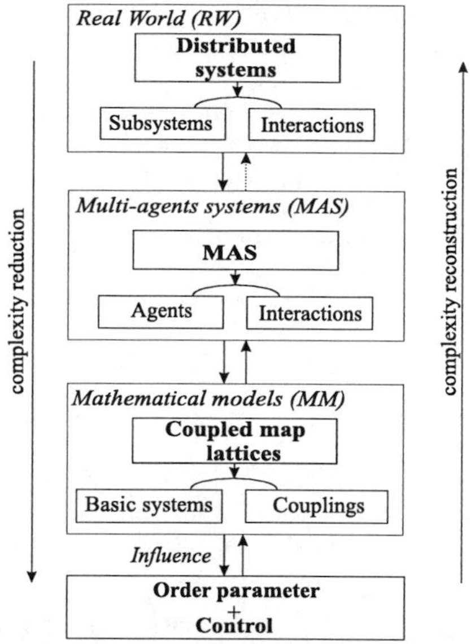

Figure 6.1: Systematic approach for the analysis and control of collective phenomena in distributed systems using the synergetic methodology (from [192]).

cedure of systematic complexity reduction from the initial system to OP is the first focus of the approach proposed here. The rigor of this procedure guarantees that we are able to reconstruct the complexity through the reverse transformation from the OP to the initial system. The dynamics of a CML-model can be viewed as a process of negotiations, and therefore the initial map determines the local negotiation strategy for every agent. In this case, the order parameter demonstrates whether the overall system is able to achieve a mutual agreement in a finite time.

Correspondingly, we can modify the behavior of the order parameter equation so as to guarantee a desired collective decision with the chosen evaluation criterion. Then, return to the CML, we can determine the changes of the couplings between maps caused by a modification of the order parameter. Our suggestion is to incorporate the analytical system directly into the structure of the multi-agent system, see Fig. 6.5. This dynamical system is denoted as an analytical component of an agent, or as the analytical agent. The realization of the decisions made is executed by the algorithmic component, denoted accordingly as an algorithmic agent. Since this dynamical system, that is, the analytical agent, is not modified by the transition from the MM level to the MAS level, it is expected that it can be also implemented in the robot swarm on the RW level.

Thus, we can begin a detailed consideration of these topics. In Sec. 6.2 we introduce the robots on the RW level and formulate the problem of collective homeostasis. Sec. 6.3 considers this problem on the MAS level and explains the main simplifications undertaken. In Sec. 6.4, we carry out a systematic complexity reduction up to the level of mathematical models. The approach is completed by returning to the MAS and RW-levels in Sec. 6.5, where we describe a step-wise reconstruction of complexity, the implementation of this approach, and several experiments performed.

6.2 RW level: Collective energy homeostasis in a robot swarm

The distinctive property of living organisms is their autonomous homeostasis. The sensor-actor "equipment" of these organisms, their behavioral strategies, even their reproduction, depends on the maintenance of their internal systems within certain parameters. The most important is energy balance and, closely connected to it, foraging behavior and strategies [195].

Homeostasis in technical systems differs from biological organisms principally in the non-autonomy of energy balance. Technical systems such as mobile robots depend on human intervention for their energy supply. When robots have responsibility for their energy homeostasis, their behavior no longer depends only on the behavioral goals defined by a designer. The robots should care about their energy state and look for ways to supplement their energy. The robots could save energy by choosing suitable behavior. Several analogies to biological organisms [19] can be noted: the robots can feel hungry, can look for a food source and can die when no food is found. The robots can even demonstrate a kind of "free will" in their behavior, in avoiding energetic death. Thus, energy homeostasis changes in several ways the traditional concept of autonomous behavior in mobile robotics. A generic experimental setup for experiments on collective energy homeostasis is shown in Fig. 6.2; a more detailed description is given in Sec. 6.5.

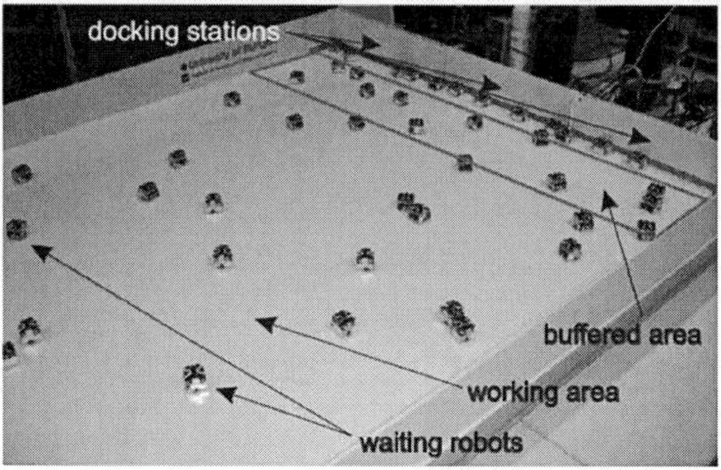

Figure 6.2: General setup for the experiments with collective energy homeostasis. Shown are robots Jasmine, different areas of robot arena and docking stations (from [196]).

The problem of obtaining energy becomes especially interesting when many robots take part in cooperative foraging for energy. Depending on their behavioral goals, the robots can collectively choose different foraging strategies: when collective energy consumption is high, the robots with a low energy level should "eat", otherwise they will not survive; when energy consumption in the robotic group is average, all the robots should recharge for a short time (and thus remain a little hungry) so that the common energetic level of all the robots is maximized. In critical cases, the robots can decide to forage individually, competing with other the robots for resources. Analogies exist with the behavioral strategies of animals where food resources are distributed [14].

6.2.1 Collective strategies in energy foraging

Robots in a swarm solve tasks collectively [6], including foraging for energy. Such cooperation not only involves functional cooperation, where robots of different capabilities and equipment work together, but also faster cooperation, where the task is done in parallel by similar robots. The expression of this cooperation is the swarm efficiency, calculated as the sum of individual robots' efforts. This effort can be estimated either as the time needed to execute a particular task (within the whole solution), or as the amount of energy required to complete the task. Obviously, swarm efficiency depends on the cooperative strategy chosen, and must be optimized. **Thus, the need for a collective strategy, that is, collective decision-making, in energy foraging emerges from the optimization of swarm efficiency.** As we will demonstrate, collective decision making allows almost a doubling of energetic efficiency.

When the robots recharge individually, certain undesired effects can appear:

- the docking station can be a bottleneck that essentially decreases the swarm efficiency;
- robots with a high energy level can occupy the docking station and block robots with low energy levels, which then energetically die and so decrease the swarm efficiency;
- the robots can crowd' around the docking station and essentially hinder the docking approach for other robots, increasing the total recharging time and reducing the energetic balance of the swarm.

Regarding energetic homeostasis, the robots have to take only two primary individual decisions: to execute a current collective task (Job A) or to move to recharging (Job B). Thus, a cooperative strategy should ensure the correct timing and the correct combination of the individual decisions of each robot. In other words, the robots should synchronize their individual decisions.

There are two methods for synchronizing individual decisions. First, the decision-making procedure can be executed collectively, and individual robots need only execute the decision. Examples of such a decision-making process are bargaining, and auctioning, where the final collective activity is not fixed and must first be negotiated (see Sec. 2.1). Second, the decision can be made individually, however the information feeding into this decision is prepared collectively. This type of decision-making process is usually applied to "switching decisions", where the final collective activity represents a sequence of predefined sub-activities. To some extent, this dynamic is similar to the voting procedures described in Sec. 6.4 or in Sec. 6.5.2 in the form of randomly changing neighbors.

For energy foraging, the second method is the most suitable. In this case, the robots can exchange information about their individual sensor inputs. After the inputs are collectively processed, the robots receive a value that defines whether each will perform Job A or Job B, see Fig. 6.3. Obviously, the individual state of a robot also influences its decision, so that the collective behavior represents a complex mix between collective "needs" and individual "desires".

We introduce two input values that the robots can exchange for collective processing: the individual energy level E_l^i, representing the digitalized voltage within the robot's accumulator, and individual energy consumption E_c^i, calculated as the energy difference per time.

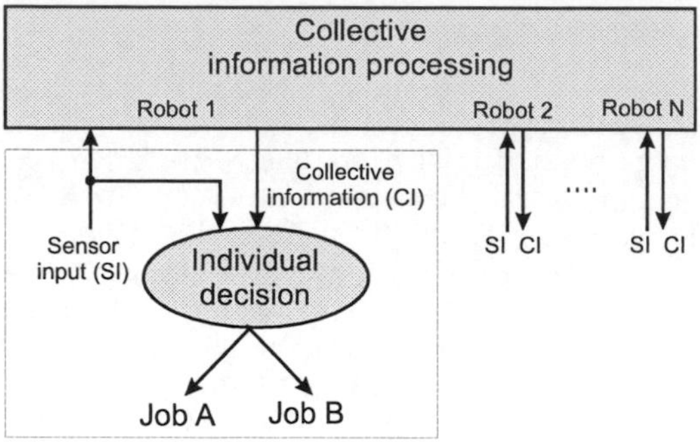

Figure 6.3: The structure of decision making based on the collective information processing.

$$E_c = \frac{\Delta E_l^i}{\Delta t} = \frac{E_l^i(t1) - E_l^i(t2)}{t2 - t1}. \tag{6.1}$$

Additionally to individual values, we introduce E_l^c and E_c^c, representing the corresponding collective value. The value of E_c^c is very useful in estimating the level of collective activity. When E_c^c is high, all the robots move, meaning that the swarm is very busy. In contrast, when E_c^c is small, the activity level of the swarm is low. The individual decision making has the form shown in Fig. 6.4.

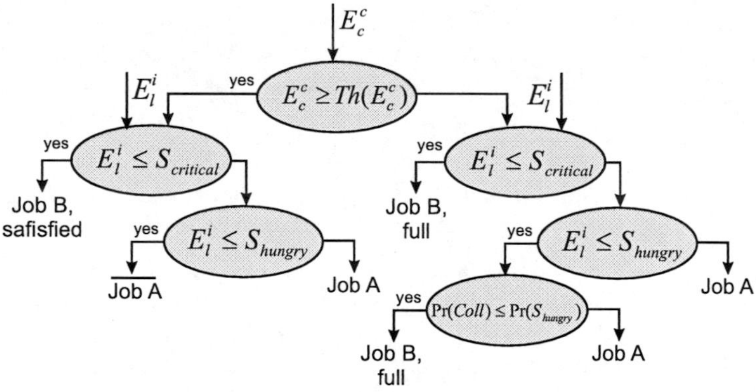

Figure 6.4: The structure of individual decision making by using collectively calculated E_c^c and individual E_l^i. "Job A" means executing current collective activity, "not Job A" – do not participate in collective activity, "Job B satisfied" – recharging until the satisfied state, "Job B full" – recharging till fully recharged.

When E_c^c is larger than a determined threshold, the swarm is very active. This means that robots with low energy should look for a docking station and recharge. This recharging period lasts only until the energy state S_s (Job B satisfied) is satisfied, because full recharging takes more time, and other robots may have reached a critical energy state. Moreover the recharging robot should not participate in any collective activities, because these can be distorted whilst the robot is recharging (not Job A). When E_c^c is smaller than a determined threshold (the swarm is passive), robots with a critical energy level can recharge until full S_f (Job B full). Hungry robots compare the priority of collective tasks with the priority for recharging. This collective strategy manages the order in which the robots recharge, depending on the collective energy consumption. This strategy is comparable with "normal" voting, as seen in social systems, where global information, such as the total number of votes, and the number of "pro-" and "anti" votes making up that total enables the making of individual decisions. The difficulty is that voting must be performed without reference to any central instances. In Sec. 6.5.2.1 we consider the question of how to perform such operations.

In considering the issue of collective homeostasis, we can see the need for collective decision-making approaches that balance individual decisions towards "Job A", or "Job B" as well as different degrees of "satisfied" and "full". In their nature, these decisions are very similar to the distributed voting and bargaining approaches described in distributed and multi-agent system theory. This section is merely a short introduction to collective energy management' we will return to this topic in Sec. 6.5, after we have derived the analytical approaches required for the decision-making procedures.

6.3 MAS-level: Collective homeostasis and decision making in simulation

In this section we consider the next step in simplifying the problem of collective energy homeostasis and deriving analytical approaches for decision making. There are three topics which have to be taken into account when considering simulative approaches: stricture of elementary agent (Sec. 6.3.1), interactions between agents (Sec. 6.3.2) and specific coordination approaches (Sec. 6.3.3). A systematic consideration of these issues allows minimizing the "gap effect", well-known in robotics, i.e., a difference between real and simulative systems.

6.3.1 Using dynamical systems in software agents and real robots

Autonomous agent reflects to some extent the real robot, mentioned the previous section. Similarly to robots, a rational elementary agent has the possibilities of perceiving and acting in the environment [58]. Moreover, each agent behaves independently of other agents but is able to interact with them. The architecture and functional properties of rational elementary agents have been already studied enough (e.g. [122], [69]). Here we mention only two essential propositions as a concept of autonomy and the autonomy cycle.

Concept of autonomy in our context implies that each agent determines decisions on the base of own knowledge (experience) and interactions with other agents. The rational elementary agent in own behaviour cyclically iterates some steps such as collecting of the sensors information, communication with other agents, planning of next state, actions and so on. This cyclical iteration is called the modified autonomy cycle of an agent, as sketched in Fig. 6.5.

Building the further model of an agent, we make some assumptions concerning the autonomy cycle. Firstly, agent intends to solve some problems, whose result is expected to be presented by some action or sequence of actions performed by the agent. The particular case of such a problem being solved is to design a plan [5], that we will briefly discuss in Sec. 7.4. In the case of energy homeostasis, such a plan consists of different degrees of “Job A” or “Job B”. In our further consideration we do not distinguish between predefined plan and forming a plan and focus mainly on the process of negotiation.

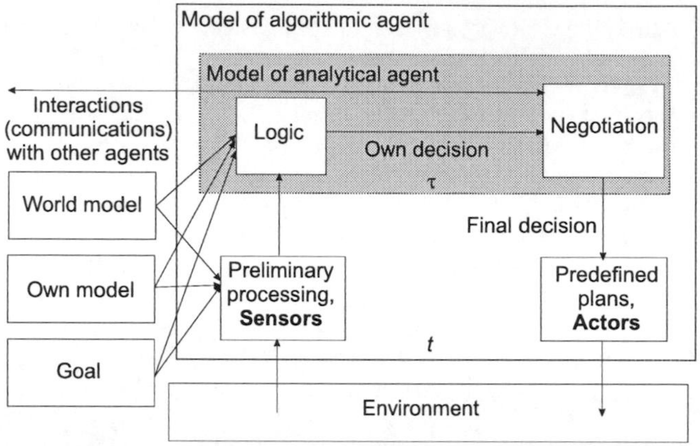

Figure 6.5: Modified autonomy cycle of a rational elementary agent that consists of an analytical decision module, denoted as the analytical agent, and algorithmic part, supporting and execution the made decisions (from [192]). This part contains a few modules and is generally denoted as the algorithmic agent. The whole agent possesses internal time τ needed for negotiations with other agent and external time t when it is acting in environment.

Secondly, it is assumed that the agent itself consists of three main units: the sensors module, the decision module and, finally, actions module. The decision module creates some decisions that are encoded by numerical values. These values are correlated with the actions in environment, that are provided by the actions module, for example ”0” corresponds to the “Job A” and “1” corresponds to “Job B”. There are some sensors which perceive external information from the environment. Forasmuch as all agents can be situated in different spatial or environmental states they will perceive accordingly different input information. This information is processed on the base of some logical rules. Let us remark that knowledge about environment, own construction and functions are collected in the form of explicit rules. These rules finally determine own decisions which will be then transmitted to other agents. Performing negotiations, every agent reaches a final (collective) decision that will be then realized by the actions module. Following the mentioned ideas of structural parting of the elementary agent, we determine the *analytical agent* that presents in fact the part of an agent making the decisions. We build the analytical model of this structure in the form of a dynamical system which is intended for the further investigations.

The *algorithmic agent* has the sensors module, actions module and supports the communication with other agents. Therefore this structure provides the concrete realization of the decision obtained by the analytical agent. It is assumed that algorithmic agent has nature of computer program. On the high abstraction level, this program either is absent or possesses the simplest structure. But the more real and complex environment the more complex is this program, i.e. the algorithmic agent. In contrast to it the analytical agent does not change on different abstraction levels.

6.3.2 Interactions among agents in simulation

Another issue about a rational elementary agent on the MAS-level concerns the communications with other agents. The main question here is what kind of information has to be transmitted to other agents in the collective decision making, taking into account the remark about privacy and security? If we assume the states of system (6.4) between "0" and "1" as being fuzzy (e.g. $x = 0.8$ as the decision "1" with probability "0.8"), we can send such fuzzy decisions from one agent to another. In this case the communication between agents can have the form shown in Fig. 6.6.

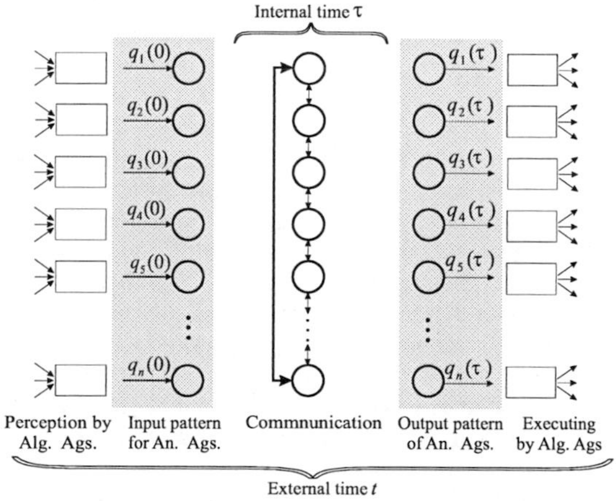

Figure 6.6: Interactions among agents. One turn of modified autonomy cycle that consists of five steps is shown. As can see after the communication phase the decisions obtained by agents are completely independent from one another. Moreover the process of negotiation is independent of a length of communication chain. This Figure is a part of the work [197].

In first phase every agent gets the local sensory information and makes its preliminary processing. In the second phase all agents are connected together (supported by algorithmic agent) and begin to send own fuzzy decisions. In this way every agent obtains the information only about its own direct neighbours. It leads firstly to the privacy and security of information (see also Sec. 7.4), secondly, because of local interactions the negotiation process does not depend of the length of the whole communication chain. This communication phase lasts the time τ. In the last phase every agent executes the made decision. Thus, the agent in Fig. 6.5 has the external time t of the whole cycle action-sensors-decision-action and internal time τ, spent for communications.

6.3.3 The hybrid coordination approach used in simulation and in real robots

The coordination approaches well-known in multi-agent communities (for example [5]) can rarely be applied to limited autonomous systems, because of their different architectures and the limits imposed on communication and computation. Another kind of coordination strategy originates in swarm-related domains [19]. These approaches are much less demanding of the capabilities of the robots. This work follows the swarm-oriented strategy. We suggest a low-complexity coordination approach for limited autonomous systems based on synergetic methodology [43]. As previously mentioned (see Sec. 6.1), from the viewpoint of synergetics, collective behavior is governed by a low-dimensional dynamical quantity, denoted as the order parameter. The OP depends, among other things, on interactions between agents. To obtain the desired collective behavior at the macroscopic level, we can modify the order parameter by adjusting the interactions. In this way, we obtain the tools which establish at least a "one-way" systematic relationship between coordinated collective behavior and the interactions in the system.

The interactions can be of varied physical nature: for example be discrete or continuous in time and space, or with underlying numeric/semantic limitations on the transmitted values. They can use different natural dynamic processes determined by chemical or physical effects (for example [210], [211]), for example light, electromagnetic fields, or chemical gradients. The required dynamics of interactions can be produced by micro-controllers, analogue circuits, (bio-) chemical oscillators or even by nonlinear functions without internal dynamics. From a mathematical viewpoint, these dynamical processes are suitable for supporting collective coordination mechanisms and can be applied to a wide range of micro,- bio-inspired and molecular robotic systems [10], [37], [212].

Though different collective phenomena can be handled in these ways (as demonstrated by physical, chemical and technical systems [28], [204]), in this section we restrict ourselves to two particular coordination problems: decision-making and spatial coordination. To differentiate between a number of agents and a group of agents, some organizational principles must be introduced. Looking to nature for examples, humans are not fascinated by birds or fish, but by their organized behavior in flocks or schools. We say therefore that agents compose a system not only when they jointly interact, but also when they are coordinated. Coordinated behavior is the most decisive element, among many other factors, in the discussion of collective behavior.

Considering micro-robotic or molecular robotic systems, as well as the domain of multi-agent systems, we focus on the problem of coordination, which represents a strong research focus (for example [72], [216]). From the viewpoint of swarm research, works on distributed coordination (for example [217]) are especially interesting. In implementing some suggested approaches in swarm robotics, we encounter a few notable difficulties: for example, Durfee in [72] identified three major issues for successful coordination:

- there must be structures which enable the agents to interact in predictable ways;
- there must be flexibility, so that agents can operate in dynamic environments and can cope with the inherently partial and imprecise viewpoint of the community;
- the agents must have sufficient knowledge and reasoning capabilities to exploit the available structures and flexibility.

Jennings in [217] pointed out that "... *the process of coordination is built upon four main structures – commitments, conventions, social conventions and local reasoning capabilities. Furthermore, it has been hypothesised that **all** coordination mechanisms can be expressed using these concepts ...* ". Micro-robots have essential problems with almost issues of sensing and actuation in the environment, and so of having sufficient knowledge and predictable interactions. When the coordination mechanism involves phases like planning, plan execution or any behavioral primitives, coordination fails, because none of these phases is "precise enough" in term of predictability, collection of sensing data, or capability of information transfer.

To exemplify this, consider an experiment (see for details [84]) in which two or three robots were assigned the task of moving a small colored Lego block in a specific direction determined by the leader robot (see Fig. 6.7). We allowed the use of one or more messenger robots to transmit messages between the leader robot and the pushing team, but a central camera or any sort of common knowledge was not allowed. To identify the object of interest, one robot was equipped with a color sensor. After the robot, through a random search, encountered the Lego block, it sent "come to me" signals and a team built around the object (see Fig. 6.7(a)). The robots surrounded the Lego block relatively quickly (Fig. 6.7(b)). The robots on one side became the pushing team, while the robots on the other side became the leader and messengers (Fig. 6.7(c)). The pushing of the Lego block by a few robots must be well coordinated, or the other robots will turn the object. We tested two approaches. In the first, each robot in turn gave the object a quick push on one side. The leader robot scanned the position of the object, and the messenger robots coordinated the pushing team: "left/right robots push more/less". Though the leader robot perceived the orientation of the object, it knew neither the positions of the pushing robots nor the timing of messages (when and whether the messenger robots transmitted coordination commands to the pushing team).

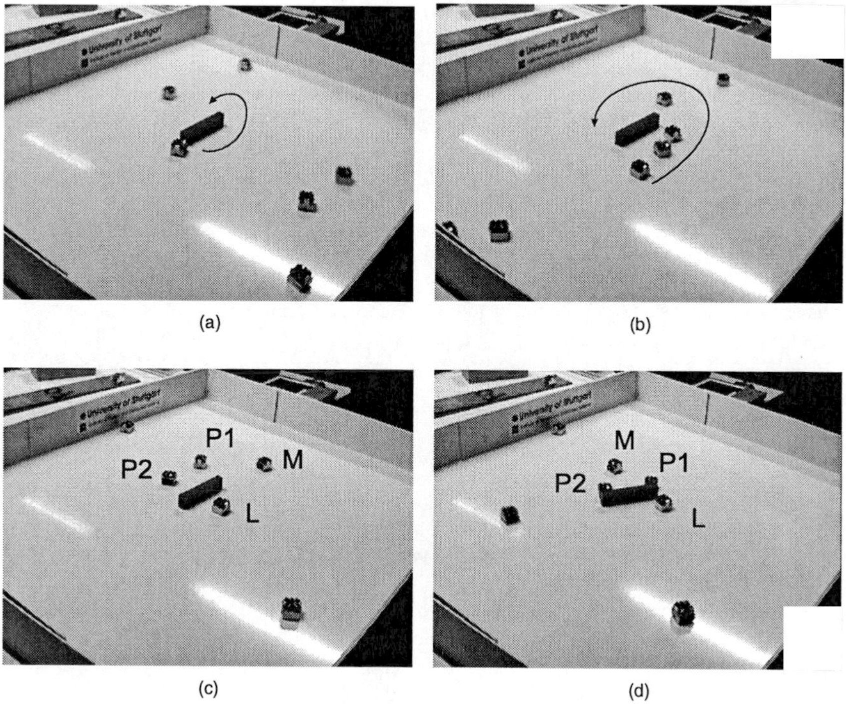

(a) (b) (c) (d)

Figure 6.7: Experiment with cooperative actuation. **(a)** The robot, equipped with the color sensor encounter the blue object; **(b)** It forms a team that surrounded the object; **(c)** Leader "L" scan the orientation of the object, messenger "M" is in contact with leader the pushing team "P1" and "P2"; **(d)** Displacement of the object, where the leader is displaced from the central axis and the messenger lost communication contact with leader. These Figures are a part of the work [218].

In the second strategy, the pushing team continuously pushed the object. The pushing robots measured their position relative to each other, and tried always to be in one line. Messengers were placed to left and right on the other side of the object and measured their distance to the leader. By maintaining a predefined distance to the leader, the messengers tried to block wrong turnings of the object. In this strategy, there was no communication at all.

In implementing these strategies with "Jasmine" micro-robots, we failed in both cases to reproduce the moving of the object. The robots surrounded the object, and split into two groups, relatively reliably. In the first approach, the leader robot was not always aligned with the middle of the object. The orientation of the object was [1] not measured in relation to the axis of symmetry. Similarly, the pushing robots were also placed at arbitrary points and at arbitrary angles to the object. When the leader robot corrected the orientation of the object, the side that appeared as a larger "lever" was displaced less (Fig. 6.7(d)). Moreover, since the leader did not know whether the pushing team had received its commands, it sent the same commands several times. Ultimately, objects were more rotated than translated. In the second approach, in addition to being in random locations around the object, the robots estimated their relative positions to each other and to the leader robot fairly inaccurately. This also led, in many cases, to rotation of the object. We consider the results of this experiment can be generalized to common cases of limited sensors and "piecewise" communication.

In analyzing this experiment, we are interested in why the robots in the first phase could reliably and reproducibly surround the object, and why in the second phase they failed to create reproducible collective behavior. Coordination in the first phase was not connected with the planning or the execution of the plan. Each robot, when receiving new information, took a decision and executed a fixed plan. When the corresponding plan was finished, it took a further decision and executed a new plan. To coordinate the robots, information circulated among robots should be "synchronized". In the second phase, coordination involved actuation, sensing, and planning, integrated into the behavioral primitives of the robot. For example, coordination depended on sensory data obtained by the pushing team or the leader robot. Unreliable communication and sensory data led to incorrect cooperative behavior. Even introducing "social conventions" (in Jenning's terminology), that is, imposing social laws on the robots (for example "make one line"), were not enough for stable coordination. Therefore, for successful coordination, we suggest making coordination independent of the sensory data and any planning activities.

Formalizing this discussion, we can say that an agent has a set of activities. Transition between activities depends on certain conditions of the environment or on interactions with other agents. These conditions can be represented by a "branching" point, as shown in Fig. 6.8.

When two or more agents are coordinated, we can say they have a synchronization between their branching points. It does not mean that all agents execute the same activities in the same time. This synchronization primarily means that agents' transition between different stages takes each other into account.

This approach allows us to discount consideration of behavioral primitives A_i and focus only on coordination. We assume that agents exchange some signals locally. This can take the form of semantics-free interactions (simple exchange of particular values) or more complex communications (with exchange of semantic messages). In molecu-

[1] The object's orientation was measured by sending a narrow IR beam and receiving reflected light. When the robot rotates and scans the object, the form of the IR images indicates the orientation of the object, see [218].

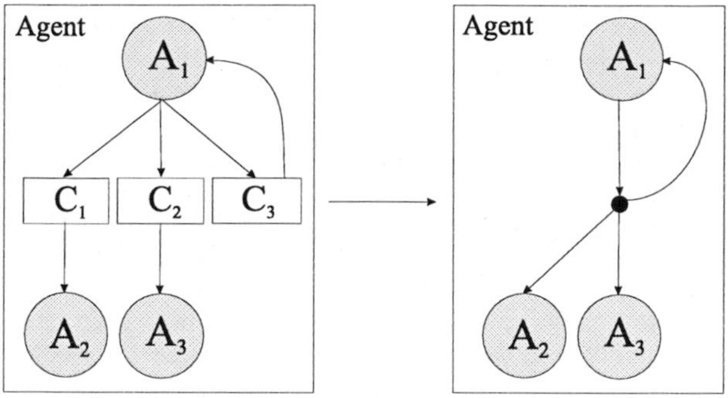

Figure 6.8: Fragment of agent's activity in the form of Petri network (left), transformed form with the "branching point" (right). These Figures are taken from [48].

lar systems, this would be an exchange of chemical elements, messenger molecules, or membrane potentials [219]. In micro-robotic systems such as the "Jasmine" robots, they are IR and RF communications [218]. As we shall show, this signal exchange is not critical for many parameters. Moreover we assume that these signals are "created" by simple chemical, optical or electric oscillators, by non-linear transfer functions, or by digital micro-controllers. In contrast to other well-known approaches, which use synchronization of chaotic oscillators, these signal generators are not chaotic. Obviously, the result of signal exchange must be "understandable" so as to be able to take decisions at branching points. With these two assumptions in mind, we can begin considering the suggested approach at the next level, of mathematic models.

6.4 MM level: Using dynamical systems for collective decision-making

In this section we deal with the mathematical models used to generate decision-making procedures. In contrast to the previous sections, which dealt primarily with systematic complexity reduction, here we focus on the second issue associated with an analytical representation of negotiation strategies. Since any collective decision is based on an evaluation criterion, a few such criteria are considered in Sec. 6.4.1. The evaluation criteria chosen enable us to derive the mathematical models underlying the desired negotiation. Examples of such models are shown in Sec. 6.4.4 - 6.4.6, for which the macroscopic order parameters are represented. Finally, the dimension scaling is discussed for each of the derived models.

6.4.1 Evaluation criteria for negotiation

In our representation, the agents will make a binding deal regarding coordination to achieve a collective goal. This case refers to cooperative games, in contrast to non-cooperative games where agents cannot make a binding deal [74]. On the one hand, cooperative games mean every agent is self-interested and tries to maximize its own good, using its own local strategy. On the other hand, cooperative games also mean that every agent accepts some social laws that guarantee, for example, that if the agent's desired local strategy is best for other agents, the agent will use it (the social welfare criterion [26]). In the context of negotiations this implies that every agent accepts a solution for cooperative games, that is, the principles of calculation and the division of social outcomes in the agent coalition. Moreover, these principles are given from outside that is, by the designer or the agents' environment. In many cases, a solution for cooperative games can simply be viewed as an evaluation criterion that is a determining factor in collective decision-making. The choice of negotiation strategies, implemented in every agent, directly depends on these criteria.

For the first criterion, we assume every agent is trying *to carry its point, and not being relied upon by other agents.* If the agent coalition has a finite number of possible decisions, set by us, in the case where the majority of agents votes for a decision, this decision is the best and has to be adopted by the whole group. Moreover, it is assumed that the quota needed for the collective decision may be varied (for example 50%, 75%, and so on). This procedure is very similar to distributed voting in the particular case when each agent does not know how many other agents are taking part [5].

When the possible decision has a continuous (infinite) nature, for example the optimization of a monetary value (provided that the required value can calculated by a mathematical or logical relationship) the *agents have to bargain until they achieve the required value.* The negotiation strategy determines the calculation rule for this value.

In exceptional cases, for example where the environment is dangerous or the event is very useful for the overall system (for example a fire or the finding of a new resource), it is expected that *the agents will be confident in the one that first detects this situation.* It is assumed that every agent is able to detect such events with high reliability. Therefore we adopt this as the third evaluation criterion, denoted as "voting with confidence" among agents.

6.4.2 Analytical agent I: low-dimensional case

Negotiations between agents can be viewed as a dynamical process whose result is presented by decisions. Considering the dynamical system that models the process of negotiation, we associate different decisions made by agents to distinguish stationary states of this system. The state variables of the dynamical system influence one another and undergo the evolution with the determined evolution law from some the initial states to the final ones [87]. These final stationary states, that all state variables will be attracted to, are ultimately determined by interactions between state variables. The dynamical system that determines the evolution law of these state variables has to satisfy the following requirements:

1. all state variables of the dynamical system at long time dynamics get the distinctly distinguished states such as "-1", "0", "1", ... that will be understood by the algorithmic agent as decisions in favour of one from the predefined plans;

2. dynamical system obtains from the algorithmic agent the local sensory information encoded by distinctly distinguished numerical values of some parameter;

3. the dynamical system is coupled with other dynamical systems, presenting other analytical agents, moreover such a coupling has to possess a local nature and all state variables of this coupled system at long time dynamics simultaneously get the states mentioned in the point 1;

4. the stationary states of this coupled system do not depend on a number of the composing dynamical systems;

5. the dynamical system is discrete in time because the algorithmic agent is also discrete in time.

The needed dynamics of states variables can be achieved in different way, e.g. by means of the selection dynamics, long time dynamics, bifurcation dynamics and so on. The main requirement in this case is that the chosen type of dynamics has to be built on the local interactions, i.e. it does not possess the centralized elements such as e.g. a black box needed for global interactions in selection dynamics.

The bifurcation dynamics of CML suggested by Kaneko [119] satisfies all above mentioned requirements. In this case the variation of bifurcation parameter (over the critical value) changes the stability of stationary states (i.e. attractors) and can be interpreted as change of decisions. Such a bifurcation parameter presents some function of sensory information, local rules and couplings with other analytical agents. It means that the coupling between analytical agents will be performed by means of this bifurcation parameter. The simplest dynamical system that undergoes the desired bifurcation type is given by the normal form of transcritical bifurcation

$$x_{n+1} = ax_n - x_n^2, \tag{6.2}$$

where $x_n \in \mathbb{R}$ is a state variable and a is the bifurcation parameter. Stationary states $x_{st_{1,2}}$ have the following form

$$x_{st_1} = 0, \quad x_{st_2} = a - 1. \tag{6.3}$$

Varying the bifurcation parameter a in the vicinity of the critical value $a_{cr} = 1$, the transcritical bifurcation occurs, thereby the stationary state x_{st_1} loses the stability and the stationary state x_{st_2} becomes stable as shown in Fig. 6.9(a). Being interested in the distinctly distinguished stationary states, we perturb (6.2) by means of the coordinate change so that x_{st_2} will equal to some constant value, e.g $x_{st_2} = 1$. In this case substituting $x_n = y_n(a-1)$ into the equation (6.2), we finally get

$$y_{n+1} = a(y_n - y_n^2) + y_n^2. \tag{6.4}$$

Bifurcation diagram of this system is shown in Fig. 6.9(b). Thus, the value of the bifurcation parameter a in the region [-1,1) leads to the stationary state $y_{st_1} = 0$ and in the region (1,3] to $y_{st_2} = 1$. These values can be viewed as a logical "0" and logical "1" for the further processing in the algorithmic agent. Other types of dynamical systems for analytical agent can be obtained in a similar way. For example the system with three states "-1", "0", and "1" can be derived by perturbing the normal form of the pitchfork bifurcation

$$x_{n+1} = ax_n - x_n^3. \tag{6.5}$$

The perturbated system has the following form

$$y_{n+1} = a(y_n - y_n^3) + y_n^3, \tag{6.6}$$

whose bifurcation diagram is shown in Fig. 6.9(c). The coefficients at nonlinear terms (y_n^2, y_n^3 correspondingly) in Eqs. (6.4), (6.6) in the bifurcation point $a_{cr} = 1$ are equal to zero. This is the distinctive attribute of nonlinear terms in such rescaled normal forms that will be used in further calculations. In case we need more potential decisions it is reasonable to use approaches that sequentially browse all possible decisions. Now let us consider the logical module shown in Fig. 6.5. Analytical agent, at the start of the negotiation process, transmits its own decision to other agents. This decision is in fact some logical function of sensory information and it contains in implicit form the knowledge about world model, own goals and ect. We assume, that such an information processing is mainly performed by the algorithmic agent.

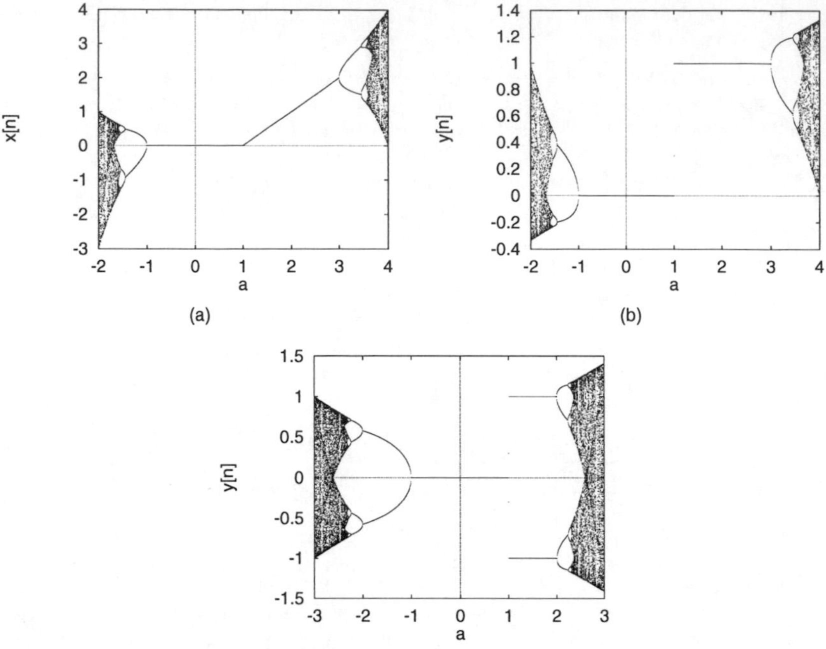

(c)

Figure 6.9: (a) The bifurcation diagram of the transcritical bifurcation (6.2); **(b)** The bifurcation diagram of the perturbated transcritical bifurcation (6.4); **(c)** The bifurcation diagram of the perturbed pitchfork bifurcation (6.6)(with positive and negative initial conditions). The stationary states "0", "1", "-1" can be easily distinguished by analytical agent and therefore can be used for encoding of decisions. These Figures are taken from [192].

6.4.3 Analytical agent II: selection dynamics

As already mentioned in the previous section, the signal exchange between agents can be viewed as a dynamical process whose result allows us to coordinate activity of agents in branching points. This dynamical process can be modelled by a system of coupled nonlinear equations, known also as coupled map lattices.

Since we are going to demonstrate the approach for bio-inspired swarm systems, we choose the basic system from bio-world, for example those which represent evolution of biological macromolecules [213], [215] in the form

$$\dot{\zeta}_i = \zeta_i \left(\alpha \zeta^{p-1} - \frac{1}{\tau} \sum_{k=1}^{N} \alpha_k \zeta_k^p \right), \dim \forall i \in N, \tag{6.7}$$

where ζ_n is a state variable and α, τ are parameters. This system generalizes equations (6.5) and (6.6) considered in the previous section. Setting $\tau = 1$, $p = 1$ and $i = 1$, as well as re-scaling the stationary states ζ_{st} of the system (6.7) so that to obtain only $\zeta_{st} = \{0, 1\}$, we derive the following time-discrete system

$$\zeta_{n+1} = \alpha \left(\zeta_n - (\zeta_n)^r \right) + (\zeta_n)^r, \quad \zeta_n \in \mathbb{R}, \tag{6.8}$$

where α is a parameter and r is the determinancy order of the corresponding normal form[2]. For example, setting $r = 2$ and varying the bifurcation parameter α in the vicinity of the critical value $\alpha_{cr} = 1$, the transcritical bifurcation occurs, thereby the stationary state $\zeta_{st_1} = 0$ loses the stability and the stationary state $\zeta_{st_2} = 1$ becomes stable as shown in Fig. 6.9(b). These values can be viewed as a logical "0" and logical "1" for the further processing in the algorithmic parts of the system. Setting $r = 3$ and varying α, we obtain the system with three stationary states "-1", "0", and "1" as shown in Fig. 6.9(c). In following we use the case $r = 2$.

From the mathematical viewpoint, interactions among robots are represented by couplings between corresponding dynamical systems. Changing the coupling among basic dynamical systems, we can correspondingly influence the output results, and so to guarantee that the system will achieve the desired coordinated activity. The coupling can be performed locally i.e. among neighbor robots. Generally the choice of the neighbors for coupling has the random nature. Considering a nature of initial dynamical system ($\zeta_{st} = \{0, 1\}$), remark that the coupling $\Phi(\zeta^{random})$ between analytical agents can be performed by means of the bifurcation parameter

$$\zeta^i_{n+1} = \Phi(\zeta^{random}_n) \cdot (\zeta^i_n - (\zeta^i_n)^2) + (\zeta^i_n)^2, \quad i = 1, ..., m, \tag{6.9}$$

where m is a maximal number of robots and $\Phi(\xi^{random}_n)$ is a coupling function with a random neighbor. The function $\Phi(\xi^{random}_n)$ is selected as the following polynomial

$$\Phi(\zeta^{random}_n) = k_0 + k_1\zeta^{random}_n + k_2(\zeta^{random}_n)^2 + O^3(\zeta^{random}_n), \tag{6.10}$$

where k_j are coefficients. From the initial system (6.8) we impose the boundary conditions for the coupling function $\Phi(\zeta^{random}_n)$

$$\begin{aligned} k_0 + k_1(\zeta^{random}_n)^{max} + k_2((\zeta^{random}_n)^{max})^2 + ... < 3, \\ k_0 + k_1(\zeta^{random}_n)^{min} + k_2((\zeta^{random}_n)^{min})^2 + ... > -1, \end{aligned} \tag{6.11}$$

where $(\zeta^{random}_n)^{max}$, $(\zeta^{random}_n)^{min}$ are the maximal and correspondingly minimal values of state variables ζ^{random}_n. We impose on the coupling function additional requirements, that originate from the nature of real communication between robots. Namely we require that communication between robots is performed only by means of one byte message, i.e.

$$\Phi(f(\xi^{random}_n)), \quad f(\xi^{random}_n) = 0, 1, 2, ...255. \tag{6.12}$$

In other cases, the limitations (6.12) can have another nature, e.g. two-byte, delayed and so on. The system (6.9) possesses one interesting property.

Property 1. *When the coupling function $\Phi(\zeta^{random}_n)$ is chosen as the first order polynomial ($k_2 = k_3 = ... = 0$ in (6.10)), the system (6.9) has $2^m + 1$ stationary states. The first 2^m stationary states are the m-permutations of the set $\{0, 1\}$ with repetition, the last state is $\zeta_{st_1} = \zeta_{st_2} = ... = \zeta_{st_m} = \frac{1-k_0}{k_1}$.*

This property is extremely useful for construction of the needed dynamics. In the Sec. 6.5.2.3 we consider two different dynamics of the system (6.9), which can be directly applied in micro-robots.

[2]The system (6.8) can be seen as a perturbed normal form with the determinancy order r.

6.4.4 Distributed voting

In this case we assume every agent has two decisions that may be individually made. During negotiation the whole group has to reach the common agreement about kind of decision that has to be collectively executed. This problem, taking into account the requirement about local couplings, involves the theory of coupled map lattice (CML) [119], where the basic systems given by (6.4) are coupled with two direct neighbours

$$q^i_{n+1} = f(q^{i+1}_n, q^{i-1}_n)(q^i_n - (q^i_n)^2) + (q^i_n)^2, \quad i = 1, \dots, m, \tag{6.13}$$

with the boundary conditions for q^{i+1}_n, q^{i-1}_n. The coupling function $f(q^{i+1}_n, q^{i-1}_n)$ is selected as the following linear polynomial

$$f(q^{i+1}_n, q^{i-1}_n) = k_0 + k_1 q^{i+1}_n + k_2 q^{i-1}_n + O((q^{i-1}_n)^2, (q^{i+1}_n)^2), \tag{6.14}$$

where k_i are coefficients. Solution of this problem is expected to be given by coefficients k_i of coupling function (6.14). In case the solution cannot be found we will increase the order of (6.14) until the solution will be obtained.

As mentioned above we use a priori information provided by the macroscopic order parameter for the solution of interactions problem. Forasmuch as we would like to show several approaches leading to the desired negotiation strategy we start with the most evident case when macroscopic dynamics possesses the coexisted attractors (e.g. [204]). Such a kind of dynamics can be illustrated by the following normal form

$$\varphi_{n+1} = \lambda_u \varphi_n + \mu_3 \varphi^3_n + O(\varphi^r_n), \tag{6.15}$$

where $\lambda_u > 1$ and $\mu_3 < 0$ are coefficients. The system (6.15) possesses three stationary states $\varphi_{st_{1,2,3}} = \{0, \sqrt{\mu_3(1-\lambda)}/\mu_3, -\sqrt{\mu_3(1-\lambda)}/\mu_3\}$ and for $\lambda_u > 1$ the state φ_{st_1} is unstable and $\varphi_{st_{2,3}}$ are stable. The attractor that the state variable φ_n will be attracted to is determined by initial conditions namely $\varphi_0 > 0$ for φ_{st_2} and $\varphi_0 < 0$ for φ_{st_3}. In other words the long time dynamics of the system with coexisting attractors is determined by the initial conditions, i.e. by their so-called attraction basins (e.g. [142]). Generalizing this consideration for the case of m state variables, we remark that the state variables, starting from one of these basins, will be attracted to the appropriate attractor. If the state variables start from different basins they will compete for the attractor that these variables will be attracted to. The negotiation (here a voting procedure) can be based on this competition where the initial conditions in CML (6.13) play a role of agent initial proposals.

Being motivated by macroscopic assumption (6.15) we require:

- the eigenvalues of the system (6.13) evaluated on the stationary states $\underline{\mathbf{q}}_{st=0} = (0, 0, \dots, 0)^T$ and $\underline{\mathbf{q}}_{st=1} = (1, 1, \dots, 1)^T$ have to be stable;
- the eigenvalues of the system (6.13) evaluated on all other stationary states have to be unstable;
- the coefficients at nonlinear terms in system (6.13) should provide the supercritical type of postbifurcation dynamics like the condition $\lambda_u \mu_3 < 0$ for Eq. (6.15).
- the initial conditions $\underline{\mathbf{q}}_0$ have to be so selected that to guarantee the reliable choice between the states $\underline{\mathbf{q}}_{st=0}$ and $\underline{\mathbf{q}}_{st=1}$ at long time dynamics of system (6.13).

$\underline{\mathbf{q}}_{st}$	x_{st}	y_{st}	z_{st}
$\underline{\mathbf{q}}_{st_1}$	0	0	0
$\underline{\mathbf{q}}_{st_2}$	0	0	1
...	...	...	...
$\underline{\mathbf{q}}_{st_8}$	1	1	1
$\underline{\mathbf{q}}_{st_9}$	0	m_1	m_1
$\underline{\mathbf{q}}_{st_{10}}$	m_1	0	m_1
$\underline{\mathbf{q}}_{st_{11}}$	m_1	m_1	0
$\underline{\mathbf{q}}_{st_{12}}$	m_1	m_1	m_1
$\underline{\mathbf{q}}_{st_{13}}$	1	m_2	m_2
$\underline{\mathbf{q}}_{st_{14}}$	m_2	1	m_2
$\underline{\mathbf{q}}_{st_{15}}$	m_2	m_2	1

$m_1 = -(k_0 - 1)/k_1$
$m_2 = -(k_2 + k_0 - 1)/k_1$

Table 6.1: Stationary states of three-dimensional system (6.13).

Now, we start with the three-dimensional CML (6.13) where $\underline{\mathbf{q}}_n = (x_n, y_n, z_n)^T$ is the state vector. The stationary states of this system are shown in the Table 6.1. The Jacobian of the three-dimensional system (6.13) is given by

$$\begin{pmatrix} f_1(1-2x)+2x & k_2(x-x^2) & k_2(x-x^2) \\ k_1(y-y^2) & f_2(1-2y)+2y & k_2(y-y^2) \\ k_2(z-z^2) & k_1(z-z^2) & f_3(1-2z)+2z \end{pmatrix}, \tag{6.16}$$

where $x = x_{st}, y = y_{st}, z = z_{st}$ are the stationary states shown in Table 6.1 and $f_1 = f(z,y), f_2 = f(x,z), f_3 = f(y,x)$ are the functions given by $f(q_n^{i+1}, q_n^{i-1})$ in (6.14). We are mainly interested in the following eigenvalues

$$\lambda_{1,2,3}^{\underline{\mathbf{q}}_{st_1}} = \lambda_{1,2,3}^{1} = k_0, \tag{6.17a}$$
$$\lambda_{1,2,3}^{\underline{\mathbf{q}}_{st_8}} = \lambda_{1,2,3}^{8} = -k_0 - k_1 - k_2 + 2, \tag{6.17b}$$

where subindex denotes the number of the eigenvalue in (6.16), superindex shows the stationary state that the eigenvalue is evaluated on. In accordance with the macroscopic assumptions we require

$$|\lambda_{1,2,3}^{1}| < 1, \quad |\lambda_{1,2,3}^{8}| < 1, \tag{6.18a}$$
$$|\lambda_j^i| > 1, \tag{6.18b}$$
$$i = 2, ..., 7, 9, ..., 15, \quad j = 1, 2, 3.$$

In addition the boundary conditions for initial map (6.4) are

$$k_0 + k_1 (q_n^{i+1})^{max} + k_2 (q_n^{i-1})^{max} < 3, \tag{6.19a}$$
$$k_0 + k_1 (q_n^{i+1})^{min} + k_2 (q_n^{i-1})^{min} > -1, \tag{6.19b}$$

where $(q_n^i)^{max}$, $(q_n^i)^{min}$ are the maximal and correspondingly minimal values of state variables q_n^i. Conditions (6.18) can be simplified by the following assumption:

- We suppose some of stationary states (eigenvalues (6.18b)) can be also stable. The arisen thereby simultaneously stable stationary states can be absorbed by appropriate choice of initial conditions.

- From three eigenvalues (6.18b) (with index j), evaluated on stationary state, only one has to be unstable.

- Forasmuch as all initial systems (6.4) are identical and moreover they are coupled in (6.13) by symmetrical coupling it is expected that their eigenvalues are also equal. Therefore it needs to consider instead of all $\lambda_{1,2,3}$ only one from them for the condition (6.18a).

Moreover a solution of the inequalities (6.18) and (6.19) is simplified by the stationary states that are equal to 0 or 1. Finally we get the linear system of inequalities that can be solved by programs of symbolic manipulations like Maple or Mathematica.

Now taking into account our assumptions, we get finally for coefficients k_i:

$$k_0 = 0, \;\; k_2 = k_1, \;\; \frac{1}{2} < k_1 < \frac{3}{2}. \qquad (6.20)$$

The calculated coefficients k_i enable us to determine a stability of stationary states, but there are two further macroscopic requirements that have to be also fulfilled. The first from them concerns the dimension of coupled system (6.13) where we will treat the problem of nonlinear terms. Second one is the boundary of attraction basin. Both points are the focus of two following sections.

6.4.4.1 Nonlinear terms and dimension scaling

The essential question arisen often in praxis concerns the dimension of the system (6.13). The problem is that the complexity of real applications is much higher than it can be treated analytically. In order to get round this problem we assume that some properties of the regular built CMLs are independent of their dimension [192]. The regular or homogeneous CMLs are such coupled maps that possess equal basic systems coupled by means of the polynomials like (6.14). In such cases the low-dimension systems can be expanded in arbitrary high dimension without a change of their linear and some nonlinear properties.

Now we would like to show that the CML (6.13) possesses the mentioned properties denoted further as the dimension scaling. For that we consider separately the linear and nonlinear parts of (6.13). Moreover it is also important to show the possible changes in the basins of initial conditions. Linear part of (6.13) is given by eigenvalues of Jacobian (6.16). Evidently that eigenvalues (6.17) evaluated on $\underline{\mathbf{q}}_{st_1}, \underline{\mathbf{q}}_{st_8}$ (the "0" and "1" stationary states are independent of the dimension of (6.13)) are always equal and therefore independent of the dimension change. To show, that nonlinear part is also independent of the dimension (6.13), is more difficult. One possible way is to assume the eigenvalues (6.18a) to be parameters and then to derive the normal form of system (6.13) like Eq. (6.15). Considering the derived normal form, one can prove whether the nonlinear terms are independent from the dimension m. But forasmuch as all eigenvalues are equal we will get thereby the normal form of m state variables and a treatment of this topic outsteps the framework of the given work [114]. But we can simplify the problem of nonlinear terms, if to consider only postbifurcation dynamics of system (6.13). For that let $\xi_n^i = q_n^i - q_{st_j}^i$ and then substituting it into (6.13) and taking into account (6.20), we get the following system for the case $q_{st}^i = 0$

$$\xi_{n+1}^i = \lambda^1 \xi_n^i + (\xi_n^i)^2 + k_1 \xi_n^i(\xi_n^{i-1} + \xi_n^{i+1}) - k_1(\xi_n^i)^2 \xi_n^{i+1} \tag{6.21}$$

and for the case $q_{st}^i = 1$

$$\xi_{n+1}^i = \lambda^8 \xi_n^i + (1 - 2k_1)(\xi_n^i)^2 - k_1 \xi_n^i(\xi_n^{i-1} + \xi_n^{i+1}) - k_1(\xi_n^i)^2 \xi_n^{i+1}, \quad i = 1, \ldots, m, \tag{6.22}$$

where m is the dimension and taking into account the boundary condition for ξ_n^{i-1}, ξ_n^{i+1}. These equations are usually denoted as mode equations (e.g. [161]).

Forasmuch as λ^1, λ^8 are coefficients and $|\lambda^1| < 1$, $|\lambda^8| < 1$ we conclude that the systems (6.21), (6.22) do not undergo any local bifurcations, moreover the stationary state ξ_{st_i} is stable in the linear approximation for both systems.

Therefore we can narrow down the proving and to verify that the state variables of systems (6.21), (6.22) at long time dynamics will always get zero state. The idea is that these systems because of the local couplings possess the homogeneous structure and if their state variables at dimension $i = 3, 4, ..., j$ are really getting zero state, then in systems with $i = j+1, j+2, ..., m$ they will also get zero state. This step has been performed numerically. In this way we have in fact proved that nonlinear terms of system (6.13) built supercritical type of postbifurcation dynamics. Therefore the linear and nonlinear terms satisfy the macroscopic requirements and we can consider the last point about attraction basins.

6.4.4.2 Initial conditions

In the procedure of simple voting the agents's initial proposals are encoded by initial conditions, i.e. every proposal gets separate numerical value. Forasmuch as every agent can vote only for one from two decisions, it needs to determine only two numerical values. Their choice is dependent on a few factors, the first from them is the quota needed for the taking of decision. In this work we would like to consider the most problematical case with the quota 50% because it effects differently systems with even and uneven dimension. We set, if the dimension m of initial system (6.13) is even, then exactly $m/2$ voices are sufficient for the taking of decision "1", but if m uneven, then the system needs only $m/2 - 1$ voices for the same decision. The last remark concerns the spatial allocation of initial values, because the system (6.13) is sensitive to the order of introduced initial conditions. For example at $m = 10$ in the sequences of initial values "1111100000" and "1010101010" five agents voice for decision "1". But, as shown in Fig. 6.10 (a),(b), the attractions basins are different for these sequences. Therefore the last requirement is that the chosen initial conditions have to guarantee the determined quota independently of the spatial allocation of input values.

Secondly, as it turned out, modifying the value m, the areas of initial conditions undergo small perturbations. Therefore it makes sense to divide the values of m on regions where the change of initial conditions still satisfies the determined quota. The important practical consequence of this effect is that the number of agents in group can be varied and that does not influence the voting procedure. From the practical viewpoint we are interested in the region up to 10 agents that is reasonable for the small group of moving robots. The boundaries of basins are investigated numerically. For that we continually change the values of initial conditions for the proposals "0" (axis P_1) and "1"

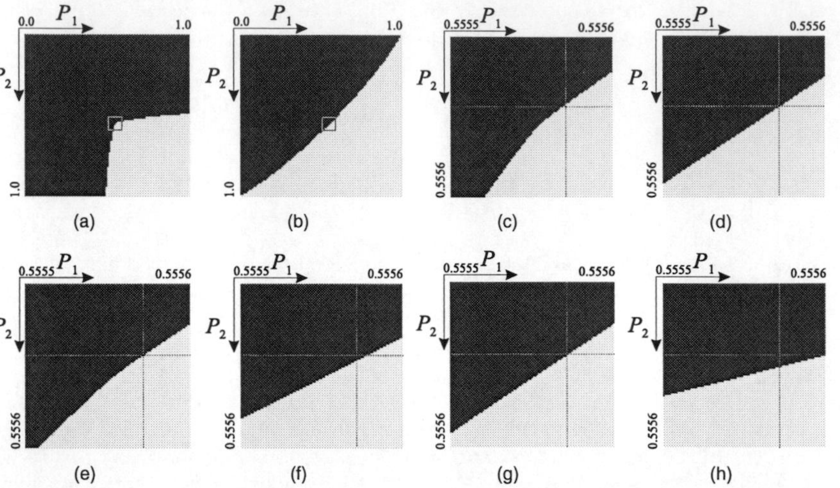

Figure 6.10: Attraction basins of ten-dimensional system (6.13). Axis P_1 on all figures shows numerical values for the initial proposal "0", the axis P_2 shows numerical values for the initial proposal "1". The dark areas in all figures show where all state variables of this system achieve the state "0", the bright areas - the state "1"; **(a)** *m=10* The case when five state variables get the initial condition "1", five - "0" in the spatial allocation "1111100000"; **(b)** *m=10* The same case as in (a) but the initial values are in spatial allocation "1010101010". Different areas of attraction basins can be observed. The small quadrate in the centre of two figures shows the area of initial conditions; **(c)** *m=10* The case when five state variables get initial condition "1", five - "0" in spatial allocation "1111100000". Required state lies in bright area; **(d)** *m=10* The case when four state variables get initial condition "1", six - "0" in spatial allocation "1010100100". Required state lies in dark area. Two last cases are decisive for the determination of initial conditions with quota 50% that are shown as an intersection of dashed lines corresponding e.g. to the numerical value $q_0 = 0.555544$ for the initial proposal "0" and $q_0 = 0.555572$ for the initial proposal "1", where $k_1 = 0.9$ and perturbation of fixed points 1E-5; **(e)** *m=9* Initial conditions in the spatial allocation "111100000"; **(f)** *m=9* Initial conditions in the spatial allocation "100001000"; **(g)** *m=5* Initial conditions in the spatial allocation "11000"; *(h) m=5* Initial conditions in the spatial allocation "00100"; As can see the attraction basins in figures (d) and (g) are identical, this restricts the values m in the area $6, ..., 10$.

(axis P_2). The dark areas in Fig. 6.10 (a)-(d) show where the whole system achieves the state "0", the bright areas show where the whole system gets the state "1". The area of initial conditions, that guarantees the desirable quota 50 % for the group $m = 6, \dots, 10$ ($m = 4, 6, 8, 10$ with quota $m/2$, $m = 7, 9$ with quota $m/2 - 1$ and $m = 3, 5$ with quota $m/2 + 1$) is shown in Fig. 6.10 as an intersection of dashed lines. That corresponds e.g. to the numerical value $q_0 = 0.555544$ for the initial proposal "0" and $q_0 = 0.555572$ for the initial proposal "1", where $k_1 = 0.9$ [3]. Selecting other numerical values for initial proposals, we can accordingly achieve other quota needed for the decision making.

[3] In order to accelerate an achievement of the stable fixed point the dynamics of (6.13) in the neighborhood of points 0 and 1 is slightly perturbated on the value 1E-5.

6.4.5 Voting with bargaining procedure

In previous section it was assumed that all agents are making a decision in favor of one of predetermined plans. But in a general case the decision can possess a form of some non-predetermined value that is calculated from the propositions of agents taking part in a cooperative game. The most simple result of such a bargaining is the arithmetical mean value calculating from the agents's initial propositions. Therefore in this section we firstly intend to construct the coupled maps that will distributively calculate this value. In this case we assume that every agent is accumulating the weighted propositions of other agents circulating around network. Our first task is to determine this weighted coefficient.

Secondly, we would like to expand the scheme (6.13), setting that agents first perform the voting of whether they will perform the further bargaining procedure. In case the agent coalition decide to carry out the bargaining, every agent will obtain the calculated arithmetical mean value otherwise it will get zero. The meaning of this scheme is on the one hand to show that an analytical agent can consist of a few equations, on the other hand that increasing the number of equation, we hereby increase the functionality of analytical agent.

The system, that accumulates the propositions of neighbor agents, has the form of linear two-way coupled maps

$$q^i_{n+1} = aq^i_n + a(q^{i-1}_n + q^{i+1}_n), \quad i = 1, \dots, m, \tag{6.23}$$

where a is a weighted coefficient. This system is first considered in the low-dimensional case (m=4), then the dimension scaling will be performed.

In order to determine the weighted coefficient a we intend to find the solution of these difference equations, using the Jordan normal form approach [205]. Rewriting the system in the following form

$$\underline{\mathbf{q}}_{n+1} = \underline{\underline{\mathbf{A}}}\,\underline{\mathbf{q}}_n, \tag{6.24}$$

where the matrix $\underline{\underline{\mathbf{A}}}$ is the two-way coupling matrix [131], we introduce new state variables $\underline{\xi}_n$ determined by $\underline{\mathbf{q}}_n = \underline{\underline{\mathbf{V}}}\,\underline{\xi}_n$, thereby the system (6.24) yields $\underline{\underline{\mathbf{V}}}\,\underline{\xi}_{n+1} = \underline{\underline{\mathbf{A}}}\,\underline{\underline{\mathbf{V}}}\,\underline{\xi}_n$. Multiplying from left side with the inverse matrix $\underline{\underline{\mathbf{V}}}^{-1}$, we get finally the following diagonalized system

$$\underline{\xi}_{n+1} = \underline{\underline{\mathbf{J}}}\,\underline{\xi}_n, \tag{6.25}$$

where matrices of the eigenvectors $\underline{\underline{\mathbf{V}}}$ and the inverse one $\underline{\underline{\mathbf{V}}}^{-1}$ are given by

$$\underline{\underline{\mathbf{V}}} = \begin{bmatrix} 1 & -1 & -1 & 0 \\ 1 & 1 & 0 & -1 \\ 1 & -1 & 1 & 0 \\ 1 & 1 & 0 & 1 \end{bmatrix}, \quad \underline{\underline{\mathbf{V}}}^{-1} = \frac{1}{4}\begin{bmatrix} 1 & 1 & 1 & 1 \\ -1 & 1 & -1 & 1 \\ -2 & 0 & 2 & 0 \\ 0 & -2 & 0 & 2 \end{bmatrix}. \tag{6.26}$$

Solution of the system (6.25) can be easily obtained as

$$\underline{\xi}_n = \underline{\underline{\mathbf{T}}}\,\underline{\xi}_0, \tag{6.27}$$

where diagonal matrices $\underline{\underline{\mathbf{J}}}$ and $\underline{\underline{\mathbf{T}}}$ are given by

$$\underline{\underline{\mathbf{J}}} = \begin{bmatrix} 3a & 0 & 0 & 0 \\ 0 & -a & 0 & 0 \\ 0 & 0 & a & 0 \\ 0 & 0 & 0 & a \end{bmatrix}, \quad \underline{\underline{\mathbf{T}}} = \begin{bmatrix} (3a)^n & 0 & 0 & 0 \\ 0 & (-a)^n & 0 & 0 \\ 0 & 0 & a^n & 0 \\ 0 & 0 & 0 & a^n \end{bmatrix}. \tag{6.28}$$

The solution (6.27) for the initial state variables $\underline{\mathbf{q}}_n$ has the following form

$$\underline{\mathbf{q}}_n = \underline{\underline{\mathbf{V}}}\,\underline{\underline{\mathbf{T}}}\,\underline{\underline{\mathbf{V}}}^{-1}\,\underline{\mathbf{q}}_0 \tag{6.29}$$

or setting $\underline{\mathbf{q}} = \{x, y, z, v\}^T$, we get explicitly

$$\begin{bmatrix} x_n \\ y_n \\ z_n \\ v_n \end{bmatrix} = \frac{1}{4} \begin{bmatrix} \beta^+ x_0 + \gamma y_0 + \beta^- z_0 + \gamma v_0 \\ \gamma x_0 + \beta^+ y_0 + \gamma z_0 + \beta^- v_0 \\ \beta^- x_0 + \gamma y_0 + \beta^+ z_0 + \gamma v_0 \\ \gamma x_0 + \beta^- y_0 + \gamma z_0 + \beta^+ v_0 \end{bmatrix}, \tag{6.30}$$

where

$$\gamma = ((3a)^n - (-a)^n), \quad \beta^{\pm} = ((3a)^n + (-a)^n \pm 2a^n). \tag{6.31}$$

In order to obtain the arithmetical mean value

$$q^i_{n\to\infty} = \frac{1}{N} \sum_{j=1}^{N} q^j_0 \tag{6.32}$$

we set $a = 1/3$. Then in the long time dynamics the coefficients γ and $\beta^{\pm}$ will equal to $\gamma = \beta^{\pm} = 1$ and then every of state variables of system (6.23) will equal to (6.32).

6.4.5.1 Dimension scaling

The dimension scaling for the system (6.23) is based on the properties of eigenvalues and eigenvectors of two-way coupling matrix $\underline{\underline{\mathbf{A}}}$ in (6.24).

The eigenvalues of this symmetric band $m \times m$ matrix (see e.g. [131], [192]) are given by

$$\lambda_i = a + 2a\cos\left(\frac{2\pi i}{m}\right), \quad i = 1, \ldots, m. \tag{6.33}$$

Apparently, that eigenvalues (6.33) take values, maximal of that in absolute magnitude is determined by $\lambda_m = 3a$. Setting $a = 1/3$ and moreover $|\lambda_{i=1,\ldots,m-1}| < 1$ we get at long time dynamics for matrix $\underline{\underline{\mathbf{T}}}$ where $T_{[1,1]} = 1$ and $T_{[k\neq 1, l\neq 1]} = 0$. It means that result of $\underline{\underline{\mathbf{V}}}\,\underline{\underline{\mathbf{T}}}\,\underline{\underline{\mathbf{V}}}^{-1}$ in (6.29) is determined only by the first column of $\underline{\underline{\mathbf{V}}}$ and first row of $\underline{\underline{\mathbf{V}}}^{-1}$.

Considering the linear problem of eigenvectors for $\underline{\underline{\mathbf{A}}}$, we ascertain that all components of the first column in $\underline{\underline{\mathbf{V}}}$ (eigenvector corresponding to the maximal eigenvalue λ_m) will always equal to 1 independently of dimension of (6.23). However the difficulty here is that we are not always able to determine analytically the inverse matrix of eigenvalues. In order to get round this problem we have numerically calculated this matrix. It was hereby confirmed that for dimension $m < 30$ we will always get $1/m$ for all elements of first row.

6.4.5.2 Implementation remarks

As already mentioned, we would like to incorporate the bargaining procedure given in analytical form by (6.23) into the voting procedure with equation (6.13). For that we first modify the coordinate change of Eqs.(6.2),(6.4), rewriting it as $y_n = \frac{x_n(a-1)}{r}$, $r \neq 0$. Then the equation (6.4) gets the following form

$$y_{n+1} = a(y_n - y_n^2/r) + y_n^2/r \tag{6.34}$$

with stationary states $y_{st_{1,2}} = 0, r$. Our idea is to replace the parameter r by the state variable q_n from (6.23). In this case every agent in the group possesses two state variables q_n and y_n with the following evolution low

$$q^i_{n+1} = aq^i_n + a(q^{i-1}_n + q^{i+1}_n), \tag{6.35a}$$

$$y^i_{n+1} = k_1(y^{i-1}_n + y^{i+1}_n)(y^i_n - (y^i_n)^2/q^i_n) + (y^i_n)^2/q^i_n + \eta[y^i_{n-1} - y^i_n], \tag{6.35b}$$

$$i = 1, \dots, m.$$

where $a = 1/3$ and k_1 is given by (6.20). The delayed function in the square brackets stabilizes the behavior of state variables y^i_n and moreover accelerates the achievement of stable fixpoint. The value of coefficient η is experimentally selected and for $m = 10$ is equal to -0.25. Initial values q^i_0 corresponding to the initial propositions of analytical agents are first normalized in interval $[1, 1.01]$. Initial values y^i_0 are chosen as suggested in Sec. 6.4.4.2. The behavior of system (6.35) is shown in Fig. 6.11 (a)-(d). The last important question concerns the transient time from the start to the time point when the state variables get the values of the stable fixed point. For the $m = 10$ the typical time lies within 50 time steps both for variables q^i_n and y^i_n and shown in Fig. 6.11 (a)-(c). But for some spatial allocations of initial conditions for state variables y^i_n the transient time is increased and lies within 200-300 time steps (Fig. 6.11 (d)). Evidently for the smaller values m the transient time will be shorter. Moreover one can influence the transient time, modifying the values of k_1 (i.e. eigenvalues of system (6.13)) and correspondingly changing the initial conditions for y^i_n.

6.4.6 Voting with confidence between agents

6.4.6.1 Construction of initial system

In order to implement this strategy of negotiation, we need the initial map that possesses some special properties. Namely we require at least three distinctly distinguished stationary states which are stable at different values of bifurcation parameter. Forasmuch as we did not found such a map, we would like to construct required dynamics, based on one of the known maps.

Creation of the suitable dynamics is complex problem that consists of the following points:

- creation of suitable stationary states;
- linear stability of the stationary states;
- an influence on the dynamics caused by nonlinear terms;
- areas of attraction basins;
- external control mechanisms.

Taking into account these points, we choose the NF of pitchfork bifurcation (6.5) (mentioned in Sec. 6.4.2) for such an initial map. Firstly, we create an unfolding of this NF, inserting the perturbated term x^2

$$x_{n+1} = ax_n + bx^2_n + cx^3_n. \tag{6.36}$$

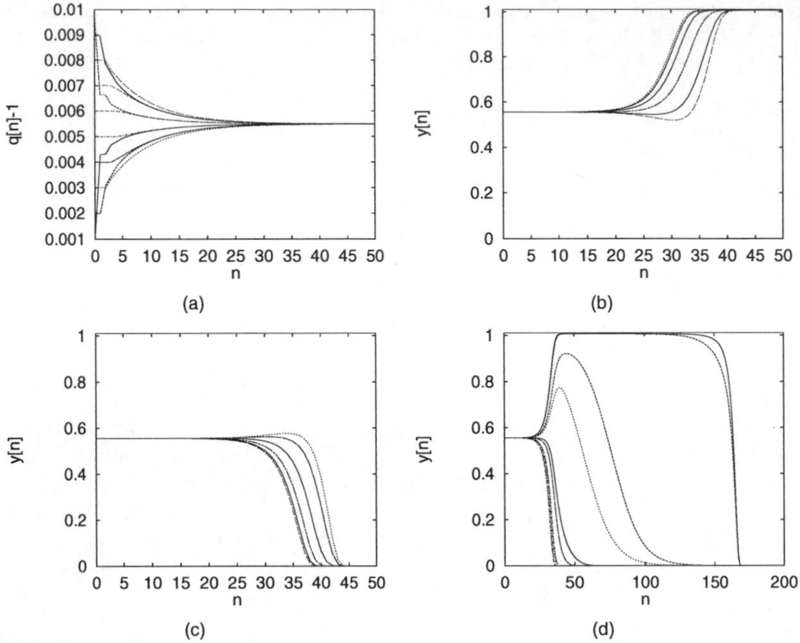

Figure 6.11: (a) The behavior of state variables q_n^i of (6.35), where $m = 10$ with the initial values $q_0^{1,2,\ldots,10} = \{0.1, 0.2, 0.3, \ldots, 0.9, 0.99\}$. The calculated arithmetical mean value equals 0.549; **(b)** The behavior of state variables y_n^i of (6.35), where $m = 10$ in case the initial values are spatial allocated as "1111100000". The state variables y_n^i achieve the attractor with the numerical value 1.0549; **(c)** The behavior of state variables y_n^i in the case of spatial allocation "1010100100" of the initial values. The state variables y_n^i achieve the attractor with the numerical value 0. The typical transient time is shown in these figures; **(d)** One of the longest transition time at the spatial allocation "1110000010" of initial values.

Stationary states of system (6.36) are given by

$$x_{st_0} = 0, \quad x_{st_{1,2}} = \frac{-b \pm \sqrt{b^2 - 4ca + 4c}}{2c}. \tag{6.37}$$

(1) Now, according to the first point, we intend to modify the dynamics of (6.36) so that to get the stationary states that will be independent of coefficients a, c. In this case the coefficient c from $b^2 - 4ca + 4c = (b-2)^2$ yields the value $c = \frac{b-1}{a-1}$. Then the equation (6.36) gets

$$x_{n+1} = ax_n + bx_n^2 + \frac{(b-1)}{a-1}x_n^3 \tag{6.38}$$

with the stationary states $x_{st_0} = 0$, $x_{st_1} = 1 - h$, $x_{st_2} = \frac{1-a}{b-1}$. Rescaling the state variable x in the following way $x = \frac{x(1-a)}{r}$, where r is a parameter and $r \neq 0$, we get finally the perturbated normal form of the pithfork bifurcation

$$x_{n+1} = a\left(x_n - \frac{x_n^2 b}{r} + \frac{x_n^3 b - x_n^3}{r^2}\right) + \frac{x_n^2 b}{r} - \frac{x_n^3 b + x_n^3}{r^2}, \tag{6.39}$$

that has the following stationary states

$$x_{st_0} = 0, \quad x_{st_1} = r, \quad x_{st_2} = \frac{r}{b-1} \tag{6.40}$$

(2) For the second point we perform the linear stability analysis of system (6.39). The eigenvalue λ evaluated on the stationary states $x_{st_{1,2,3}}$ (6.40) is shown below

$$\lambda(x_{st_0}) = a, \tag{6.41a}$$
$$\lambda(x_{st_1}) = ba - 2a - b + 3, \tag{6.41b}$$
$$\lambda(x_{st_2}) = -\frac{-2b + ba + 3 - 2a}{b-1}. \tag{6.41c}$$

The stability of stationary states is independent of coefficient r. Therefore these states can be scaled and that does not influence their linear stability. Now we intend to set the coefficient b so that to determine stability of $x_{st_{1,2,3}}$ at different values of the bifurcation parameter a. The eigenvalues (6.41) at $b = 6$ are shown in Fig. 6.12. Setting $r = 1$, we

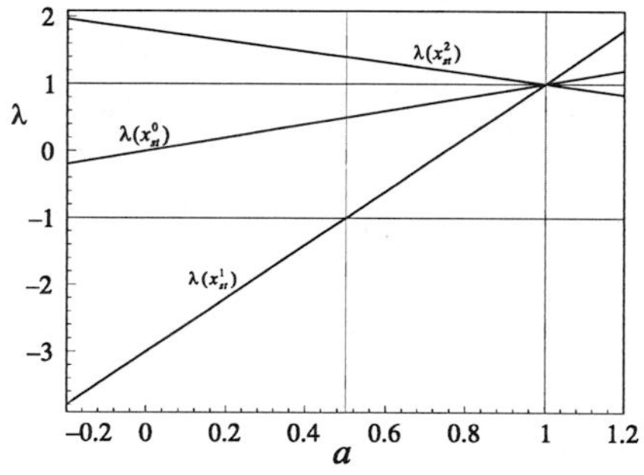

Figure 6.12: Eigenvalues of (6.41) at $b = 6$.

get the following stationary states $x_{st_{0,1,2}} = 0, 1, \frac{1}{5}$. As we can see from Fig. 6.12 there are two instability points at $a_{cr} = 0.5$ with $\lambda_u = -1$ and at $a_{cr} = 1$ with $\lambda_u = 1$. In the regions $[-1, 0.5]$ and $[1, 2]$ of the parameter a the stationary states $x_{st_0} = 0$ and $x_{st_2} = 0.2$ are correspondingly stable. In the region $[0.5, 1]$ of the parameter a both x_{st_0} and x_{st_1} are stable, therefore x_{st_1} will be chosen by initial conditions. In this way we have obtained three stationary states that are stable at different values of the bifurcation parameter a.

(3) Determining the influence of nonlinear term, we take into account mainly the sub- and supercritical conditions and determinancy order [192]. In order to determine these conditions we have to calculate the normal form of local instabilities at $\lambda_u = 1$ and $\lambda_u = -1$. Introducing the nonlinear coordinate transformation up to the fourth order denoted as near identity transformation (e.g., [42], [45]).

$$x_n = \varphi_n + g_2\varphi_n^2 + g_3\varphi_n^3 + O(\varphi_n^4), \tag{6.42}$$

where g are coefficients at state variable φ_n and substituting it into the equation (6.39), we get two normal forms. In the case $\lambda_u = 1$ the normal form is

$$\varphi_{n+1} = \lambda_u\varphi_n + \mu_2\varphi_n^2 + \mu_3\varphi_n^3 + O(\varphi_n^4), \tag{6.43}$$

where $\lambda_u = 4a - 3$, $\mu_2 = 9a - 9$, $\mu_3 = 5a - 5$ with $x_{st} = x_{st_1}$, that corresponds to the perturbated pitchfork bifurcation [4] with determinancy order three. At the bifurcation point $a_{cr} = 1$ the coefficients μ_2, μ_3 are getting zero therefore we determine sub-supercritical condition. In this case if the bifurcation branch at $a > a_{cr}$ exists and moreover is also stable this bifurcation refers to the supercritical type. Investigating the behavior of (6.43), we can ascertain that bifurcation branches at $a < a_{cr}$ and $a > a_{cr}$ both exist and are stable.

In the case $\lambda_u = -1$ we get the normal form

$$\varphi_{n+1} = \tilde{\lambda}_u \varphi_n + \tilde{\mu}_3 \varphi_n^3 + O(\varphi_n^4), \tag{6.44}$$

where $\tilde{\lambda}_u = 4a - 3$, $\tilde{\mu}_3 = \frac{40a^2+11a-51}{8a-6}$. As we can see from the condition

$$\tilde{\lambda}_u \tilde{\mu}_3 < 0, \tag{6.45}$$

this period-doubling bifurcation at $a_{cr} = 0.5$ is also supercritical with the determinancy order three. Thus the influence of nonlinear terms fulfils our requirements.

Now, with the defined coefficient b, c, r the system (6.39) has finally the following form

$$x_{n+1} = a(x_n - 6x_n^2 + 5x_n^3) - 5x_n^3 + 6x_n^2, \tag{6.46}$$

where a is a bifurcation parameter. The bifurcation diagram of (6.46) is shown in Fig. 6.13 (a).

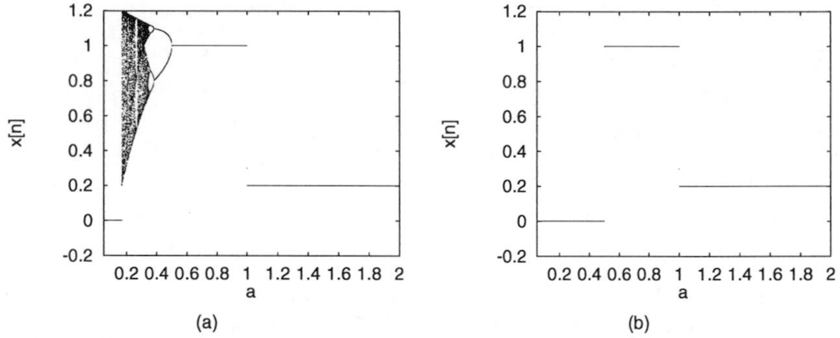

Figure 6.13: a) Bifurcation diagram of the system (6.46); b) Bifurcation diagram of the system (6.46) with the introduced control elements (6.47), (6.48).

(4),(5) The determination of the boundaries of attraction basins does not require a special techniques and is performed numerically. But in contrast to that to change these boundaries is sometimes impossible because it needs to modify the whole dynamics of (6.46). To get round this problem we first look for the boundaries that are critical for the dynamics e.g. where the system loses a dissipation. Then we insert some external control elements introducing the needed corrections, assuming, that they will be executed by the algorithmic agent.

In our case the numerical investigations show, that for the stationary state $x_{st_1} = 0.2$ such critical boundaries are $x_{init}^{low} = 0$ and $x_{init}^{upper} = 1$ and for the stationary state

[4] The determinancy order of this normal form is increased in order to achieve a topological equivalency between the original bifurcation problem and the reduced one. See for details [192].

$x_{st_1} = 1$ the critical boundary is $x_{init}^{upper} = 1.2$. In case the state variable x_n possesses through these boundaries the system will lose a dissipation. Consequently the first control condition is

$$if\ a > 1\ and\ x_n < 0 \rightarrow x_n = 0.01, \tag{6.47a}$$

$$if\ a > 1\ and\ x_n > 1 \rightarrow x_n = 0.99. \tag{6.47b}$$

This control is destined to restrict the motion amplitude of state variable x_n. It introduces a small perturbation into the dynamics but does not change the linear stability of (6.46). In the following section we discuss the introduced influence once more and construct the coupling system so that to minimize it.

Secondly, in the region $[0.18, 0.5]$ of the parameter a the initial condition $x_{init} > 0.2$ leads to the stationary state x_{st_1} and correspondingly to the period-doubling cascade. In order to avoid this undesirable state we require

$$if\ a < 0.5 \rightarrow a = 0.15. \tag{6.48}$$

Let us remark that the condition (6.48) also does not influence the linear stability (6.41). The bifurcation diagram of the system (6.46) with these control conditions is shown in Fig. 6.13.

6.4.6.2 Construction of the coupled system

Constructing a coupled system, we use a priori knowledge about the dynamics given by the order parameter equation (or the normal form). In this case we require the defined behavior of the order parameter and the coupling has to be introduced into an initial system so that the order parameter really achieves this behavior.

Returning to the evaluation criteria, we assume that during negotiation all agents are performing the voting in favor of one of two decisions encoded by stationary states $x_{st_{1,2}}$. This type of dynamics is similar to the nonbifurcation dynamics considered in Sec 6.4.4. But if one of agents perceives exclusive local information, introduced into analytical agent by means of e.g. the parameter k_{0i}, it causes the bifurcation that turns over the whole system into a new state $x_{st_{ex}}$. The plan that has to be executed in this situation is encoded by this stationary state. Therefore the initial system (6.46) had to possess at least three stationary states.

The simultaneous change of the stationary states $\underline{\mathbf{x}}_{st_1} \rightarrow \underline{\mathbf{x}}_{st_{ex}}$ (or $\underline{\mathbf{x}}_{st_2} \rightarrow \underline{\mathbf{x}}_{st_{ex}}$), occurred by a change of the parameter k_{0i}, points to the transcritical bifurcation on the macroscopic level given by the following normal form [42]

$$\varphi_{n+1} = \lambda_u \varphi_n + \mu_2 \varphi_n^2 + O(\varphi_n^r) \tag{6.49}$$

that represents the order parameter for this kind of decision making. Our requirements to the coupling methods is in fact based on this one-dimensional equation with the following linear term

$$\lambda_u(k_{0i}^{cr}) = 1, \quad dim(\underline{\lambda}_u(k_{0i}^{cr})) = 1. \tag{6.50}$$

Moreover the coupling method should not modify the stationary states (that equal constants), attraction basins and finally should not increase the determinancy order of (6.49). This can be expressed by

$$\mu_2(k_{0i}^{cr}) = 0, \quad \lambda_u(k_{0i}^{subcr})\mu_2(k_{0i}^{subcr}) < 0, \quad r = 3, \tag{6.51}$$

where k_{0i}^{cr}, k_{0i}^{subcr} are correspondingly critical and subcritical values of parameter k_{0i}.

Generalizing (6.50), (6.51) and taking into account the similar conditions given in Sec. 6.4.4 for the procedure of simple voting, we require:

- in order not to modify the stationary states and to guarantee the (6.51) the coupling between initial systems has to be performed by means of the parameter a of system (6.46);
- all stationary states $\underline{\mathbf{x}}_{st_{1,2}}$, $\underline{\mathbf{x}}_{st_{ex}}$ have to be stable if the parameter k_{0i} is not activated. In this case the initial conditions have not to reach an attraction basin of the stationary state $x_{st_{ex}}$;
- if the parameter k_{0i} is activated then $\underline{\mathbf{x}}_{st_{1,2}}$ become unstable and only $\underline{\mathbf{x}}_{st_{ex}}$ still remains stable. In this case the overall coupled system is forced to jump on this state. The condition (6.50) has to be fulfilled.

The stationary states of initial map (6.46) will be arranged in the following way $x_{st_{1,2}} = 1, 0.2$, $x_{st_{ex}} = 0$.

Now, basing on these conditions we construct the CML, first for low-dimensional case with $m = 3$, $\underline{\mathbf{q}}_n = (x_n, y_n, z_n)^T$

$$x_{n+1} = f_1(z_n, y_n, k_{01})(x_n - 6x_n^2 + 5x_n^3) - 5x_n^3 + 6x_n^2, \quad (6.52a)$$
$$y_{n+1} = f_2(x_n, z_n, k_{02})(y_n - 6y_n^2 + 5y_n^3) - 5y_n^3 + 6y_n^2, \quad (6.52b)$$
$$z_{n+1} = f_3(y_n, x_n, k_{03})(z_n - 6z_n^2 + 5z_n^3) - 5z_n^3 + 6z_n^2, \quad (6.52c)$$

then consider the dimension scaling for this system. In (6.52) the coupling functions f_i are determined as the following polynomials with respect to x_n, y_n, z_n and k_{01}, k_{02}, k_{03}

$$\begin{aligned} f_1 = (&k_{01} + (k_{01}k_{11} + k_{12})z_n + (k_{01}k_{21} + k_{22})y_n + (k_{01}k_{31} + k_{32})z_n y_n + \\ &+(k_{01}k_{41} + k_{42})z_n^2 + (k_{01}k_{51} + k_{52})y_n^2), \end{aligned} \quad (6.53a)$$
$$\begin{aligned} f_2 = (&k_{02} + (k_{02}k_{11} + k_{12})x_n + (k_{02}k_{21} + k_{22})z_n + (k_{02}k_{31} + k_{32})x_n z_n) + \\ &+(k_{02}k_{41} + k_{42})x_n^2 + (k_{02}k_{51} + k_{52})z_n^2), \end{aligned} \quad (6.53b)$$
$$\begin{aligned} f_3 = (&k_{03} + (k_{03}k_{11} + k_{12})y_n + (k_{03}k_{21} + k_{22})x_n + (k_{03}k_{31} + k_{32})y_n x_n) + \\ &+(k_{03}k_{41} + k_{42})y_n^2 + (k_{03}k_{51} + k_{52})x_n^2). \end{aligned} \quad (6.53c)$$

The $k_{0i} = \{k_{0i}^0, k_{0i}^1\}$ denote the parameters that correspond to the exclusive information obtained from outside. The coefficients k_{0i}^0 correspond to nonactivated values (there is no exclusive local information) in contrast to k_{0i}^1 that denote the activated values.

The system (6.52) coupled by means of the functions (6.53) possesses a set of stationary states $\underline{\mathbf{q}}_{st}$. According to the second requirement

$$\left|\lambda_{1,2,3}^{\underline{\mathbf{q}}_{st}=0}(k_{01}^0, k_{02}^0, k_{03}^0)\right| < 1, \ \left|\lambda_{1,2,3}^{\underline{\mathbf{q}}_{st}=1}(k_{01}^0, k_{02}^0, k_{03}^0)\right| < 1, \ \left|\lambda_{1,2,3}^{\underline{\mathbf{q}}_{st}=0.2}(k_{01}^0, k_{02}^0, k_{03}^0)\right| < 1. \quad (6.54)$$

where a notation for the sub- and superindices is similar to (6.17) in Sec. 6.4.4. The third requirement and the condition (6.50) determine that

$$\left|\lambda_{1,2,3}^{\underline{\mathbf{q}}_{st}=0}(k_{01}^1, k_{02}^0, k_{03}^0)\right| < 1, \ \lambda_{1,2,3}^{\underline{\mathbf{q}}_{st}=1}(k_{01}^1, k_{02}^0, k_{03}^0) > 1, \ \lambda_{1,2,3}^{\underline{\mathbf{q}}_{st}=0.2}(k_{01}^1, k_{02}^0, k_{03}^0) > 1. \quad (6.55)$$

In expression (6.55) k_{01}^1 is activated only as an example. In common case every of coefficients k_{0i} has to cause the transctritical bifurcation in the system (6.52).

Additionally to conditions (6.54), (6.55) we set that other stationary states $\tilde{\underline{q}}_{st}$ (except for $\underline{q}_{st} = 1$, $\underline{q}_{st} = 0$, $\underline{q}_{st} = 0.2$) are unstable

$$\left|\lambda_j^{\tilde{\underline{q}}_{st}}\right| > 1, \quad j = 1, 2, 3. \tag{6.56}$$

Now simplifying the coupling functions (6.53), setting

$$k_{41} = k_{42} = k_{51} = k_{52} = 0, \ k_{21} = k_{11}, \ k_{22} = k_{12}, \tag{6.57}$$

we get finally the following system of inequalities

$$|4k_{01}^0 + 8k_{01}^0 k_{11} + 8k_{12} + 4k_{01}^0 k_{31} + 4k_{32} - 3| < 1, \tag{6.58a}$$

$$|\frac{9}{5} - \frac{4}{5}k_{01}^0 - \frac{8}{25}k_{01}^0 k_{11} - \frac{8}{25}k_{12} - \frac{4}{125}k_{01}^0 k_{31} - \frac{4}{125}k_{32}| < 1, \tag{6.58b}$$

$$(4k_{01}^1 + 8k_{01}^1 k_{11} + 8k_{12} + 4k_{01}^1 k_{31} + 4k_{32} - 3) > 1, \tag{6.58c}$$

$$(\frac{9}{5} - \frac{4}{5}k_{01}^1 - \frac{8}{25}k_{01}^1 k_{11} - \frac{8}{25}k_{12} - \frac{4}{125}k_{01}^1 k_{31} - \frac{4}{125}k_{32}) > 1, \tag{6.58d}$$

$$|(4k_{01}^1 + \frac{8}{5}k_{01}^1 k_{11} + \frac{8}{5}k_{12} + \frac{4}{25}k_{01}^1 k_{31} + \frac{4}{25}k_{32} - 3)| > 1. \tag{6.58e}$$

The last condition has to be imposed on the functions (6.53) and is connected with the areas of attraction basins. If the parameter k_{0i} is nonactivated, the system (6.52) possesses the nonbifurcation dynamics. In this case the state, that the system achieves in finite time, is determined by competition between the state variables. Using the initial conditions, we can influence this competition. But if the area of initial conditions that leads to the desired attractor is too small we are not able to affect effectively the dynamics. Shift of the boundaries of attraction basins can be performed e.g. by the change of eigenvalues, in the following way

$$(k_{01}^0 + (k_{01}^0 k_{11} + k_{12})l_1 + (k_{01}^0 k_{11} + k_{12})l_2 + (k_{01}^0 k_{31} + k_{32})l_1 l_2)) < 1, \tag{6.59}$$

where the values of l_1 and l_2 are estimated experimentally (in this case they are equal to 0.89).

Now using the program of symbolic manipulations, we solve the systems (6.58), (6.59) and get the following values of coefficients k

$$k_{01}^0 = 0.15, \ k_{01}^1 = -0.15, \ k_{11} = 2.4, \ k_{12} = 2.245, \ k_{31} = -7.25, \ k_{32} = -3.7. \tag{6.60}$$

6.4.6.3 Dimension scaling

Dimension scaling is in this case more complex mainly because a dynamics of the system (6.52) consists of the bifurcation and nonbifurcation components, which the dimension scaling has different form for. The nonbifurcation dynamics is already discussed in Sec. 6.4.4.1. The bifurcation dynamics is governed by the low-dimensional OP equation (6.49) and if this equation derived from the original system (6.52) is dimension invariant, the original system will also dimension invariant.

In order to confirm that the Eq. (6.49) is independent of the dimension of the initial system (6.52) we first derive the mode equation similarly to (6.21), (6.22) that has the following form

	Case $\underline{\mathbf{q}}_{st} = 1$
λ	$k_{0i}(4 + 8k_{11} + 4k_{31}) + 8k_{12} + 4k_{32} - 3$
μ_2	$k_{0i}(9 + 18k_{11} + 9k_{31}) + 18k_{12} + 9k_{32} - 9$
μ_3	$k_{0i}(4k_{11} + 4k_{31}) + 4k_{12} + 4k_{32}$
μ_4	$k_{0i}(5 + 10k_{11} + 5k_{31}) + 10k_{12} + 5k_{32} - 5$
μ_5	$k_{0i}(9k_{11} + 9k_{31}) + 9k_{12} + 9k_{32}$
μ_6	$k_{0i}4k_{31} + 4k_{32}$
	Case $\underline{\mathbf{q}}_{st} = 0.2$
λ	$k_{0i}(-4/5 - 8/25\, k_{11} - 4/125\, k_{31}) - 8/25\, k_{12} -$ $-4/125 k_{32} + 9/5$
μ_2	$k_{0i}(-3 - 6/5k_{11} - 3/25k_{31}) - 6/5k_{12} - 3/25k_{32} + 3$
μ_3	$k_{0i}(-4/5\, k_{11} - 4/25\, k_{31}) - 4/5\, k_{12} - 4/25\, k_{32}$
μ_4	$k_{0i}(5 + 2\, k_{11} + 1/5\, k_{31}) + 2k_{12} + 1/5\, k_{32} - 5$
μ_5	$k_{0i}(-3k_{11} - 3/5\, k_{31}) - 3k_{12} - 3/5\, k_{32}$
μ_6	$-4/5 k_{0i} k_{31} - 4/5\, k_{32}$

Table 6.2: Coefficients for the mode equation (6.61).

$$\xi_{n+1}^i = \lambda \xi_n^i + \mu_2(\xi_n^i)^2 + \mu_3(\xi_n^i)(\xi_n^{i-1} + \xi_n^{i+1}) + \mu_4(\xi_n^i)^3 + \mu_5(\xi_n^i)^2(\xi_n^{i+1}) + \mu_6 \xi_n^{i-1}\xi_n^i\xi_n^{i+1}, \quad i = 1, \ldots, m, \tag{6.61}$$

where coefficients μ are summarized in Table 6.2. Introducing the near identity transformation up to the third order (e.g., [42], [45]).

$$\underline{\boldsymbol{\xi}}_n = \underline{\boldsymbol{\varphi}}_n + \underline{\mathbf{p}}^{(2)}(\underline{\boldsymbol{\varphi}}_n) + O(\underline{\mathbf{p}}^{(3)}(\underline{\boldsymbol{\varphi}}_n)), \tag{6.62}$$

where $\underline{\mathbf{p}}^{(r)}$ are nonlinear functions of the state vector $\underline{\boldsymbol{\varphi}}_n$ of the order r and substituting it into the mode equation (6.61), we get the normal form (6.49). For $k_{0i}^{cr} = -0.1448275862$ in the Table 6.2 for both $\underline{\mathbf{q}}_{st}$ are $\lambda(k_{0i}^{cr}) = 1$, $\mu_2(k_{0i}^{cr}) = 0$, $\lambda(k_{0i}^{subcr})\mu_2(k_{0i}^{subcr}) < 0$ as it was required by (6.50), (6.51). Moreover, neither the linear part given by $\lambda\varphi_n$ nor the nonlinear part $\mu_2\varphi_n^2$ is dependent on the dimension of the system (6.52).

Generalizing the main results of this section, we point to the OP equation (6.49). Beginning our consideration, we made some assumptions about the behavior of the order parameter. Then based on this assumption, we derived the restrictions imposed further on the coupling method. Performing the required calculations we obtained the coupled system (6.52). Now, deriving the order parameter equation from this system, we obtain the initially assumed equation (6.49). Moreover, this equation is independent of the dimension of the coupled system (6.52).

6.4.6.4 Implementation remarks

Combining the coupling functions given by (6.53) (taking into account (6.57) and (6.60)) and the initial system (6.46) (with the control (6.47), (6.48)), we get dynamical system in the following form

$$
\begin{aligned}
q_{n+1}^i = & (k_{0i} + (k_{0i}k_{11} + k_{12})(q_n^{i-1} + q_n^{i+1}) + (k_{0i}k_{31} + k_{32})q_n^{i-1}q_n^{i+1})(q_n^i - \\
& -6(q_n^i)^2 + 5(q_n^i)^3) - 5(q_n^i)^3 + 6(q_n^i)^2 + \eta[q_{n-1}^i - q_n^i], \qquad (6.63) \\
i = & 1, \dots, m, \qquad (6.64)
\end{aligned}
$$

taking into account the two-way coupling neighborhood for q_n^{i-1}, q_n^{i+1}. The time-delay function in the square brackets ($\eta = 0.01$ for $m = 8$) accelerates the achievement of the stable fixpoint (likewise (6.35)). Moreover, this control together with the condition (6.59) restricts the motion amplitude of the state variable q_n^i in the area $[0, 1]$. In this case the external control (6.47) is very seldom turned on therefore does not introduce in fact any essential perturbation into the dynamics of the initial system (6.46). Forasmuch as the dynamics of (6.63) is intricate enough, its most stable behavior is achieved for $m \leq 10$ (see Fig. 6.14). These can be further composed into the hierarchical structures.

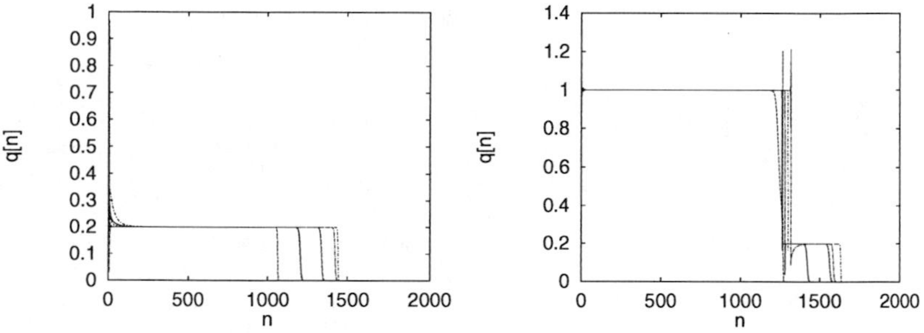

(a) (b)

Figure 6.14: The temporal behavior of state variables of the system (6.63) with $m = 7$. (a) In the first case the overall system, using nonbifurcation dynamics (voting procedure) with all coefficients $k_{0i} = 0.15$, achieves the attractor with numerical state 0.2. At $n = 1000$ one of the coefficients gets the value $k_{0i} = -0.15$ and then the whole dynamics undergoes the transcritical bifurcation after that all state variables of system (6.61) achieve the attractor with numerical values 0; (b) The second case is similar to first one, but in the first phase all variables achieve the attractor with numerical value 1.

6.4.7 Distributed voting dynamics

As mentioned in Sec. 6.3.3, a coordination between robots can be achieved by "attracting synchronization" between branching points in individual plans of robots. In the simple case we assume every robot has two alternative plans (A and B) that can be individually executed. When robots are not coordinated, the whole group will execute A and B randomly (e.g. $AABABBABBA$, $m = 10$). We require that during coordination the whole group has to reach the common agreement about kind of activity being executed collectively, i.e. either $AAAAAAAAAA$... or $BBBBBBBBBB$.... We can say that all robots vote either for A or for B. Remember, that in distributed voting none of agents knows neither the decisions of other participants nor even the number of participants in group, therefore this goal is not so trivial as it may seem.

The collective agreement about A or B represents a qualitative change in the group behavior which can be described by macroscopic order parameters. This kind of macroscopic dynamics is very similar to the case of coexisting attractors (e.g. [101]) illustrated by the following one-dimensional normal form

$$\varphi_{n+1} = \lambda_u \varphi_n + \mu_3 \varphi_n^3 + O(\varphi_n^4), \tag{6.65}$$

where $\lambda_u > 1$ and $\mu_3 < 0$ are coefficients. The system (6.65) possesses three stationary states $\varphi_{st_{1,2,3}} = \{0, \sqrt{\mu_3(1-\lambda)}/\mu_3, -\sqrt{\mu_3(1-\lambda)}/\mu_3\}$ and for $\lambda_u > 1$ the state φ_{st_1} is unstable and $\varphi_{st_{2,3}}$ are stable. As shown in fig. 6.15(a) the attractor, that the state variable φ_n will be attracted to, is determined by initial conditions (so-called attraction basins (e.g. [42])), namely $\varphi_0 > 0$ for φ_{st_2} and $\varphi_0 < 0$ for φ_{st_3}.

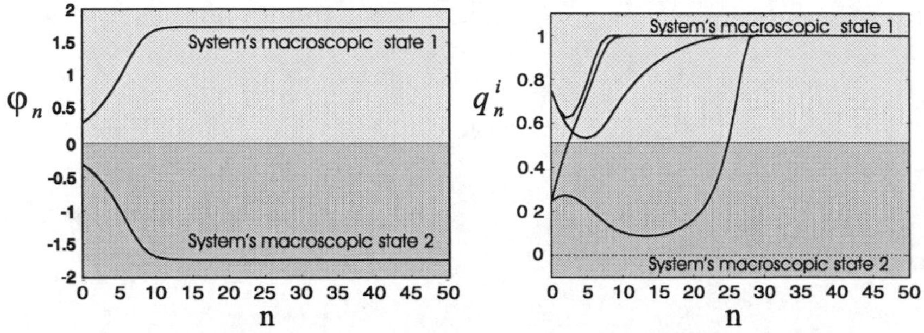

(a) (b)

Figure 6.15: (a) Macroscopic dynamics of the system (6.65) (the order parameter), where $\lambda_u = 1.3$ and $\mu_3 = -0.1$, (two times: at initial condition $x_0 = -0.3$ and at initial condition $x_0 = 0.3$); **(b)** Microscopic dynamics of the system (6.9) $m = 7$ with two different initial values $q_0 = 0.25$, $q_0 = 0.75$. These Figures are taken from [192].

In other words, according to initial states, the system will be attracted to the macroscopic states A or B that represent two different coordinated activities. Therefore the system (6.65) describes the macroscopic dynamics of distributed voting. Following synergetic tradition, we denote this system as the order parameter. Our intension now is so to transform the initial system (6.9), that it demonstrates such a macroscopic dynamics as the order parameter (6.65). This can be achieved by using the synergetic slaving principle [101] and adjusting of the coefficients k_j of coupling function (6.10) (by necessity to increase the order of (6.10)). The initial conditions $\underline{\zeta}_{st=0}$=(ζ_0^1, ζ_0^2,..., $\zeta_0^m)^T$ play a role of agents's initial proposals towards A or B. More exactly we require:

- the eigenvalues $\underline{\lambda}^{\zeta_{st_0}}$ of the system (6.9) evaluated on the stationary states $\underline{\zeta}_{st=0} = (0, 0, ..., 0)^T$ and $\underline{\lambda}^{\zeta_{st_1}}$ evaluated on $\underline{\zeta}_{st=1} = (1, 1, ..., 1)^T$ have to be stable;
- the eigenvalues of the system (6.9) evaluated on all other stationary states have to be unstable;
- the coefficients at nonlinear terms in system (6.9) should provide the supercritical type of post-bifurcation dynamics like the condition $\lambda_u \mu_3 < 0$ for Eq. (6.65);
- the initial conditions $\underline{\zeta}_0$ have to be so selected that to guarantee the reliable choice between the states $\underline{\zeta}_{st=0}$ and $\underline{\zeta}_{st=1}$ at long time dynamics of system (6.9).

We will use the following approach: first we consider the system (6.9) as low-dimensional system (e.g. $m = 5$) with fixed neighbors in coupling function $\Phi(\zeta_n^{random})$ (e.g. $\Phi(\zeta_n^{i-1})$)

and without restrictions (6.12). After deriving all needed relations, we consider the dimension scaling and effects of random coupling and restrictions. So, we start with the five-dimensional CML (6.9) where $\underline{\zeta}_n = \left(\zeta_n^{(1)}, \zeta_n^{(2)}, \zeta_n^{(3)}, \zeta_n^{(4)}, \zeta_n^{(5)}\right)^T$ is the state vector. This system has five eigenvalues λ_{1-5}, which can be evaluated on $2^5+1 = 33$ stationary states. We are interested in the following eigenvalues of Jacobian of the system (6.9)

$$\lambda_{1-5}^{\underline{\zeta}_{st=0}} = \lambda_{1,2,3}^{1} = k_0, \tag{6.66}$$

$$\lambda_{1-5}^{\underline{\zeta}_{st=1}} = \lambda_{1-5}^{2} = 2 - k_0 - k_1, \tag{6.67}$$

where subindex denotes the number of the eigenvalue, superindex shows the stationary state that the eigenvalue is evaluated on. In accordance with the macroscopic assumptions, defined by (6.65), we require (additionally to (6.11) and (6.12))

$$|\lambda_{1-5}^{1}| < 1, \quad |\lambda_{1-5}^{2}| < 1, \tag{6.68}$$

$$|\lambda_j^{i}| > 1, \quad i = 3, ..., 33, \quad j = 1 - 5. \tag{6.69}$$

Conditions (6.68),(6.69) can be simplified by the following assumption:

- some of stationary states (eigenvalues (6.69)) can be also stable. The arisen thereby simultaneously stable stationary states can be absorbed by appropriate choice of initial conditions;
- from five eigenvalues (6.69) (with index j), evaluated on stationary state, only one has to be unstable;
- Forasmuch as all initial systems in (6.9) are identical and moreover they are coupled in a symmetrical way, the eigenvalues are expected to be also equal. Therefore it needs to consider instead of all λ_{1-5} only one from them for the condition 6.68).

Finally, we get the linear system of inequalities that can be solved by programs of symbolic manipulations like Maple or Mathematica. Now taking into account our assumptions, we get for coefficients k_i in (6.10):

$$k_0 = 0, \quad 1 < k_1 < 3. \tag{6.70}$$

The calculated coefficients k_i enable us to determine a stability of stationary states, but there are two further macroscopic requirements that have to be also fulfilled. They concern the dimension of coupled system (6.9) (the problem of nonlinear terms) and boundaries of attraction basin.

6.4.7.1 Nonlinear terms and dimension scaling

As known from the investigations of coupled map lattices, some properties of the regularly built CMLs are independent of their dimension. In such cases the low-dimension systems can be expanded into arbitrary high dimension without a change of their linear and some nonlinear properties (so-called dimension scaling). To prove a dimension scaling for (6.9) we consider separately the linear and nonlinear parts of (6.9). Moreover, it is also important to show possible changes in attraction basins.

Linear part of (6.9) is given by eigenvalues of Jacobian. Evidently that eigenvalues (6.66) evaluated on $\underline{\zeta}_{st=0}, \underline{\zeta}_{st=1}$ are independent of the dimension of (6.9). To show, that the nonlinear part is also independent of the dimension (6.9), we derive the normal form of the system (6.9) and prove whether the nonlinear terms are independent of the dimension m. Since all eigenvalues are equal, we get the normal form of m state variables and a treatment of this topic outsteps the framework of the given work [114]. However, we can simplify the problem of nonlinear terms, when to consider only a postbifurcation dynamics of system (6.9).

For that let $\xi_n^i = \zeta_n^i - \zeta_{st_j}^i$ and then substituting it into (6.9) and taking into account (6.70), we get the following system for the case $\zeta_{st}^i = 0$

$$\xi_{n+1}^i = \lambda^1 \xi_n^i + (\xi_n^i)^2 + k_1 \xi_n^i \xi_n^{i-1} - k_1 (\xi_n^i)^2 \xi_n^{i-1} \tag{6.71}$$

and for the case $\zeta_{st}^i = 1$

$$\xi_{n+1}^i = \lambda^2 \xi_n^i + (1 - k_1)(\xi_n^i)^2 - k_1 \xi_n^i \xi_n^{i-1} - k_1 (\xi_n^i)^2 \xi_n^{i-1} \tag{6.72}$$

where $i = 1, ..., m$. These equations are denoted as mode equations (e.g. [161]). Since $\lambda^1 = 0$, $\lambda^2 = 2 - k_1$ are independent of m, we focus on nonlinear terms. When representing the systems (6.71), (6.72) in a matrix form, the coefficients of nonlinear terms build always matrices with elements only on the main diagonal, i.e. the structure of nonlinear parts does not undergo any perturbations at a dimension growing. In this way we proved that the system (6.9) has supercritical type of postbifurcation dynamics independently of the dimension m, i.e. the system is scalable, at least, from the viewpoint of structural dynamics.

6.5 Returning back: Implementation and experiments

In this section, we describe an overview of the experiments performed concerning energy homeostasis, as mentioned in the previous sections, and the implementation of analytic decision-making approaches. First, we describe the hardware of the "Jasmine" robot and its individual energy decision tree. Since several experiments were performed in simulation, the structure of the simulation is briefly described. Then, the mechanisms developed for collective decision-making, implemented in simulation and in real robots, are discussed. Finally, the results are compared.

6.5.1 Implementation notes on recharging hardware, software and behavior

6.5.1.1 Recharging hardware

General design of the micro-robot is often limited by the size and geometrical configurations of used components. The "Jasmine" robot is $3 \times 3 \times 2$ cm robot, cheap and easy in assembling, the whole integration of all principal components is done on two PCBs (Printed Circuit Board) in the form of a "sandwich design" [82]. It uses two 8-bit "Atmel" MPUs, contains several sensors, such as IR-based, light, color, odometrical and other. Energy resources are given by 250mA/h Li-Po accumulator. More detailed overview about the robot can be found in [82]. To make the micro-robots capable of autonomous recharging, there are four following components, installed on the robot:

- internal energy sensors, monitoring energy level of Li-Po accumulator;
- especial recharging circuits for Li-Po process;

- reliable connectors to docking station with a low electrical resistance;
- communication with docking station.

Internal energy sensors. The internal energy sensor is a resistive voltage divider with a coefficient of 0.55. The whole resistance is 726k (402k+324k); the continuous current is about 5μA. The voltage divider is directly connected, with a non-regulated power line, directly with the Li-Po accumulator and with an ADC input for the microcontroller. The ADC conversion uses about 64μs, so that the energy level monitoring can be performed very quickly.

Recharging circuits for the Li-Po accumulators. The battery used in the "Jasmine" robot is a single-cell Li-Po accumulator. When it is fully charged, it provides the micro-robot with enough energy for 1.25 hours. Li-Po accumulators require a specific recharging process. For controlling and energy management we use the linear Li-Ion/Li-Po battery charger LTC4054-4.2 in a small ThinSOT package with low dropping voltage, installed on the Li-Po accumulator.

Mechanical connectors with low-resistance contacts. To perform autonomous recharging, the robots have connectors for the docking station that allow simple docking and reliable mechanical contact with low electrical resistance [83]. The connectors are two 0.4mm silver-plated wires, glued to the front of the robot. The connectors are installed at different heights, see Fig. 6.16(a). The docking station is a wall, on which two 5mm-wide copper strips are glued 0.2mm apart. Both copper strips are connected to a stabilized 5V source. The length of the docking station is chosen so as to allow simultaneous recharging of 5-10 robots, see Fig. 6.16(b). To connect to the docking station, a robot has to move to this wall (based on the docking signal), until it receives a positive signal from its touch sensor (see Fig. 6.16(a). After that, the robot makes a small turn, to produce a slight mechanical strain, sufficient to create a reliable mechanical contact.The resistance of such a contact is less than 0.1 Ohm (measured statistically).

Communication with the docking station. The docking station has a communication system similar to the robot. This communication system continuously sends a docking signal: "a free slot exists in the docking station" (coded numerically). This signal can be received up to 10-15cm away from the docking station. A "hungry" robot, on receiving the docking signal (it "smells food"), approaches the docking station and requests a recharge. The docking station decreases by one the number of free slots, and sends a confirmation to the requesting robot. The robot can then begin the docking and recharging process. While recharging, the body of the recharging robot blocks the signal from that slot, so that no other robot can "smell food" in that region. When recharging is finished, robot sends a "finished" signal to the docking station and moves away. The docking station then increases the number of free slots by one.

6.5.1.2 Effect of communication between robots and docking station

[5] The experiments described here are highly dependent on local IR-based communication; more reliable communication paths will result in improved behavior. The IR communication can be described in two regards: first, communication between two micro-robots and second, communication between a micro-robot and the docking station. In most cases, although not always, the transmission of a message is an important issue in the

[5]The experiments described in this subsection were performed jointly with A. Attarzadeh and are partly described in his MSc thesis. The author was responsible for the supervision of his work within the "Collective Microrobotics" project.

(a) (b)

Figure 6.16: (a) Contacts for recharging installed on the robot Jasmine. Shown is also the toch sensor in the front of the robot; **(b)** The docking station and a few robots recharging there. These Figures are taken from [197] and modified.

communication between two robots. However, in the communication with the docking station, the communicating sensor on micro-robot has the same importance as the message. In fact, the communicating sensor plays a decisive role in guiding a robot to the docking station.

A maximum and a minimum range is defined for communication. Below the minimum, the two robots are so near that the IR emitter on one cannot see the receiver on the other. This occurs below about $3cm$. Fig. 6.17 gives other valuable information; in the range between $3cm$ to $5cm$, the overlapping area of the two adjacent emitters is depicted. As shown in Fig. 6.17, the effective communication range is about $15cm$. However, communication between two micro-robots can still be established at greater distances, for example up to $25cm$.

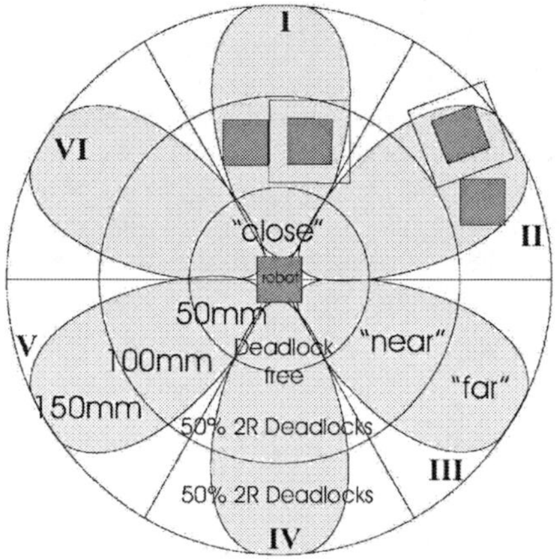

Figure 6.17: Different zones of IR-based communication (from [200]).

When a robot sends a message to another robot while they are in the overlapping region, it is possible that the receiving robot receives the message on two adjacent receiving sensors. Where the robot wants simply to hand over a message to its neighbor, the receiving sensor is not as important as the safe delivery of the message, but if this interference happens when a robot is trying to find the docking station, this might create a large error. Assume that a robot is approaching the station and is communicating with the communication point on the docking station, using its front transceiver. The robot reaches the overlapping area; suddenly one of the adjacent receiving sensors (receiver 6 or 2) receives the acknowledgment sent from the docking station. Based on the algorithm, the robot assumes that it stands at an angle of about 60 degrees with the communication point, and tries to correct its course by a stepwise rotation of 30 degrees at each step. While rotating, the robot checks if its course has been corrected. To do this, it sends a request signal to the docking station and listens to the acknowledgment. If it receives the acknowledgment on its front receiver, it stops rotating and continues to move forward. However, it was on the correct course before this confusion occurred and now has actually drifted! This confusion continues; in the end the robot is misled and looses the station.

As tested, the micro-robots have a communication gap of about $3cm$, which means when two micro-robots are closer than 3cm there have no chance to communicate with each other. This limit appears because of the physical structure of the micro-robots. Based on this value, and the overlapping range of adjacent communication signals, it was decided to approach the communication point by no more than about $6cm$. At this distance, as shown by experiment, there is a good and strong communication path between the micro-robot and the docking station and nearly no interference with communication occurs.

To lower the average time of docking, several improvements have been applied to the docking algorithm. In these improvements, every docked robot becomes a communication point that guides other robots to the station. After a robot is docked to the station, it sends the proper docking acknowledgment to other requesting robots on its rear transmitter, which is numbered as the fourth transceiver set. After applying this improvement, the results were not only significant, as expected, but also the overall complexity was increased. When two robots were docked near each other, in guiding other robots, both received the request and both sent the acknowledgment. This causes an interference in communication, which results in loss of guidance. In such cases, when the requesting micro-robot does not receive the proper acknowledgment, it halts its approach to the docking station and starts to move and search again for the communication point.

To reduce such confusion, each robot, shortly before docking with the station, sends a shutdown signal to the communication point that guided it to the station. After docking, the micro-robot itself becomes a communication point, using its rear transceiver set, number four. The results showed a considerable improvement. In this improved algorithm, the robots did not need to take note of the already docked robots, because the shutdown policy guarantees there is always a free slot next to the communication point.

As the number of docked micro-robots increases, the probability of finding the docking station with another robot decreases. The micro-robots are more in the middle of the field than in the space near the walls or the corners. Fig. 6.18 shows this dramatic effect. As seen in the end picture, although enough free space exists for a further robot to dock with the station, the final robot, because of the smaller area of communication, cannot find the communication point. This happens because the last-docked robot, which is now acting as a communication point, is so near the corner that it is rather unlikely the other robots can find it. Before they can establish communication, they detect walls and obstacles and try to avoid them. A remedy to this problem would be to change the position of the

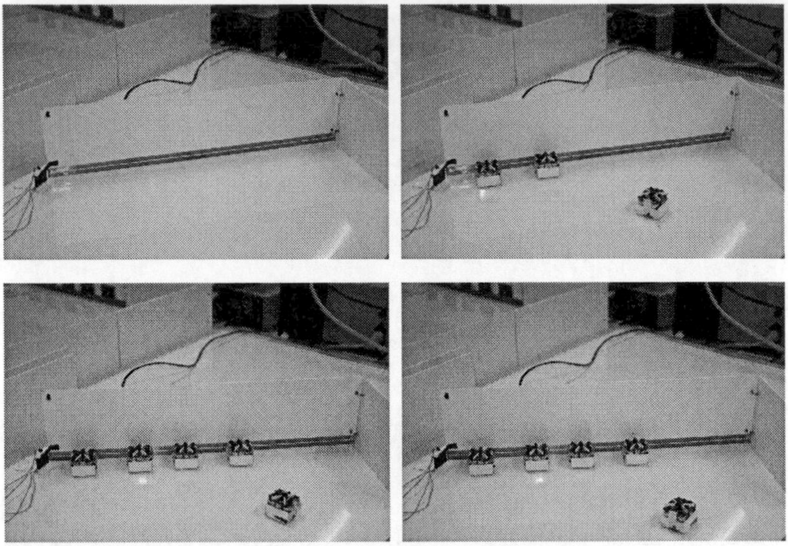

Figure 6.18: Docking approach based on the communication with the docking station. Shown are four images with step-wise docking of robots. These Figures are taken from [201].

communication point to the corner. In this case, the search area would not be limited by the adjacent wall, as previously.

6.5.1.3 Energetic homeostasis of individual robot

As described in the last sections, the robot is able to feel its own energetic level, is able to smell the position and availability of "food place". Capacity of single cell Li-Po accumulator is 250 mAh, the robot consumes about 200mA when moving and sensing, about 20mA when only sensing (communicating) and about 10 mA when only listening. In this way, the running time of the robot at least about 1,25 hour. The optimal working mode of Li-Po accumulator is discharging only till 75-80% of capacity. The critical level of accumulator is archived when the voltage dropping less than 3V, because in this case the internal power regulator is not able to stabilize voltage fluctuations and microcontroller can spontaneously reboot. The recharging current is 1C (250mA), the full recharging take about 90 minutes, the partial recharging is almost equal to discharging (15 min. motion requires about 15 min. recharging).

The energetic homeostasis of the robot is based on the diagram, shown in Fig. 6.19 and includes five different states:

S_d, **energetic death:** when voltage fails under 3.05V (ADC value 142), robot should stop and go in stand-by mode. In this state it is not able to react on external stimulus and need human assistance for recharging;

S_c, **critical state:** when voltage is under 3.2V (ADC value 150), robot should look for docking station independently of the current task. It has about 3-5 minutes to find it, other case it will energetically die;

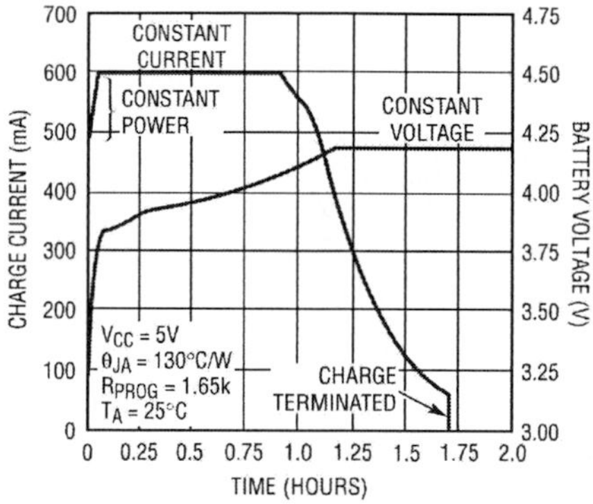

Figure 6.19: Complete charge cycle of Li-Po 750mA battery (source: the datasheet of the LTC4054-4.2 standalone linear Li-Ion/Pi-Po battery charger from the Linear Technology).

S_h, **hungry state:** when voltage is less than 3.7V (ADC value 173) but more then 3.2V, the robot has different degrees of hunger. It means it can start look for food, when there are no more important tasks. The less is energetic level, the more important is the "desire" of looking for food (the more higher priority has it). Ideally, when reaching 3.65V (ADC value 142), robot should start looking for docking station.

S_s, **satisfied state:** when during recharging, a robot achieves 4.0V (ADC value 187), this state can be characterized as satisfied - accumulator is not fully recharged (80-85%), but enough to run again and make slot free for other robots.

S_f, **full state:** the voltage increases between 4.1 and 4.2 very slow, (accumulator is already recharged up to 90-95%). When voltage is about 4.2V (ADC value 196) the electronic circuit stops recharging. In this state the robot is "full".

In Table 6.3 we have collected above-mentioned characteristics of energetic homeostasis.

State	Voltage	Reaction
S_d, energetic death	$<$ 3.05V	power down, stand-by mode
S_c, critical state	3.05-3.2V	looking for docking station
S_h, hungry state	3.2-3.7V	working or locking for recharging
S_s, satisfied state	3.7-4.0V	primarily working
S_f, full state	4.0-4.2V	stop recharging

Table 6.3: Characteristics of energetic homeostasis.

The robot can manage its own behavior in the way, shown in Fig. 6.20. Firstly, in a critical state, robots should break the currently executed collective or individual activity. This is not typically in robotics, however when a robot will execute its activity further, it can essentially distort this, when die, especially in collective case. Secondly, a robot should have the priority of a currently executed activity $Pr(Task)$ and the priority of looking for

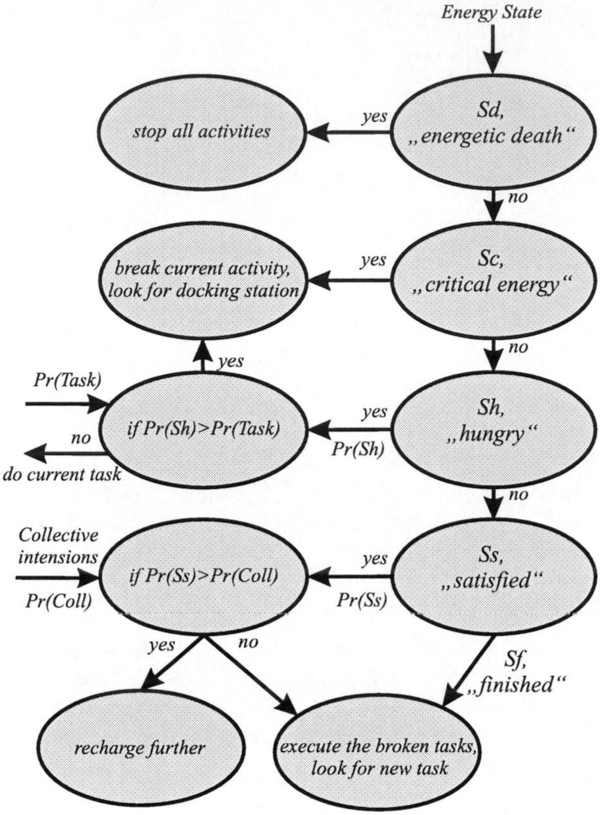

Figure 6.20: The structural scheme of the energetic homeostasis of the robots "Jasmine" (from [209]).

food $Pr(S_h)$ ("hunger feeling"). When, for example, the priority of current activity is 0.6, but hunger is 0.7, robot will look for a docking station. Finally, a robot can have so-called "collective instinct", it can recharge only until "satisfied state" (it takes less time), and makes a slot free for recharging of another robot. Generally, potential cooperativeness of individual energetic homeostasis can lie:

- in more higher priority of collectively executed tasks. For example, the tasks, where a few robots work cooperatively, get a maximal priority;
- in activities, when a swarm "permits" an express recharging "out of turn" for robots with critical energy state;
- in avoiding a full recharging when there are other robots in "hungry" states.

6.5.1.4 Recharging approach in simulation

Modern advances in robotic design use models of robots within a simulation environment in order to train the robot and to anticipate software and hardware design faults, enabling the robot to always be upgraded and redesigned quickly and efficiently. Further, using simulation environments reduces the cost of using real robots to advance robot software and to minimize the amount of destruction a robot might do if it went out of control. Thus, modeling the Jasmine robot in a simulation environment is considered essential to ensure the success of the robot design.

As already mentioned, several experiment are performed in simulation. We used an object-oriented 3D simulation environment called “Breve” (e.g. [47]). Breve is a free, open-source software package for multi-agent, distributed and artificial life simulation. Breve simulation is easy to design and program. It uses an object-oriented programming language called “Steve”. The Breve software comes with a library of already built and tested classes for using and building the simulation environment which reduces the effort to design and simulate robots. The software environment is supported by OpenGL display engine so that it provides realistic visualization of the simulated world. Breve includes physical simulation so it is possible to simulate physics for robots, vehicles or animals. This means that agents in the simulated world can be configured to behave according to the laws of physics. It uses Open Dynamics Engine (ODE) to support physics simulation. Useful class methods that come with Breve are collision detection and ray tracing which enable modeling of real robot sensors.

Program architecture of the simulation system shown in Fig. 6.21. The general sim-

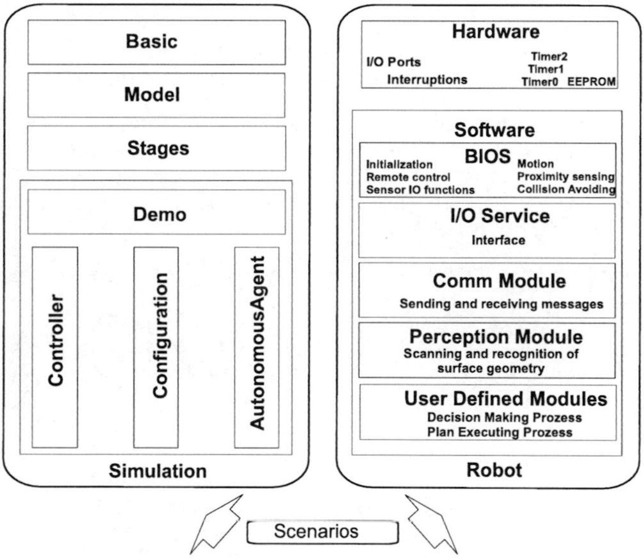

Figure 6.21: Main program architecture of the simulation.

ulation features are characterized by having a dynamic arena architecture to support different simulation scenarios. The dynamic arena architecture was designed to support a dynamic wall communication beacon creation and placement where the wall communication beacons are used to provide anti-collision signals (passive sensing). The simulation features also come with a dynamically positioned target object used by robot scenarios. These features add support to the robot passive sensor used for sensing the robot surroundings. The general simulation features are listed as following:

- easy way to create new arenas;
- arena with beacons of 360 degrees aperture that send anti-collision signals with location. It uses different wall beacon sensor range shapes (Cube, Disk);
- target object with beacons of 360 degrees aperture that sends anti-collision signals with its location.

The simulation files are easy to handle and use. They are grouped in a similar manner to those of Jasmine simulation model. They are divided into five parts which are described as following [84]:

Basic Directory contains basic simulation classes files such as: *basicAutonomousAgent3d.tz* - this class contains the basic 3D robot model with all the methods and variables, *basicController.tz* - this class contains the basic physical simulation platform and also enables and sets fast physics, *basicModel.tz*- it is the model for agents, *Stage.tz* - this class creates the simulation arena using the stage file specified in *demoAutonomousConfig.tz*, *basicSensorModel.tz* - it is the model for sensors, *basicSensorFactory.tz* - this is a factory which contains all sensors.

Model directory contains two directories that have the classes for modeling the robot and the different sensors in the simulation. These directories are: *robot directory* - contains the models of different modules of the 3D robot; *sensor directory* - contains the model of the different sensors in the simulation and of the robot.

Demo directory contains the main files of the simulation and they are: *demoAutonomous-Config.tz* - contains the main parameters of the simulation, robot, arena and target obstacle, *demoAutonomousAgent.tz* - contains the main class of the 3D robot, *demoAutonomousController.tz* - this contains the main simulation file which has to be used to run the simulation.

Arena directory contains the files for creating different arenas. These files are text files that have the wall layout of the arena.

Target obstacle directory contains the classes files that create the target obstacle and that have the methods for target relocation and anti-collision signal generation and the target beacon sensor.

6.5.2 Experiments with collective decision making

In the section we describe four different experiments, performed in simulation and with real robots, where analytical decision procedures are used.

6.5.2.1 Decision making with randomly changing neighbors based on collective energy level

This approach is based on the scheme shown in Figs. 6.3 and 6.4 from Sec. 6.2 and is implemented both in simulation and in real robots. The idea of this approach originates from the Sec. 6.4.5, in particular from the equation (6.24), which can be used for calculation of global energy consumption E_c^c in fully distributed way. As mentioned, the value of E_c^c is used in the decision making process for changing the priority of recharging $Pr(S_{hungry})$ and $Pr(S_s)$ and for decisions towards executing different variants of "Job A" and "Job B".

In the equation (6.24) we assumed that robots are connected as two-ways ring with fixed connections. However in real world, robots are moving and so the neighbors in the two-ways ring are continuously changing. Generally, this kind of systems is denoted as CML with random coupling and investigated from the viewpoint of oscillator synchronization [206], global stability [207] and so on.

We are interested here in the following observation. CML can be thought of as a 2-dimensional grid, where each node is a fixed robot (agent, equation) and couplings mean connections between these fixed nodes. When we follow this idea, then the moving robots represent the changing couplings. However, when we consider the node in this CML-grid as "place holders" for a robot, then the coupling means the connections between "place holders" and not between robots. We only require that for each "place holder" there exists a robot, when even information update in CML is asynchronous (i.e. a robot can connect and disconnect to a "place holder" any arbitrary time). In this way we can

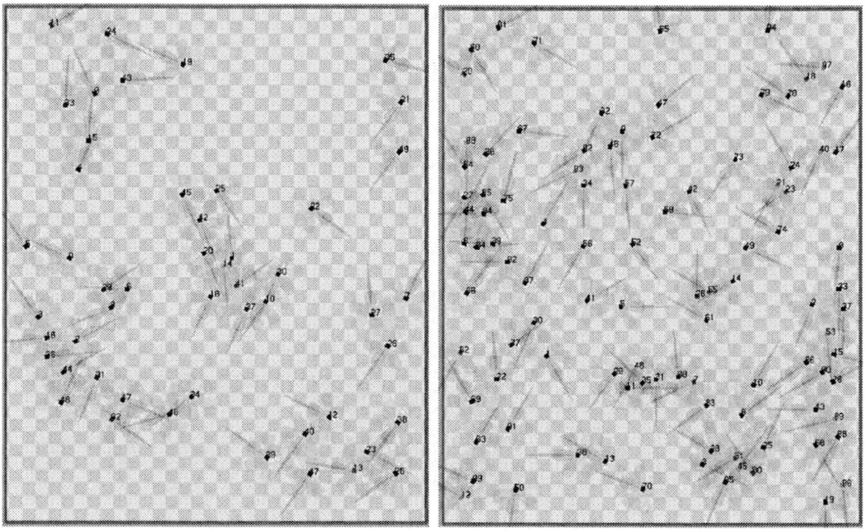

(a) (b)

Figure 6.22: Snapshots of the "Breve" simulation of **(a)** 50 and **(b)** 100 agents. Shadowed areas are covering areas of corresponding sensor (6 proximity sensing, 1 geometry perception sensors, all sensors can be used for half-duplex communication). Different clusterization degree is well visible.

perceive the structure of the equation (6.24). The change of robots in "place holders" does not influence the eigenvalues and eigenvectors, but obviously will introduce additional nonlinear effects in the dynamics of CML.

Now we try to understand the dynamics of CML, when neighbors are randomly exchanged. There are two main effects that influence this dynamics. Firstly, as known in a swarm robotics, robots often build so-called clusters, which are sometime separated from the rest of a swarm. Obviously, that this clusterization effect will be observed in CML, where we can see a separate sub-CML with their own $q^{local}_{n\to\infty}$. When robots are moving long enough (the time is primarily defined by a swarm density), they will achieve $q^{global}_{n\to\infty}$ as defined by (6.32). Secondly, the solution (6.32) with eigenvalues (6.33) represents a stable fix point. When L_j clusterized robots achieved in p-asynchronous updates the stable fix point $q^{local,j}_{p\to\infty}$ they will have this value so long as they do not enter into a new cluster

$$q^{local,j}_{p\to\infty} = \frac{1}{L_j}\sum_{i=1}^{L_j} q^i_{0,j}, \tag{6.73}$$

where j is a number of such a subcluster, $j = 0...K$ and K is a total number of subclusters. The values of $q^{local,j}_{p\to\infty}$ represent a local averaging of $q_{0,j}$ and generally have a random character, because subclusters are built more or less randomly. In this way we will observe two-steps-dynamics, where firstly we will see buildings of local fix points $q^{local,j}_{p\to\infty}$ and then their averaging into $q^{global}_{r\to\infty}$

$$q^{global}_{r\to\infty} = \frac{1}{K}\sum_{j=1}^{K} q^{local,j}_p, \tag{6.74}$$

where $r + p = n$ (n from the expression (6.32)). The value of $q^{global}_{r\to\infty}$ does not longer depend on the initial value q_0 and represent an averaging of random $q^{local,j}_{p\to\infty}$.

The relation between p and r in expressions (6.73) and (6.74) defines a dynamics of CML (6.24) when neighbors are randomly changed. In case $p > r$, the dynamics is defined by robot behavior in clusters, whereas for $p < r$ and especially for $p \ll r$, the final result is (almost) independent of clusterization effect. Both cases have their applications for swarm robotics, in this paper we are more interested when $p < r$ or $p \ll r$. This condition can be achieved when asynchronous information update in CML is slow, so that subclusters cannot be built and the $q^{global}_{r\to\infty}$ is defined mostly by q_0.

To exemplify these calculations and before making the real experiments, we performed simulations of this behavior. The simulation of the robots and especially robot-robot communication is implemented in "Breve" and have a physical character, therefore the simulated behavior matches relatively good with a real behavior of "Jasmine" robot. To investigate the clusterization effect, we simulate 50 and 100 agents, see Fig. 6.22, with fast and slow updates of information. For the case of 50 agents we expect more clusterization effects than for the case of 100 agents, primarily due to lesser swarm density. This is well visible in Fig. 6.22(a) and (b). For both cases we implement a fast and slow updates of information, so that we have 4 different situations for analysis. The algorithms for fast and slow information update in shown in Fig. 6.23. The main difference in both algorithms is the number of communication messages sent away. For slow update, the messages are sent only at the first collision contact. The averaging takes place only when two different messages are received. For the fast information update, the communication takes continuously place, not only at collision contacts but also at averaging. Moreover robot sends messages even when there is no collision. In this way we expect 3-5 times faster information propagation, than in the slow case. In the simulation we collect the q values of all robots.

In Figs. 6.24 we plot these value in dependence on the number of performed information updates. Since all updates are asynchronous (in the same time step one robot can

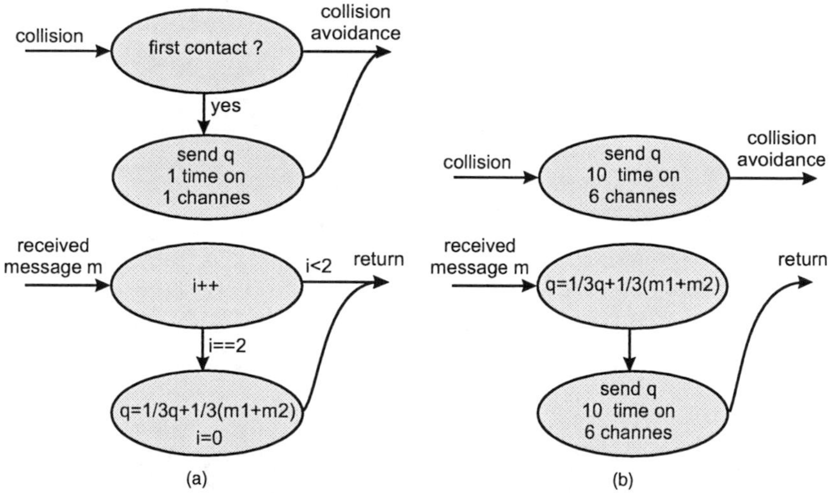

Figure 6.23: (a) The slow information update, where the message, triggering averaging, is sent only 1 time at collision with other robot; **(b)** the fast information update, messages are sent continuously.

performed 5 updates whereas another one 50 updates), these plots show a relative time-dynamics. Anyways we stop simulation only then, when all robots achieved a common value, so that finally all robots achieve the fix point.

In the Figs. 6.24(a), (b) we see the foreseen clusterization effects. The robots build many small clusters, where they archive the stable fix points $q_{p\to\infty}^{local,j}$. Finally, the global averaging is firstly relatively slow, secondly the inaccuracy of averaging is about 33% ! In the Figs. 6.24(c), (d) we demonstrate the case of slow update. Here the clusterization effect is essentially less, and therefore the accuracy of averaging is pretty better, the maximal mistake about 10%. In this way, setting the decision threshold on +20-25% and -20-25% can guarantee a stable collective decision about the collective energy consumption in a swarm. Note, that slow and fast cases differs in number of information updates, the slow update requires about 3-5 times less communication effort. However the running time (how many times agents are running) is almost the same for both cases of 50 and 100 robots, because the number of collision contacts primarily depends on swarm density, velocity of motion and other parameters of robots.

The experimental setup for real experiments is shown in Fig. 6.25. In the docking station we implemented the two-line approach, shown in Figs. 6.25(c),(d): the robots which received a direct signal from docking station navigate along this signal for docking and then start recharging. During recharging these robots in turn send secondary signal with meaning "docking station is here, but it is currently busy". This secondary signal has larger covering area than the direct signal from the docking station. When another robot receive the secondary signal from a robot, it can reduce its own velocity and slightly rotates. In this way the waiting robots perform a local search - when a position in a docking station gets free, a robot has a higher chance to find it. These waiting robots build a "recharging buffer" which allows a self-regulation in a swarm.

All robots, except for currently recharging ones, communicate and calculates the value

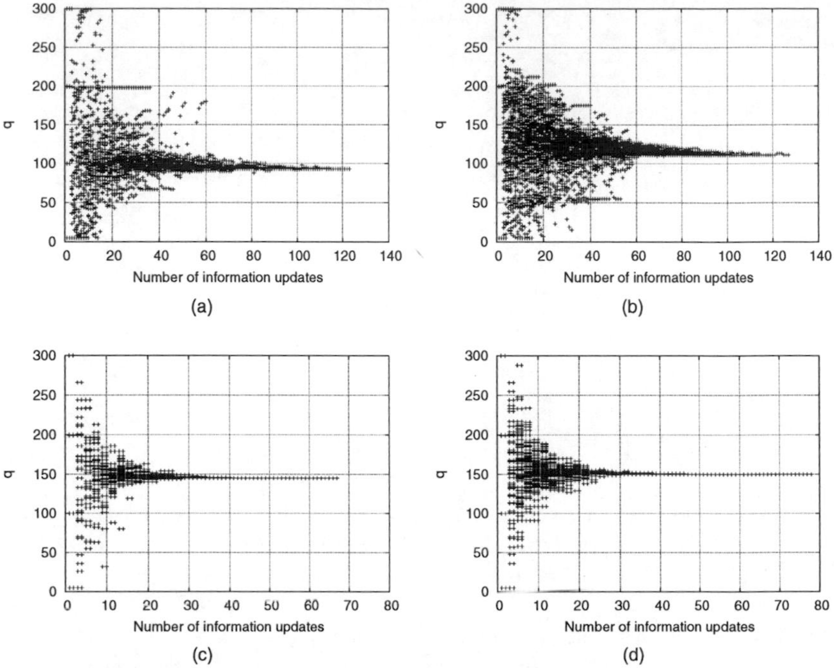

Figure 6.24: Plot of numerical q values from the simulation with 50 and 100 robots. The expected the mean value is 150. **(a)** 50 robots, the case of fast information update; **(b)** 100 robots, the case of slow information update; **(c)** 50 robots, the case of fast information update; **(d)** 100 robots, the case of slow information update;

of E_c by using the expression (6.1). For the estimation of individual energy consumption we use the following schema. The energy consumption generally depends on a relation between the time spent for motion, staying and communicating; it expresses the activity of a robot. The activity depends on the type of jobs that robots have to execute. To simulate in the experiment different types of jobs, and so different energy consumptions, each robot moves randomly but during collision avoidance has to wait some time t_{wait}. Due to the clusterization effect, the robots can have about 3-5 times difference in individual energy consumption. The general level of energy consumption depend on the t_{wait}. In this way we can experimentally test a swarm foraging behavior for different activities.

The collective behavior of robots is divided into three macroscopic states, see Fig. 6.26. In the first state all robots execute their regular collective job (in the experiment they move randomly with waiting) and calculate its own energy consumption. This state is finished by the most "hungry" robot: the first one which achieved the state S_h. It fixes the value of $q_0 = E_c$ and send the first message with q. Each other robot, which is received the q value, fixes its own $q_0 = E_c$ and calculates q by using the equation (4.49) and by using the update algorithm shown in Fig. 6.23(a). The values are limited by 8bit variables, so that only values between 0 and 255 are communicated. To have a larger range of q values, the ADC values between 140-200 (from the energetic homeostasis) are re-scaled into 0-255.

Figure 6.25: Collective energy foraging in a swarm of 50 micro-robots "Jasmine". **(a)** Initial set-up. Marked robots are landmarks with a large covering area; **(b)** The macro-state 2 with the phase of collective decision making; **(c)** The macro-state 3, where the first "hungry" robots are approaching the docking station; **(d)** The robots are docked and start recharging.

In the second macroscopic state all robots calculate the collective energy consumption, see Fig. 6.25(b). Duration of this state depends generally on the swarm density. Based on the simulation results, shown in Figs. 6.24, we set up this to 30 information updates. These 30 values are stored in the internal EEPROM and can be used for debugging purposes. After 30 updates, the collective energy consumption is calculated and robots can take a decision about going to recharging or executing their job further. This decision is based on the scheme, shown in Fig. 6.4. When E_c^c is large, only robots with low energy state move to recharge. These "hungry" robots wait (perform a local search) when receiving secondary signals from recharging robots. When E_c^c is small, a robot goes to recharge only when it received a direct signal from docking station (it ignores secondary signals from recharging robots). Corresponding to individual energetic homeostasis, all robots start actively look for docking station when their energy drops under the critical level. This is the third macroscopic state, where robots can either recharge or working on collective tasks, shown in Figs. 6.25(c)-(d).

Robots, when leaving docking station, reset their number of information updates, collective and individual energy consumption. Moreover they ignore other communication contacts with q values until they calculated the required values for energy consumption. In this way they enter into the first macroscopic state again and the whole macroscopic cycle of collective energy foraging is repeating. In Figs. 6.27 we plot the values of q read from robots EEPROM after two such experiments. Since all updates are asynchronous,

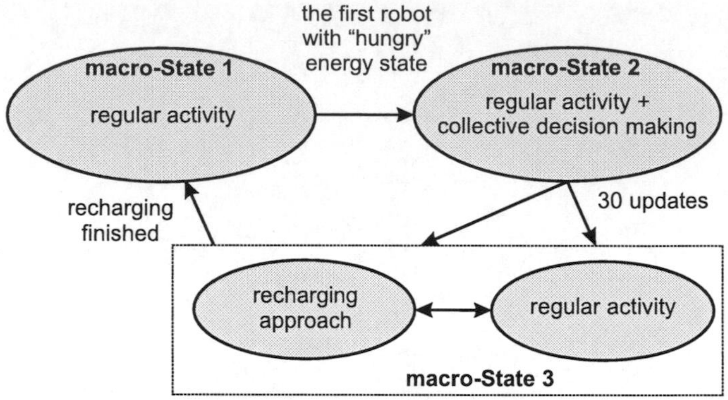

Figure 6.26: Macroscopic cycle of collective energy foraging.

the points, shown in Fig. 6.27 do not represent the same time steps. Both data indicate a good convergence and almost no clusterization effects and confirm in this way the made assumptions and theoretical calculations.

6.5.2.2 Distributed voting with random initial conditions and random couplings

This experiment intends to investigate the voting dynamics based on the system (6.9) from the Sec. 6.4.7. As demonstrated in the previous section, real systems introduce random effects, such as random couplings and s random spatial allocation of robots. The last effect has an interesting consequence for the voting dynamics. In the distributed voting the robots' initial proposals are encoded by initial conditions $\underline{\zeta}_0$. Their choice is dependent on the quota needed for taking a decision, the number of robots and a spatial allocation of initial values. For example, the sequences of initial values "AAAAABBBBB" and "ABABABABAB" ($m = 10$) have different impact on voting dynamics, despite the number of "A"- and "B"-proposals is equal in both sequences. Since, robots in real experiments can have absolutely different spatial allocations, we assume the initial proposals are allocated randomly. Thus, in this section we are interested whether random couplings and random initial conditions have any effects on the dimension scaling and on the dynamics of the system (6.9). Experiments with random initial conditions and random couplings are performed in real setup with 50 "Jasmine" robots. Scaling boundaries are selected as 100 and 1000 robots correspondingly, so that this part of the experiment was performed in the "Breve" simulation.

Random coupling is implemented (in simulation and in real robots) in the following way. We generate the float array of $\zeta[m]$. This array is randomly initialized with two numerical values a and b so that $\sum_{i=1}^{k} a$, $\sum_{j=1}^{m-k} b$, where m the number of robots. The integer parameter k shows how many robots are voting for "A". Simulation calculates four following functions

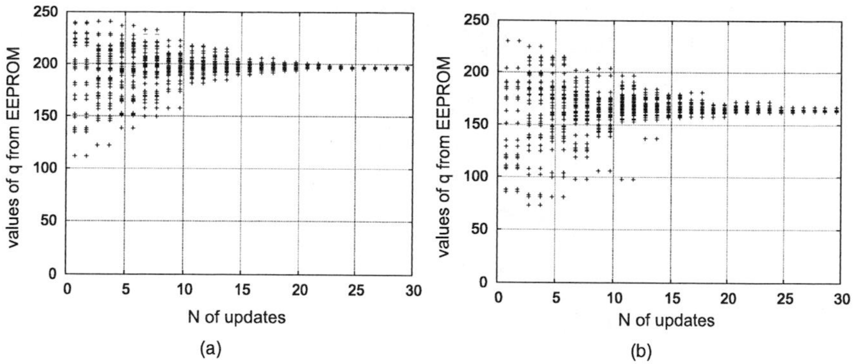

Figure 6.27: Plot of numerical q values from real experiments with 50 robots. The same experiment is repeated consequently two times, as shown in (a) and (b). Since the experiment (b) was performed after experiment (a), the values of q and E_c^c in (b) are smaller than in (a). To have a larger range of q values, the ADC values between 140-200 (from the energetic homeostasis) are re-scaled into the range of 0-255.

$$C_1 = (int)(k_0 + k_1\zeta_{[r1]})(255/k_1 - k_0), \tag{6.75}$$
$$C_2 = (int)(k_0 + k_1\zeta_{[r2]})(255/k_1 - k_0), \tag{6.76}$$
$$\vartheta_1 = C_2/(255/k_1 - k_0) \cdot (\zeta_{[r1]} - (\zeta_{[r1]})^2) + (\zeta_{[r1]})^2, \tag{6.77}$$
$$\vartheta_2 = C_1/(255/k_1 - k_0) \cdot (\zeta_{[r2]} - (\zeta_{[r2]})^2) + (\zeta_{[r2]})^2, \tag{6.78}$$

where C_1, C_2 are integers between 0 and 255, and r_1, r_2 are random integer numbers between 0 and m. The functions C simulate one-byte communication. At each calculation of these functions a counter is increased by one, meaning one information update (one communication contact between robots). Obviously, that in real robots only one C and ϑ functions are calculated. It is well-known, that when the variables ζ achieves fix points ($\zeta = 0$ or $\zeta = 1$), the system cannot leave them anymore. To prevent this effects and additionally to accelerate achieving a common decision, we perturb the dynamics so that ζ never comes to fix points, e.g. as following

$$\begin{aligned} if(\zeta_{[ri]} > 0.9999)\ \zeta_{[ri]} &= \vartheta_i - 0.001; \\ else\ if(\zeta_{[ri]} < 0.0001)\ \zeta_{[ri]} &= \vartheta_i + 0.001; \\ else\qquad\qquad \zeta_{[r2]} &= \vartheta_i; \end{aligned}$$

This function is implemented both in simulation as well as in real robots.

For this experiment we prepared a specific setup, where we try to maximize the swarm density (to have a maximal information exchange). As shown in Fig. 6.28, 50 "Jasmine" robots are randomly placed in the arena size 1100 × 1400 mm^2. Numerical values are stored in EEPROM as described in the previous section. In Fig. 6.29 we plot the dependency between the number of robots, voting for the decision "A" and the number of attempts (each experiment has 10 attempts), where all robots collectively choose the decision "B". Setting different initial conditions, we can achieve different decision quota, as for example shown 25%,50% and 75%. In Fig. 6.29(b) we plot the number of information updates (information contacts n_c) for each decision quota. The real time needed for

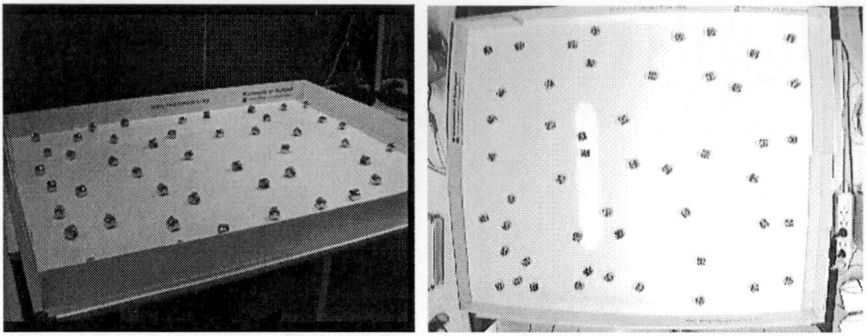

(a) (b)

Figure 6.28: Experimental setup for voting experiment with random initial conditions and random couplings. **(a)** Sidewise look; **(b)** Top-down image from the camera, installed above the arena.

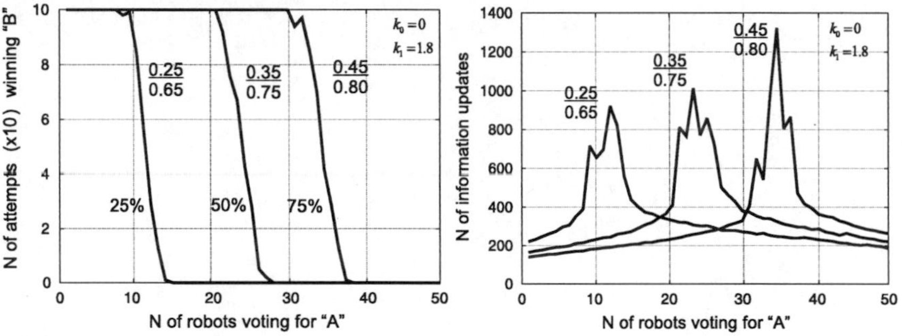

(a) (b)

Figure 6.29: **(a)** Dependency between the number of robots (50 robots), voting for the decision "A" and the number of attempts (each experiment has 10 attempts), where all robots collectively choose the decision "B". Shown are different decisions quotas, fraction shows initial conditions for "A" and "B"; **(b)** The number of information updates, required for the collective decision.

taking the decision can be obtained from

$$t = \frac{n_c}{\varphi_s N}, \tag{6.79}$$

where $\varphi_c = \frac{2\sqrt{2}R_c v}{S}$ is the swarm constant. For the values of swarm areal $S = 1400 \times 1150mm^2$, the communication radius $R_c = 150mm$ and the velocity of motion $v = 300mm/s$, the time needed for 1000 communication updates is equal to 228 sec. Facilitation of such a fast information exchange was a reason for selecting the experimental setup with a maximal swarm density, as shown in Fig. 6.28. In Fig. 6.30 we plot the same dependency for 100 and 1000 simulative robots for decision quota 33%, 50% and 66%. As visible from this plot, the mechanism is structurally stable at least in a few orders of number of robots. Another effect, encounter here, is that the number of information updates depends extremely on the choice of initial conditions. The more close initial condition are to "0" or to "1", the more large is the required time to achieve the collective decision. This

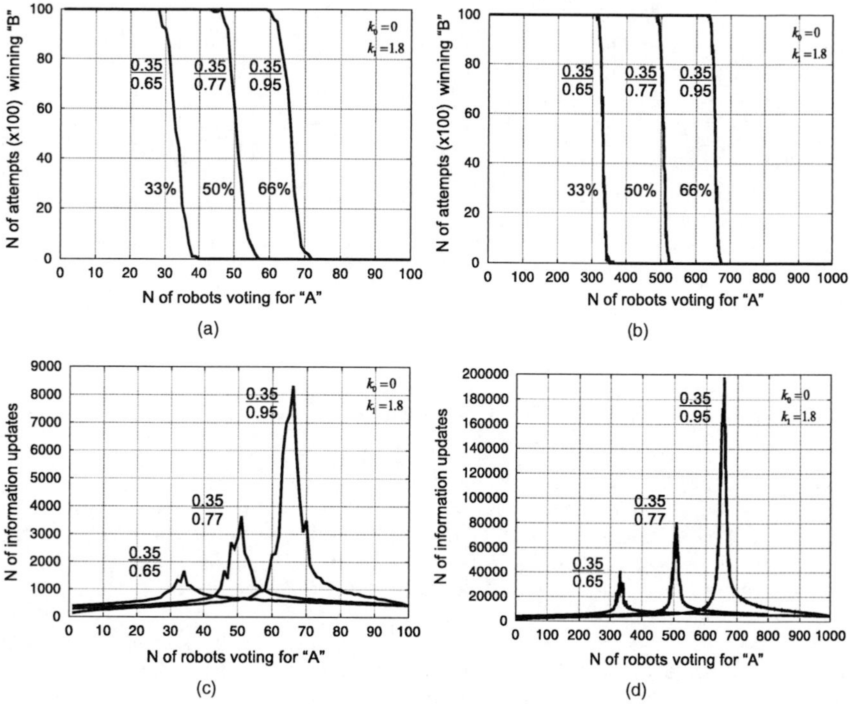

Figure 6.30: (a),(b) Dependency between the number of robots (100 and 1000 robots in simulation), voting for the decision "A" and the number of attempts (each experiment has 100 attempts), where all robots collectively choose the decision "B". Shown are different decision quota; **(c, d)** The number of information updates, required to the collective decision in the case of 100 and 1000 robots.

mechanism can underline the feedback control of swarm behavior.

Here we encounter one interesting effect. Let the dynamics of the system (6.9) to be perturbed by some external force. This force represents a changed environment, irregular process or simple some failure. As shown in Fig. 6.31 the system either absorbs this perturbation or jumps into another state, but anyway it follows the macroscopic dynamics determined by (6.65). It can be illustrated by analogy if to define the point "A" on the top of a mount and the point "B" on root of the mount. Evidently a ball (a system) will slither from "A" to "B" (it is predetermined), but the exact route can be different and depends on the landscape. The most important point is that we do not need to describe this adaptive behavior explicitly, it is already contained in the dynamics of the system.

6.5.2.3 Selection dynamics perturbed by random values

The distributed voting, considered above, represents one specific coordination mechanism. However the initial system (6.9) allows deriving different coordination mechanisms. These mechanisms are based on the Property 1, formulated in the Sec. 6.4.3. Stabilizing

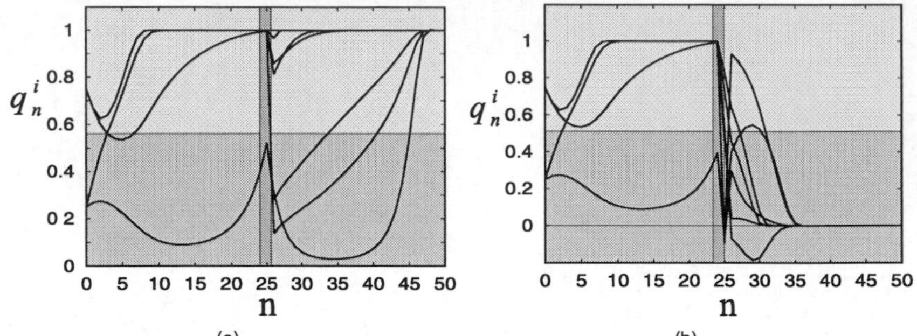

(a) (b)

Figure 6.31: (a) (b) Microscopic dynamics of the system (6.9) $m = 7$ with two different initial values $q_0 = 0.25$, $q_0 = 0.75$; **(c, d)** Perturbed microscopic dynamics of the system (6.9) $m = 7$ with the same initial conditions. Shadowed rectangle represents the perturbation.

other stationary states (taking into account nonlinear terms in corresponding mode equations, like (6.71), (6.72), we can achieve completely different dynamics. In this section we demonstrate the selection dynamics. In the selection dynamics only one (or a few) state variables ζ win the competition, i.e. receive the value "1", whereas all other variables ζ lose, i.e. receive the value "0". We show, that this approach allows a few useful swarm coordination mechanisms.

We start with the stationary states and the eigenvalues of the system (6.9). First introduce the value $\sigma = \sum_{i=1}^{m} \zeta_{st_i}$, which demonstrate how many stationary states are equal to "1". Then we order all stationery states by the value of σ. Obviously, that this order starts with the state $\underline{\zeta}_{st=0}$, finish with $\underline{\zeta}_{st=1}$ and $\sigma_{max} = m$. We introduce the repelling value $\Psi^{\underline{\zeta}_{st_i}}$ of corresponding stationery state $\underline{\zeta}_{st_i}$, defined as

$$\Psi^{\underline{\zeta}_{st_i}} = \sum_{j=1}^{m} \begin{cases} |\lambda_j^{\underline{\zeta}_{st_i}}| > 1 : |\lambda_j^{\underline{\zeta}_{st_i}}| - 1 \\ |\lambda_j^{\underline{\zeta}_{st_i}}| < 1 : 0 \end{cases} \tag{6.80}$$

When $\Psi^{\underline{\zeta}_{st_i}} = 0$, it means that the corresponding stationary state $\underline{\zeta}_{st_i}$ is stable.

Property 2. *When the stationery state $\underline{\zeta}_{st=0}$ is unstable with positive eigenvalues, e.g. at $k_0 = 2$, the repelling value $\Psi^{\underline{\zeta}_{st_i}}$ has a minimum for those stationary states whose index σ is between $m/2$ and m.*

This property originates from the observation that any eigenvalue $\lambda_j^{\underline{\zeta}_{st_i}}$, evaluated on any of the stationary states $\underline{\zeta}_{st_i}$ takes one of the following values k_0, $k_0 + k_1$, $2 - k_0$, $2 - k_0 - k_1$, (the eigenvalues for $\underline{\zeta}_{st=0}$ and $\underline{\zeta}_{st=1}$ are defined by (6.66)). When k_0 is so chosen that $|k_0| > 1$, it is not difficult to prove that stationary states at lower σ involves more unstable $\lambda = k_0$, whereas stationery states at higher σ involve more $\lambda = 2 - k_0 - k_1$. In Fig. 6.32 we plot the repelling value $\Psi^{\underline{\zeta}_{st_i}}$ for different stationary states (ordered by σ) of the system (6.9) with $m = 10$ and coupled as one-way ring (ζ^{i-1}). We see that at different values of coefficient k_1 the stationary states between $\sigma = 5$ and $\sigma = 10$ consequently are getting minimal. This effect with random coupling dynamics lead to interesting dynamics of initial system (6.9).

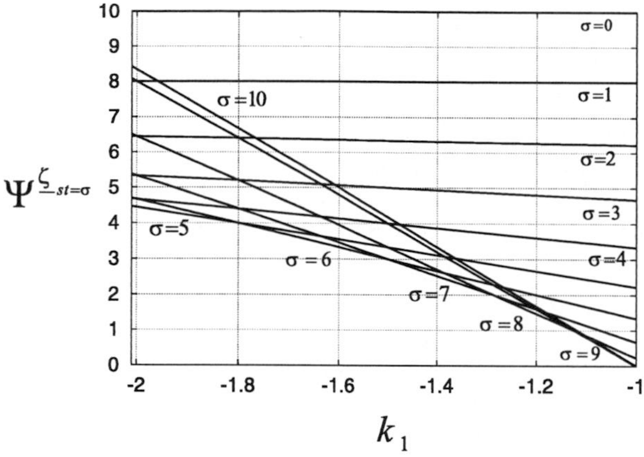

Figure 6.32: Dependency between the repelling value (ordered by σ) and the coefficient k_1 of the system (6.9), $m = 10, k_0 = 2$.

Property 3. *When all eigenvalues of the system (6.9), coupled randomly, are unstable, the dynamics of this system is attracted to those stationery state $\underline{\zeta}_{st_i}$ which possesses the lowest repelling value $\Psi^{\underline{\zeta}_{st_i}}$.*

In Fig. 6.33 we plot the number of robots attracted to the decision "A" and "B" in dependence on the value of the coefficient k_1 in the polynomial (6.10). We see that

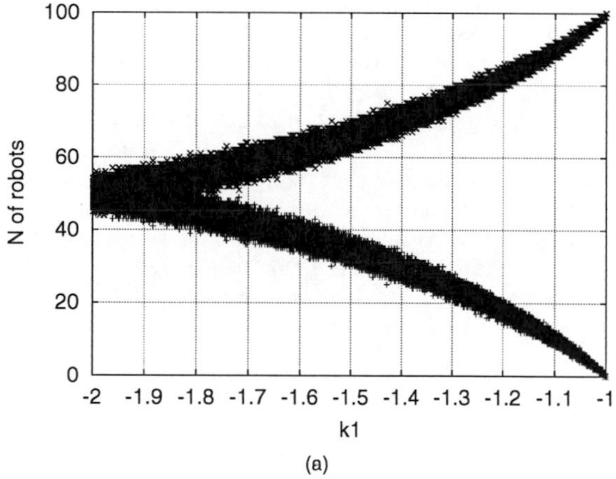

Figure 6.33: Number of robots attracted to the decision "A" and "B" in dependence on the value of the coefficient k_1 in the polynomial (6.10), $m = 100$. At each value of parameter k_1 100 attempts are performed, couplings and initial conditions are random.

swarm of robots can separate itself into two group where the relation between group can be varied from 100:0 till 50:50. The variation of random dynamics in this mechanism is about 1% - 2%. The interesting effect here is that both groups ("A" and "B" groups) are not fixed, robots perform a competition for being included into the group "A" or "B". In effect is well-visible in Fig. 6.33. The control can be executed by initial conditions and by perturbation of random dynamics.

The effect of splitting into two groups can underlie several different mechanisms in the energetic homeostasis. In particular, when the amount of energy is not enough for all robots, it is useful to separate robots with a high and low energy state from each other. Robots with a low energy stage can recharge in the first order, whereas all other robots can wait. The problem of such a split into two groups is that an energetic threshold for the decision "to include into A or B" is unknown. The group of robots can have different common energetic state and first of all robots have to calculate an average energy level. Based on the average level, each robot can decide whether it belongs to the first or to the second group. The selection dynamics provides another way for such a split. Robots map their own energy level into initial values for the selection dynamics. After interactions, those robots, which receive "0" in the selection dynamics, have a low energetic state in respect to all other robots. Thus, selection dynamics represents more efficient way for energetic grouping than procedures based on calculation of average value.

To perform experiments with this mechanism, we used the recharging setup, described before and shown in Fig. 6.35. The number of docking station is less than a number of "hungry" robots, so that splitting into two groups and waiting is necessary to survive in this scenario. The implemented approach is described in the following steps below.

1. Initially all robots are randomly distributed on the surface of $1100 \times 1400\ mm^2$, see Fig. 6.35(a). Before experiment, robots are run some time so that they have different energy state. All robots are involved into the "Job A" activity, which is a random motion on the arena. During executing "Job A", robots consume energy, so that after achieving the "hunger" threshold, robots start looking for docking station. The number of access slots in the docking station is less than the number of robots, therefore robots have to manage the collective energy foraging. Since we use the function (6.9) in this scheme, we can combine voting and selection dynamics simply by varying the coefficients k_0 and k_1, as shown in Fig. 6.34. In this way we can control the collective behavior of all robots without any central instances.

2. Every 20 minutes the robotic swarm periodically receive the external triggering event "start". For this we use signals either from the remote control or from lamp on the top of arena. This triggering signal can be thought as of some periodical events, like day-night change, that happens due to natural laws. Alternatively, the most hungry robot can send the signal "start". The meaning of this external event is to trigger the start of exchanging numerical values. Robots can setup first the voting dynamics with coefficients $k_0 = 0$ and $k_1 = 1.8$ as described in Sec. 6.4.7. During the voting robots decide whether the number of hungry robots is enough for simultaneous recharging. We performed several experiments based on the scheme in Fig. 6.34. The problem was to achieve a synchronous transition between the voting dynamics in the state 32 (st=32) and the selection dynamics in the state 34 (st=34) in all robots. It seems that a swarm needs an external synchronization mechanism for such a transition. Thus, in a final version of this experiment we start directly the selection dynamics.

3. After setting the coefficients k_0 and k_1, robots enter into the selection dynamics state. They first map the ADC values of Li-Po accumulator from the range 140-200 into the initial values between 0,3-0,7. When two robots meet each other, they exchange values of the function (6.9). Finally, achieving the values $< 0,1$ or $> 0,9$ indicates corre-

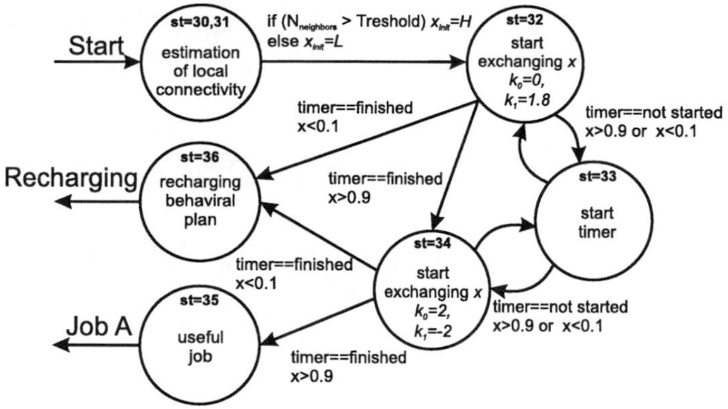

Figure 6.34: Scheme of switching between voting and selection dynamics in the energetic scenario.

sponding energetic state and can be used for the decision making (in this state to go the docking station or to wait aside the docking station). The dynamic of this process can be observed in Fig. 6.35.

The first cluster is built closely to the docking station, the position of the second cluster is random; it depends on the first meeting between two robots with high values in their dynamics, see Fig. 6.35(f). When entering into the cluster, robots stop the motion; the second group of robots primarily waits until the first group finishes the recharging procedure. To prevent the effect of clusterization, firstly, all staying robots still participate in the communication, secondly, we run the experiment longer than it is requited to achieve a stationary state (one experiment from Fig. 6.35 takes around 10 minutes). Evaluation of results is undertaken by a visual control of building these two groups and by the number of robots still running on the robot arena. Thus, we do not store the numerical values in the EEPROM (reading these values from all robots represent very time consuming operation). We repeated this experiment several times with the number of robots between 10 and 30 and always observed 3-5 running robots after 10 minutes. This corresponds to 85%-95% of robots taking part in the selection dynamics. It can be improved in different technical ways, primarily by improving a local communication between robots. As we observed many times, robots even meeting each other do not transmit numerical values.

6.5.2.4 Comparison of energetic homeostasis with and without collective decision-making

[6] In this section we compare the different strategies considered so far in relation to robots which are not synchronized in any way. Thus, we are able to draw conclusions concerning the initial proposition of improving energetic efficiency by collective decision-making. Robots that do not perform decision-making are denoted "not-knowing". In this experiment, we also used two other types of robots, "inertial" and "spontaneous", which used two different bio-inspired approaches for their behavior [209]. Both types of robots

[6]The experiments described in this subsection were performed jointly with T. Kancheva within the "Collective Microrobotics" project, supervised by author and partially published in [199]. The results of that work as well as other results obtained in a collaboration between the author, S. Kernbach, and V. Nepomnyashchikh are published in [209].

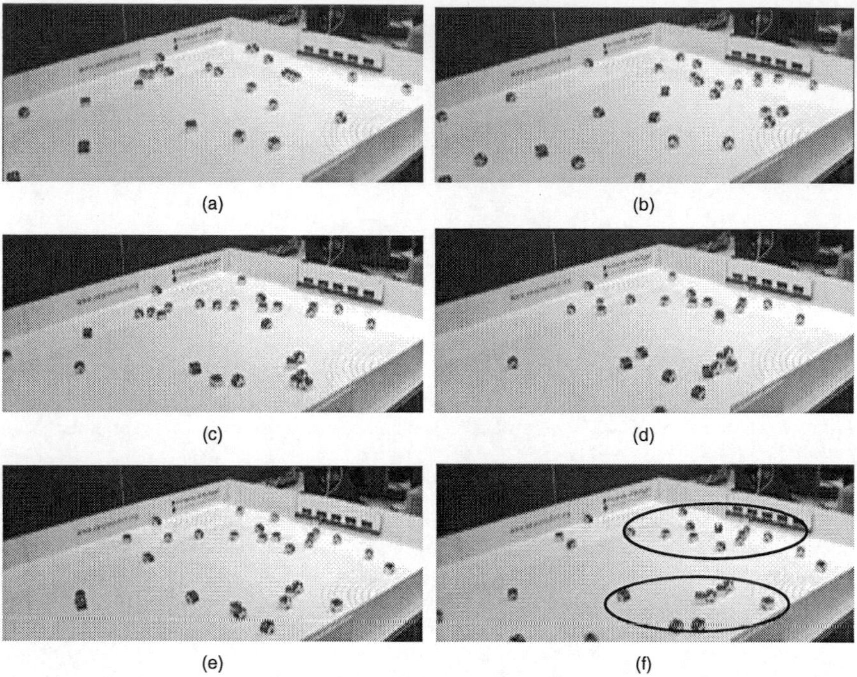

Figure 6.35: The recharging setup used in experiments with selection dynamics. **(a)** Initial random distribution of robots in the arena; **(b, c, d, e)** Dynamics of building two groups: one – closely to the docking station and another – far away from the docking station. Robots, when entering the cluster stop their movement, duration of the whole experiment is about 10 minutes; **(f)** Final clusters.

have integrated voting decision-making mechanisms with random neighbors, described in Sec. 6.5.2.1 and use the effect of random perturbations, described in Sec. 6.5.2.3. "Inertial" and "spontaneous" robots differ only in how they react to voting dynamics and are not the focus of this section, which concentrates on the difference between "not-knowing" and all other robots.

We introduced a priority $Rr(Task)_t$ for each task ("Job A" and "Job B"), which represents a local voting result after random interactions with other robots. While working, the robot starts to "feel hungry" as soon as its energy is below the "hunger" threshold. From this moment, the "hungry feeling" increases the priority of the recharging "Job B" in steps. In our model, inertial robots take the decision, where the number of working robots decreases the priority of the task "Job A", and the number of waiting robots increases it. As a final decision, the inertial robot switches to the recharging role "Job B" and starts searching for the docking station. As soon as a docking slot is available it docks, otherwise it conserves energy in the buffer zone. While recharging, the robot starts to "feel full" when its energy level exceeds a predefined recharged energy threshold. Again as above, "feeling full" increases the need to change the role. The number of waiting robots contributes to the voting dynamics and increases the need for change, whereas the number of working robots positively influences the priority of recharging. Spontaneous robots use a simple threshold-based model to react to collective decisions. If, during voting decision-making with random neighbors, it encounters a waiting robot, the voting procedure changes its local decision to "Job A", that is, it goes back to its working task. However, it remembers that it is hungry and when it reaches a free docking station, abandons its work and recharges.

Thus, this experimental setup allow us not only to estimate the qualities of spontaneous and inertial robots (this represents the bio-inspired part of this experiment) but also to compare the efficiency of decision-making mechanisms in relation to collective energy foraging.

The first experiments were performed in a small arena. The available docking slots were reduced to 3 and the experiments were executed with 3, 6 and 10 robots. For the small arena, one waiting station was used (otherwise, the search time for finding the waiting station would not be comparable). In order to deliver comparable results, many of the common parameters were set to equal for both types of experiments: maximum energy 185 marks the energy value of an accumulator up to 90% recharged, $Th_{dead} = 120$ the "dead" energy threshold, and $Th_{crit} = 150$.

After a comparison between the performances of two foraging strategies, the experiments were extended to the large arena, with 30 robots. Two waiting stations were installed, on each side of the docking station. All experiments were designed to take 10 minutes to complete. The speed of energy reduction had to be controlled, in order to allow several recharge phases in any one experiment. Therefore, in addition to measuring the physical energy value, a variable *energyValueSim* was introduced that could be used to simulate a recharge and discharge cycle. In all experiments with real robots, the $energyValueSim$ was updated every 4 seconds. A full discharge cycle of a moving robot from maximum energy to the "dead" threshold took around 4.4 min. The full recharge cycle takes just as long. In the small arena each experiment was repeated 10 times, in the large arena 5 times. Since all the experiments demonstrated good repeatability, Table 6.4 shows only mean values. The parameters and results of these experiments are collected in Table 6.4.

"Not-knowing" strategy in a small arena. A sample run in the "not-knowing 10" experiment is shown in Fig. 6.36.. The influence of energy reduction on the performance of "not-knowing" robots was studied by comparing the results of "not-knowing 3", "not-knowing 6" and "not-knowing 10" experiments. In the case where 3 docking slots provide energy for 3 robots, there is sufficient energy available, and the swarm retains a high collective energy. Doubling the number of agents reduces both the collective energy and swarm efficiency, but the swarm manages to stay alive even though acting in an environment with limited resources. In the "not-knowing 10" experiment, where energy is very constrained, the efficiency stays almost equal to the "not-knowing 6", but the collective energy is further reduced and falls below the critical threshold. Also, three to four of the 10 robots die, which reduces the swarm size back to six or seven and leads to an efficiency comparable to the "not-knowing 6" run. In summary, "not-knowing" robots offer the best performance when enough energy is available, shown by a very good collective energy level, but an efficiency of only 38.56%. Reducing the energy available leads to an extreme breakdown in the energy efficiency, whereas collective energy falls slowly.
"Limited robots" strategy. Three types of experiments were conducted to examine the performance of the "limited robots" strategy: inertial robots only, spontaneous robots only, and a mixed society of spontaneous and inertial robots. The following energetic thresholds were used: for inertial robots: $Th_{hungry} = 170$, $Th_{recharged} = 178$; for spontaneous

Experiments	N	R_i	R_s	Collective energy	Efficiency Φ^s [%]	Deaths
not-knowing 3	3	—	—	174,47	36,58	0
not-knowing 6	6	—	—	157,10	16,23	0
not-knowing 10	10	—	—	144,14	15,57	3,4
inertial 3	3	3	0	166,40	41,49	0
inertial 6	6	6	0	153,23	24,07	0,8
inertial 10	10	10	0	131,08	19,66	4,6
spontaneous 3	3	0	3	177,80	39,33	0
spontaneous 6	6	0	6	147,97	47,80	0,2
spontaneous 10	10	0	10	132,06	36,30	3,4
mixed 6	6	3	3	154,87	37,80	0,6
mixed 10	10	5	5	140,14	29,56	4
not-knowing 30	30	—	—	136,3	13,4	14
mixed 30	30	15	15	136,5	23,04	15

Table 6.4: Parameter and results for experiments with 3, 6, 10 and 30 robots, $Th_{hungry} = 171$, $Th_{recharged} = 181$ and initial energy value is equal to 173 (from [209]).

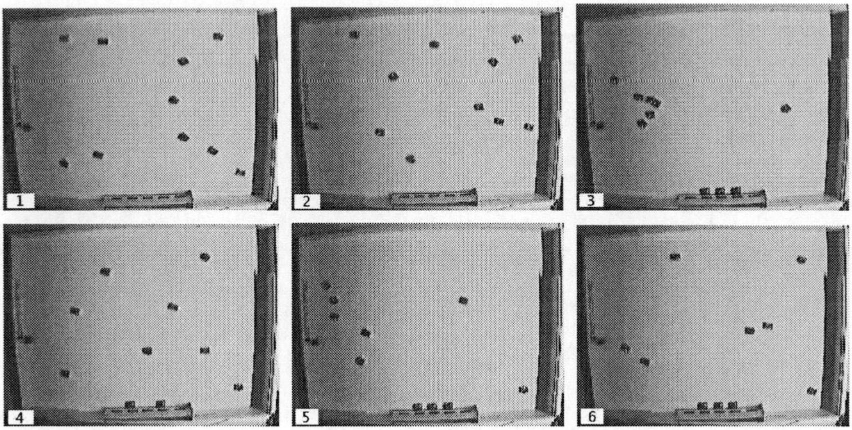

Figure 6.36: Experimental run for the "not-knowing" strategy with 10 robots. First, the robots execute regular work (1), then the Th_{hungry} is exceeded and all robots "get hungry" (2). After a while, all slots are busy and the rest of the swarm clusters around the waiting station (3). Then some robots are recharged and the waiting ones start to search for a slot again (4). As all slots are occupied, the "hungry" robots collect once again at the buffer zone (5). This goes back and forth until some of the robots "die" while waiting or searching (6). Images are from [209].

robots: $Th_{hungry} = 180$, $\Delta_{recharged} = 5$. The inertness coefficient was set in accordance with the simulation value of medium inertness. In the "not-knowing" experiments, three docking slots and one waiting station were made available to the swarm. Table 6.4 gives an overview of the experiments with "limited" strategy.

Limited strategy I: experimental run of an inertial swarm. Inertial robots achieve almost the best possible efficiency where no restrictions apply to the energy input. As more work is done, the collective energy level is not especially high, but that does not particularly affect the swarm's behavior in the "inertial 3" experiments. Increasing the number of swarm members leads to a reduction in efficiency and collective energy. However, the feedback from the swarm and the inertness allows the robots work for longer, which leads to robot "exhaustion" and "death" even when the swarm has only 6 members. Further reduction of the available energy leads to a collective energy below the critical threshold. Many robots "die" and the achieved efficiency is around 20%. Therefore, inertial robots offer very good efficiency where energy resources are adequate. Constraining energy resources leads to reduction of both efficiency and collective energy.

Limited strategy II: experimental run of a spontaneous swarm. Where sufficient energy is available to the swarm, spontaneous robots can recharge whenever they wish, so they recharge very often and maintain their energy at a high level. Respectively, frequent recharging means that less work is done, so the efficiency is only 39.33%. Increasing the number of agents greatly affects the results. Agents cannot recharge as much as they want to and work until real "exhaustion" (Th_{crit}) sets in. Thus, an efficiency of 47.80% is achieved, but the collective energy falls severely. Further constriction of available energy again reduces the collective energy, but the efficiency stays comparably high.

Limited strategy III: experimental run of a mixed swarm. As the swarm consists of two different subclasses, it can benefit from both. A swarm of six robots achieves an efficiency of 37.80%, at a collective energy level above the critical threshold. Within a more constrained environment, both parameters decrease slowly, and some robots "die".

Experimental run of a "not-knowing" swarm in a large arena. To test behavior in a heavily energy-restricted environment, two experiments were performed with 30 robots in the large arena. A sample run of the "not-knowing" scenario is shown in Fig. 6.37. The experimental run was characterized both by long periods of little or no movement and by intervals of considerable movement, when the agents essentially hindered each other.

Experimental run of a mixed swarm in a large arena. For the experiments with large mixed swarms, an equal proportion of inertial and spontaneous robots was used. The collective energy and swarm efficiency obtained exactly reflected the behavior described above. As shown in Table 6.4, half the swarm was "dead" at the end of both experiments and the resulting collective energy was likewise equal. A difference occurred in the efficiency obtained, which represents the time for which the swarm was free for useful activities. "Limited" robots worked for 23% of the time, whereas "not-knowing" robots managed to work for only 13%. Since the recharging role consists of waiting, recharging or searching, the "not-knowing" robots spent most of their time halted in clusters, either at the waiting station or the the recharging station.

In summary, experimental observations show different energy dynamics in each group of robots. None the less, when we compare the 30 robot "not-knowing" and "mixed" swarms, in other words with and without decision-making mechanisms, we observe an almost doubling of efficiency from 13.4% to 23.04%, with similar numbers of dead robots; 14 and 15 respectively. Comparing 10 robot "not-knowing" and "mixed" swarms, we observe efficiencies of 15.57% and 29.56% respectively, with three to four dead robots. This indicates that synchronizing collective behavior, even with simple analytical voting procedures, enables collective systems to behave much more efficiently.

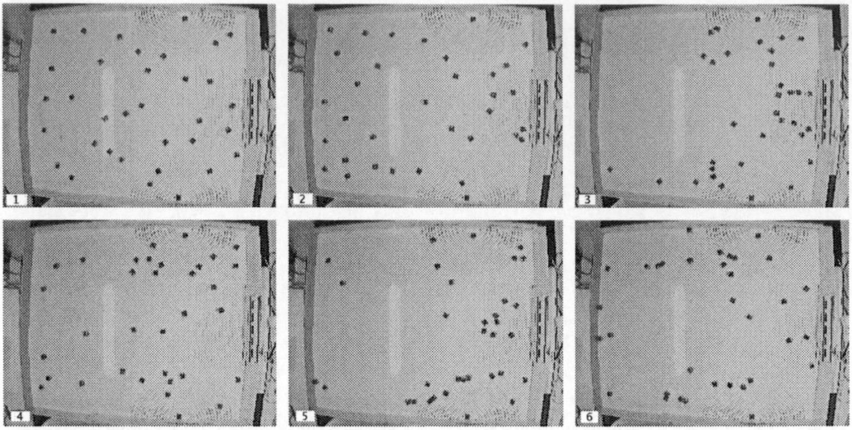

Figure 6.37: Experimental run for the "not-knowing" case strategy with swarm of 30 robots. All robots start the experiment with equal energy autonomy and execute some useful activities (1). When the individual energy falls below the Th_{hungry}, the robots switch to recharge role (2). Since all robots switch almost simultaneously, the recharge slot is soon blocked by recharging robots and the rest cluster around the two waiting stations (3). After a while, the recharging robots are ready and dissolve the cluster around the slot. At the same time, the waiting robots start to search again and the clusters at the waiting stations are being melted too (4). Now, new recharging robots stop around the slot and block it. The remaining "hungry" ones collect again at the waiting station (5). Images are from [209].

6.6 Summary

In this chapter we have considered several applications of analytical mechanisms of collective decision-making to real robot swarms. We have demonstrated a systematic method of complexity reduction from the level of real systems, to multi-agent systems (simulation) and finally to mathematic models.

We began the discussion at the real world level, because only here can we formulate the general problem, the expected results of analysis performed, and derive the common restrictions that will be borne in mind in all further steps. At this level, a huge number of parameters influence the whole distributed system, for example hardware and software components, electrical circuits, and so on. In order to simplify the models of real systems, we assume that all elements are working in a specified way, that is, they are not exerting any influence on the behavior of the overall system. Moreover, real systems possess plenty of functional features, not all of which are equally important for the issue being considered. Therefore, we assume that the functional properties of real systems can be considered independently of each other. In this way, we can take into account only essential functional features, often replacing them with an idealized description. In collective decision-making we assume that the actual type of decision is not relevant, but it is important that the whole system reaches a mutual agreement. It is expected that in the results of analysis, every elementary subsystem will implement a negotiation strategy.

At the level of mathematical models, synergetic methods can be applied to the problem of coordination in collective systems. In terms of synergetic methodology, the system is considered separately at the macroscopic and microscopic levels. On the macroscopic level, we can observe the global behavior of the whole systems. Usually, we do not consider details of individual agents at this level, focusing primarily on the collective behavior of the group. In contrast, the microscopic level represents separate agents, with individual behavior. This level has a large number of behavioral elements and, due to this high complexity, we cannot say anything about global behavior of the group.

The abstracted behavior at the macroscopic level is described by low-complex order parameters. We can say that this is responsible for coordination. By defining the order parameters, we can create a desired collective behavior in the group. There is no strong direct transformation from macroscopic to microscopic levels. However we can adjust systems at the microscopic level so that we demonstrate the desired order parameter and so allow the desired coordination to emerge.

We have shown a return path from the level of mathematic models to the level of the real world, and demonstrated an increase in complexity, which was removed during the simplification step. In particular, this is related to the effect of randomly changing neighbors, the appearance of random influences and a large number of technological details, such as docking approach or collective dynamics around the docking station. Finally, we compared the energetic efficiency of swarms with and without decision-making: in the particular experiments described, with collective energy recharging, collective decision-making almost doubled collective efficiency.

Chapter 7

Collective decision making: Agent-based manufacturing

Things should be made as simple as possible, but not any simpler.
Albert Einstein

The previous section dealt with the problem of collective energy homeostasis in distributed environment. The needed in that case collective decision-making had several different forms of distributed voting with variable votes quota. The example in this chapter treats the problem of collective decision making for distributed assignment problem. The approach developed in the last sections is expected to be expanded for a high dimensional selection dynamics, implemented in analytical as well as in algorithmic parts of autonomous agents.

7.1 Introduction into autonomous production systems

Competition among international suppliers and the globalization of national markets requires production systems that can successfully operate in a global and quickly- changing market [222]. Several requirements must be satisfied by these systems: first, the time needed between development of a product and production should be as short as possible. Second, manufacturing systems should be oriented to clients' needs, even small clients. This means that a product will be fabricated in small runs for different consumer requirements, such as color, specification, and so on. Furthermore, a product could be fabricated on demand, with an unique specification [223]. These requirements can be satisfied only by flexible and quickly reconfigurable production systems. Moreover, not only should physical fabrication be flexible (for example equipped with reconfigurable machinery), but also the operational, executive and developing processes of the modern factory. For such factories, completely new structures, organizational principles, and software and hardware instruments should be developed [53].

Taking into account the spatial distribution of manufacturing facilities and the requirement for flexibility in the whole system, the concept of autonomous agents has found some applications in this field [5], in particular:

- the activity of agents is a result of the group behavior, based on different forms of negotiation among agents. Because of this specific form of "programming", the problem-solving (decision- making, planning, etc.) in multi-agent systems (MAS) has essentially more degrees of freedom than traditional centralized systems. In this way MAS are stable in different and (sometimes unpredicted) manufacturing circumstances, for example machine failures, technological changes, and so on;

- there is a trend to equip processing elements (processing machines) with some degree of autonomy and "intelligence", allowing them to react to short-term turbulence, to self-maintain, to integrate with autonomous manufacturing. This trend corresponds to the agent concept, in which the processing element becomes an agent;

- in some situations (for example hazardous and dangerous environments) a human worker may not be suitable. The replacing technology should behave similarly to a human; that is, it should be autonomous, able to take decisions, and communicate with human or non-human workers.

The application of agents in manufacturing (see Fig. 7.1) also requires the development of new approaches towards the typical problems of multi-agent technology, such as distributed problem-solving, planning or collective decision-making (CDM) [74].

Figure 7.1: Example of agent-based manufacturing systems. Simulation by using Delmia © Quest^{TM}

The following example of CDM is addressed to the lowest level of manufacturing architecture, where low-level jobs (such as "produce one piece of work with a defined specification") should be assigned to the machines available. The goal is to generate this assignment through agents that represent different factory departments as well as processing elements.

7.2 Assignment of jobs to machines

The assignment problem is often encountered in manufacturing, it is a part of Operations Research and Management Science, where the flow-shop and job-shop problems with deterministic, stochastic, one-step or many-steps character are distinguished [225]. Generally, the assignment problem can be classified into scheduling, resources allocation and planning of operations order (see [226]). This is a classical NP-hard problem, there are known solutions by combinatorial optimization [227], dynamical optimization [238], evolutionary approaches [228], constraint satisfaction and optimization [229], discrete dynamic programming [230]. However these methods are developed as central planning approaches, the distributed or multi-agents planning for the assignment problem is in fact not researched (overview e.g., in [72]).

The assignment problem is a typical task of a short-term planning. It is assumed that all organizational, supply, storage, etc. tasks (that belong to long-term and middle-term planning) are already resolved. In a consequence of these planning steps there is a lot of available machines that are able to perform all needed technological operations. A n-step sequence of these operations in a specific order represents a processing of one workpiece. The following scenario exemplify the distributed assignment problem.

There are 5 types of workpieces (5-20 workpieces in each type) that have to be manufactured on available machines. Processing of each workpiece consists of several working steps (defined by a technological process), all these working steps cannot be processed on one machine. Each from the working steps has different length and cost. Moreover, each type of the workpieces has own technology, i.e. processing consists of different working steps.

piece type	number of pieces of each type	technology/ number of steps	number of avail. machines	optimization
A	5	table A/11	3	cost, time, cost/time
B	15	table B/11	3	cost, time, cost/time
C	10	table C/7	4	cost, time, cost/time
D	20	table D/10	2	cost, time, cost/time
E	5	table E/5	3	cost, time, cost/time

Table 7.1: Types, workpieces and machines in the assignment problem.

For simplification it is assumed that available machines are of the same type, therefore the cost and length of a working step do not differ on these machines (in general case they are different). The goal is to generate a plan of how to manufacture these workpieces with minimal cost, minimal time and minimal cost at determined delivery time, taking into account restrictions summarized in Tables 7.1 and 7.2. We can see, this problem belongs to so-called constraints optimization, where, firstly, optimization landscape is discrete (small islands on the landscape), secondly, there is no continuous gradient. Therefore, as the first step, it needs to formalize the constraints.

Let us denote a working step as WS^i_j, where i is a type of workpieces and j a number of the working step, an available machine is denoted as M_k, where k is a number of machine. We need also to introduce a workpiece P^m_n, where m is a priority of production and n is number of this workpiece. In this way $st(P^m_n(WS^i_j))$, $fn(P^m_n(WS^i_j))$ denote a start and a final position of corresponding working step that belongs to corresponding workpiece ($st(WS^i_j)$, $fn(WS^i_j)$ for all workpieces). We start with the definition of these values

working step	length/machine 1	length/machine 2	length/machine 3	order
1	1	0	1	1
2	2	0	2	2
3	0	1	1	2
4	3	0	3	2
5	1	1	0	2
6	2	2	2	2
7	0	1	1	3
8	3	0	3	4
9	2	0	2	4
10	1	1	0	4
11	1	1	1	4

Table 7.2: Technological table A for the workpiece type A. Zero in a length (of a working step) at corresponding machine means that this machine cannot perform the requested operation. Order of working steps means, that e.g., the steps 2,3,4,5,6 should be produced after the step 1 and before the step 7. It is natural to assume that these steps cannot be performed at the same time on different machines.

$$P_{n\in[1-20]}^{m\in[1-20]}(WS_{j\in[1,\ldots,11]}^{i\in\{A,B,C,D,E\}}) = o \in operation, \tag{7.1}$$

$$M_{k\in\{1,2,3,4\}} = \{o \in operation\}, \tag{7.2}$$

$$st(P_n^m(WS_j^i)) = \{t \geq 0, t \in R\}, \tag{7.3}$$

$$fn(P_n^m(WS_j^i)) = \{st(P_n^m(WS_j^i)) + length(P_n^m(WS_j^i))\} \tag{7.4}$$

The first constraint determines correspondence between operations of working step and of the k-machine

$$C_1 = \{(o_1, o_2) | o_1 \in P_n^m(WS_j^i), o_2 \in M_k, o_1 = o_2\} \tag{7.5}$$

The technological restrictions given by the Table 7.2 (for all workpieces of type A) can be rewritten in the following form

$$C_2 = \{(fn(WS_{[1]}^A) < st(WS_{[2-6]}^A)) \subset WS_j^A \times WS_j^A\}, \tag{7.6}$$

$$C_3 = \{(fn(WS_{[2-6]}^A) < st(WS_{[7]}^A)) \subset WS_j^A \times WS_j^A\}, \tag{7.7}$$

$$C_4 = \{(fn(WS_{[7]}^A) < st(WS_{[8-11]}^A)) \subset WS_j^A \times WS_j^A\}, \tag{7.8}$$

where $WS_{[2-6]}^A$ (for all workpieces) cannot be performed at the same time

$$C_5 = \{(j \in [st(WS_{[w]}^A), \ldots, fn(WS_{[w]}^A)] \neq j \in [st(WS_{[w']}^A), \ldots, fn(WS_{[w']}^A)]) \subset WS_j^A \times WS_j^A; w, w' = 2,3,4,5,6; w \neq w'\} \tag{7.9}$$

and also $WS_{[8-11]}^A$

$$C_6 = \{(j \in [st(WS_{[w]}^A), \ldots, fn(WS_{[w]}^A)] \neq j \in [st(WS_{[w']}^A), \ldots, fn(WS_{[w']}^A)]) \subset WS_j^A \times WS_j^A; w, w' = 8,9,10,11; w \neq w'\}. \tag{7.10}$$

Priority of production can be expressed by

$$C_7 = \{(m \in P_n^m > m \in P_{n'}^m | st(P_n^m(WS_j^A)) > st(P_{n'}^m(WS_j^A)))$$
$$\subset P_n^m \times P_n^m, n \neq n'\} \quad (7.11)$$

As soon as the variable, its domains of values and constraints are defined, one can start a propagation approach. The goal is to restrict the values of variables (to find such values of variables) that will satisfy all constraints. This propagation can be represented in the way shown in Fig. 7.2. All working steps that belong to the same workpiece build a sequence. Every node in this sequence gets a "finish"-position of a working step from previous node. Using this value, a current node looks for "start"-positions of the next working step that satisfy local constraints, calculates "finish"-positions and propagates them to the next node. If no position satisfying local constraint can be found, the node requests another "finish"-position from previous node. In this way the network can determine locally consistent positions of all working steps. After that, the obtained values should be tested for a global consistence.

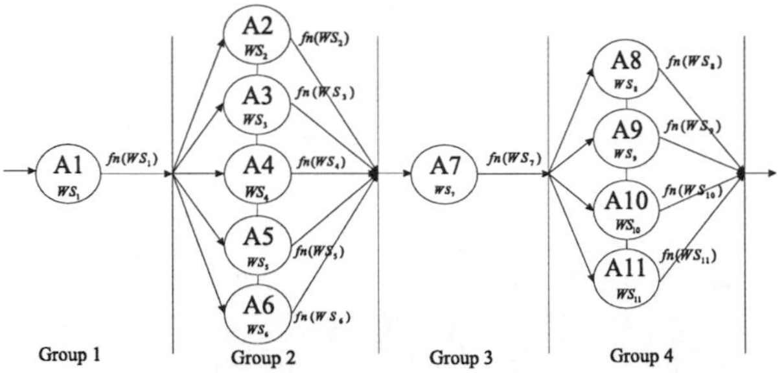

Figure 7.2: Constraint network for the problem of assignment (from [48]). Each working step WS is represented by a node, this node calculates a possible end position and sends it to the next node. If there is no possible position, satisfying all constraints, a node requires another final position from previous node.

7.3 Application of the autonomous agents

Each node in this network behaves according to some rules defined by the constraint approach. Therefore if these rules can be extracted and formalized, this network can be implemented by a multi-agent system. The question is why the constraint network should be replaced by multi-agent system performing the same operation ? The architecture of MAS is the key issue. The multi-agent system is an *autonomous* system with essentially higher degree of freedom than conventional systems. These additional degrees of freedom allow the system to react more flexibly to different perturbations. For example if a technological order of the working steps will be changed, it is also necessary to change the constraint network. In the case of multi-agent system the required changes can be performed by agents themselves. It means that the network becomes stable to different technological or organizational turbulence, occurring in a modern manufacturing.

However, a transformation of the network shown in Fig. 7.2 into a MAS consists of several steps. Firstly, the interaction among agents should be reformulated from a macroscopic to microscopic point of view. It means e.g., distribution of global information, introduction of decision making algorithms for synchronization and so on. Secondly, behavior as well as communication routines of every agent should be described by a specific language (based on the modified Petri network). This language and a software environment are developed within of the SFB 467. Details of this approach as well as of the agent software environment (Role Oriented Programming Environment) can be found e.g., in [231]. Treatment of these points here outsteps the framework of the thesis. However, the last point needs to be mentioned before discussion of results, concerns distributed optimization.

Generally, optimization consists of two steps that influence the common result. These are, firstly, an order of working steps in the group 2 and 4 (see Fig. 7.2) and, secondly, local decisions of agents concerning machine and position. Forasmuch as there are only 2881 combination between WS_1-WS_{11}, the first optimization step can be performed by distributed exhaustive search. However, the second step is more difficult because of the following reason. Some working steps can be executed on different machines as shown in Fig. 7.3. Moreover, each working step can start some time (so called forecasted positions) after the previous working step.

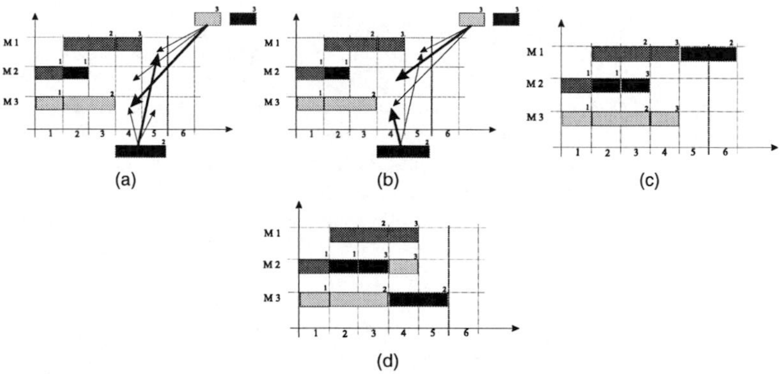

Figure 7.3: Example of the plan making for the minimal time and minimal transport cost (from [48]). There are three workpieces to be manufactured (P_1 grey, P_2 white, P_3 black). Processing of each workpiece consists of three working steps (with length 1, 2, 1 correspondingly). The working step 2 cannot be processed on the second machine M2. Each working step is represented by an agent. **(a)** Decisions of the agents, where $WS_2(P_3)$ is going to M1 in position 5, $WS_3(P_2)$ is going to M3 in position 4. Local decision is based on the criterion of minimal transport cost; **(b)** Decisions of the agents, where $WS_2(P_3)$ is going to M3 in position 4, $WS_3(P_2)$ is going to M2 in position 4. Local decision is based on the criterion of minimal length of this plan; **(c)** final plan of the situation (a); **(d)** final plan of the situation (b).

Local decisions (concerning positions) made by agents influence the time and cost of the whole assignment plan. Thus, instead of global optimization of positions, it needs to optimize the agent's local decisions towards positions. This is very characteristic for agents technology. The optimization of these local decisions can produce a plan with minimal cost or a plan with minimal manufacturing length (time). However a search space grows in this case exponentially and e.g., even for 22 agent (2 production workpieces, forecast for next 5 positions) comes to $\sim 10^{10}$. The search space can be essentially reduced if to reduce a number of forecasted positions. Forecasted positions try to compute what will happen if the next processing step will begin not immediately after the previous step (e.g., manufacturing of a workpiece will be shifted on some steps, i.e. it increases a cost of temporary storaging, see Fig. 7.4). Being argued by a minimization of the storaging cost, a number of this possible shifting steps can be reduced.

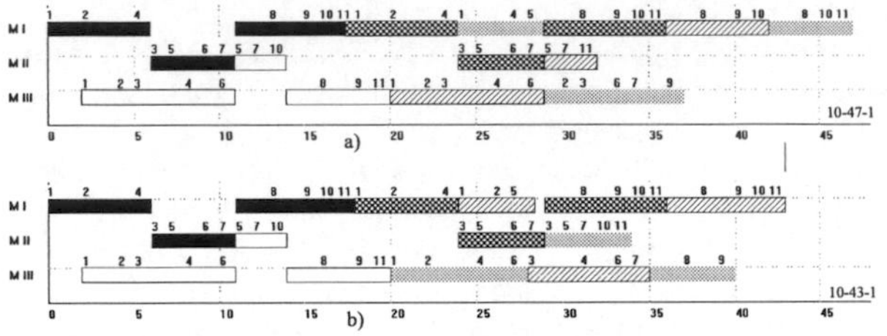

Figure 7.4: The "forecasting" effect in assignment planning (from [48]). **(a)** Each working step begins immediately after the previous step. The length of a whole plan is equal to 47; **(b)** The start of the working step 8 (3th workpiece) is delayed on one step, that allows reducing the common length to 43 steps.

For optimizing agent's local decisions we can apply some heuristic methods dealing with this kind of problems, e.g., ant colony optimization algorithms [232]. This method originated from observation of ants in the colony. Going from the nest to the food source, every ant deposits a chemical substance, called pheromone, on the ground. In the decision point (intersection between branches of a path) ants make a local decision based on the amount of pheromone, where a higher concentration means a shorter path. Similar strategy can be applied to local decisions of agents, participating in the plan making. In this case either the possibly closest position to the previous step or a position on the same machine can play a role of pheromone rate. General effect of such a strategy is that agents build some pattern of how to make an assignment plan (for the first workpiece), all next workpieces will be executed on the base of the same pattern. This approach can be modified so that agents compare different pheromone rate in order to achieve the best group effect.

The criterions of a minimal transportation cost as well as of a minimal length of a plan are chosen only as a simplified example illustrating this approach. As mentioned in the beginning of this section, generally there exist a manufacturing cost, transportation cost and storaging cost defined for each machine as well as the working step. However, this do not change the methodical approach. Results of the assignment plans are shown in Fig. 7.5.

7.3.1 A plan with a minimal cost at constant delivery date

The previous section deals with an assignment plan that is generated in distributed way by software agents. This plan can be optimized according to two criteria: minimal (transportation) cost and minimal production length. However, in real manufacturing there is the third criterion for optimization of an assignment plan. In this case the assignment plan remains optimized by the cost criterion, however there is a deadline (a delivery date) that the plan may not exceed. If deadline is yet exceeded a delayed delivery penalty should be paid.

Now the question is how to generate a plan that remains with minimal cost however does not exceed a deadline? Let us remember, that a length and cost of a plan stands in inverse relations, decreasing of cost increases a length. Therefore there is only one way

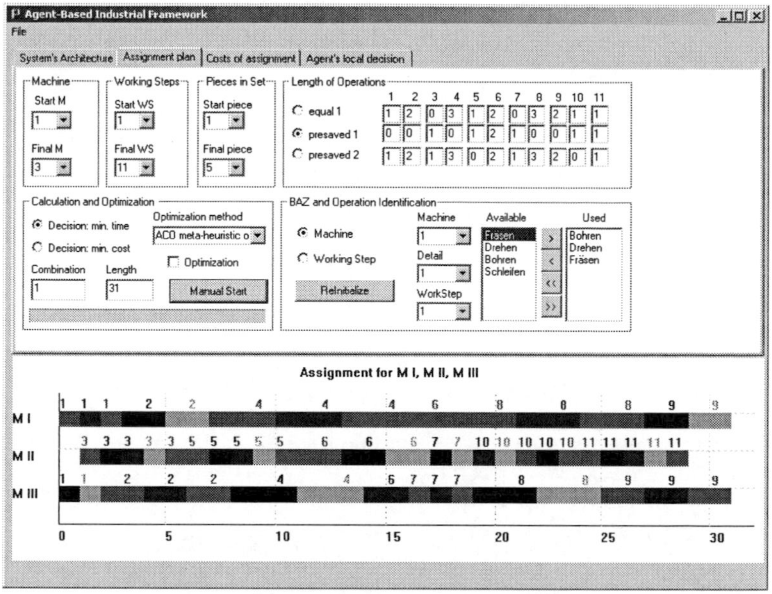

(a)

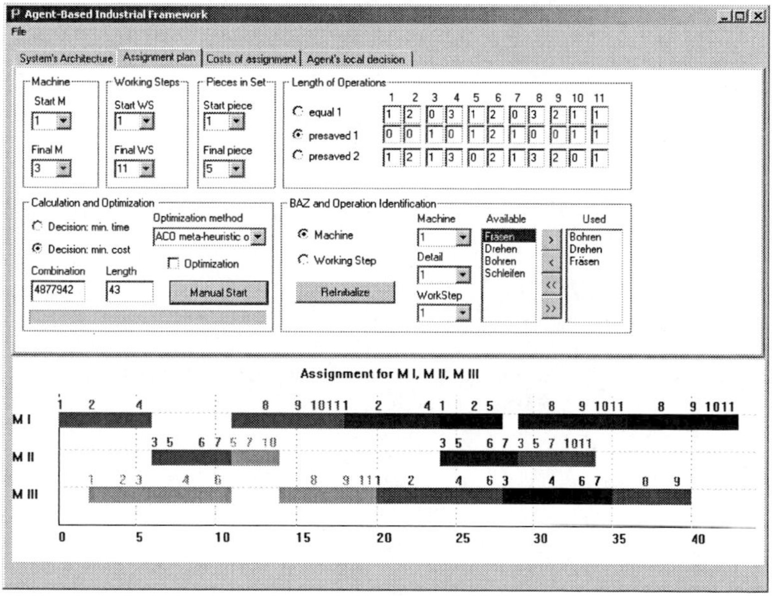

(b)

Figure 7.5: The generated assignment plans **(a)** with a minimal plan length; **(b)** with a minimal transportation cost.

to generate such a plan, namely to truncate an optimal plan. Cost of such a "distorted optimal" plan will be increased, however in real manufacturing the penalty is much higher than this growth of cost.

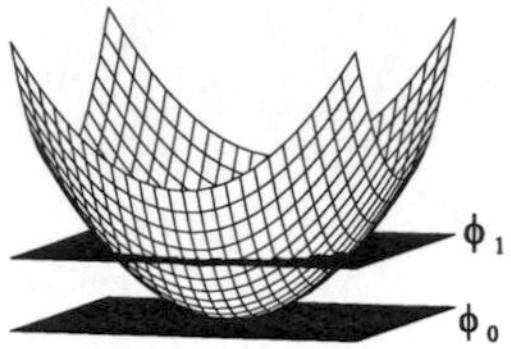

Figure 7.6: 3D plot of the function $x^2 + y^2$ that is intersected by two plains $\phi_0 = 0$ and $\phi_1 = 0.5$.

Generating a truncated plan, we encounter one problem. There is only one plan (usually) with a minimal cost, but there are a many plans with nonminimal costs. This case can be illustrated by the potential landscape in Fig. 7.6, where there is only one minimum point (intersection of $x^2 + y^2$ with the plane ϕ_0), but many nonminimal points (intersection of $x^2 + y^2$ with the plane ϕ_1).

This problem from the viewpoint of distributed agents is shown in Table 7.3. After

agent	optimal position (cost/length of plan)	nonoptimal positions (cost/length of plan)
...	...	...
A30	30.1 (10/43)	28.1, 19.1, 19.1, 20.1 (11/42) 28.1 (11/41), 28.1 (11/40) 19.1 (14/39), 21.1 (21/35)
A31	33.1 (10/43)	31.1, 22.3, 22.1, 23.1 (11/42) 31.1 (11/41), 31.1 (11/40) 22.1 (14/39), 25.1 (21/35)
A32	35.1 (10/43)	33.1, 24.1, 34.1, 18.2 (11/42) 33.1 (11/41), 33.1 (11/40) 24.1 (14/39), 24.1 (21/35)
...	...	...

Table 7.3: Nonoptimal planning from the viewpoint of agents. Numbers before and after decimal point denote position and machine.

performing constraint based approach and optimization each agent gets only one position which is optimal. In case of nonoptimal positions each agent gets several solutions for the same length of a common plan, moreover there exist several plans with different length (evidently that only the dedicated combinations of these solutions are consistent). It can happen that the considered plan is already executed in part. Therefore the agents standing in the executed part have other motivations than the agents at the end of a plan. Taking into account that agents behave autonomously (on the base of own rules and goals), it needs to coordinate their solutions for a collective plan. Forasmuch as neither central planning nor globalization of local sensors of agents are acceptable here, only a collective decision making can be applied in order to coordinate agents solutions. Therefore in case of constant delivery date, the problem of optimization gets new component - collective decision making.

An assignment plan is a result of collective activity (of self-organization) in the agents group. From this viewpoint, the ideas explained in the previous chapters can be applied to decision making in the plan generation. If the produced plan exceeds the deadline, the bifurcation occurs and the system jumps to another state. However, the problem is how to calculate this new state. This state represents a "distorted" optimal solution generated by agents, it cannot be calculated in analytical way. Therefore in this case the problem of collective decision making consists in choice of one of the prepared nonoptimal solutions. The agents coalition should recognize which of these plans is the most adequate one to the current positions of agents, taking into account, firstly, the deadline and, secondly, the agents position in already executed part of an assignment plan.

The problem of pattern recognition is well-known problem not only in the image processing, but also in nonlinear dynamics, where we refer primarily to the synergetic computer [204]. In this approach there are p prestored images (see Fig. 7.7), that are represented as a linear sequence of gray values (m values). Every gray value corresponds to one state variable $q_i, (i = 1, \ldots, m)$. There is also a distorted image on the input whose gray values determine initial values q_0. In a long time dynamics $q_0 \to q_n \to q^p_{n \to \infty}$ the synergetic computer recognizes, to which of the prestored images a distorted image belongs to (see Fig. 7.7).

(a)

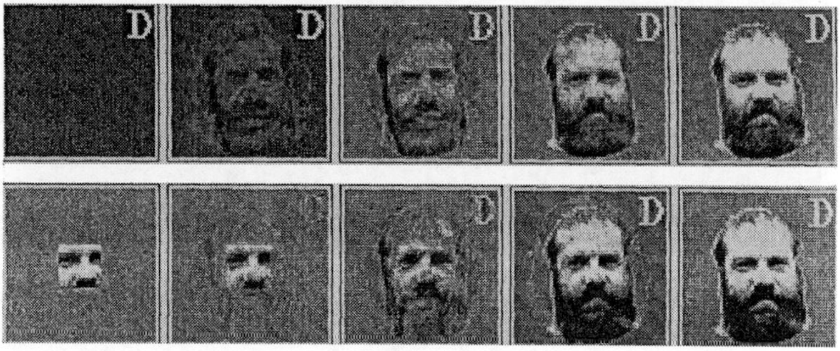

(b)

Figure 7.7: Pattern recognition using synergetic computer (Reprint from [204] with permission of Prof. H. Haken). **(a)** Prestored images using for recognition. Input image will be attracted to one of them. **(b)** Long time dynamics of synergetic computer. The first images represent initial values of state variables (distorted image on the input), the last images represent the recognized images from (a).

Recognizing of the assignment plans can be performed in a similar way. The nonoptimal assignment plans will be stored in a form of numerical positions. They represent the patterns that an input pattern will be reconstructed to. The assignment plan with minimal (transportation) cost will be distorted by cutting the part after deadline. In a long time dynamics it is expected that the distorted pattern will be reconstructed to one of the prestored assignment plans. This recognition process means that each agent gets position of the according nonoptimal plan, i.e. their decisions will be synchronized in this way.

7.3.2 Analytical agent

Although the pattern recognition based on the synergetic computer underlies the collective decision making, implementation of this approach differs from the method suggested in [204]. In the proposed here approach the decision making is based on the recalled normal form of the transcritical bifurcation shown in Sec. 6.4.2. This equation represents the order parameter for the decision making. Both stationary states of this macroscopic equation means two decisions, e.g., to use optimal plan or to use nonoptimal plan. The coupled mode equations are derived as shown in 6.4.4. The coupling function (6.14) is so chosen that to stabilize only one of the stationary states. Finally the mode equations will be transformed in the output form, where the number of equations is equal to the number of agents. Each of these equations represents analytical part of the corresponding agent, that interacts with other (analytic and algorithmic) agents.

As mentioned in Sec. 6.4.4, we start from the following p-dimensional system

$$\underline{\boldsymbol{\xi}}_{n+1} = f_c(\underline{\boldsymbol{\xi}}_n)(\underline{\boldsymbol{\xi}}_n - (\underline{\boldsymbol{\xi}}_n)^2) + \underline{\boldsymbol{\xi}}_n^2, \tag{7.12}$$

where f_c is a coupling function. The initial systems (6.4) coupled in this way guarantee that the stationary states of each state variable will be equal to 0, 1 or r ($r \neq \{0, 1\}$ that is caused by the growth of dimension). Therefore the goal is to choose this coupling function so that to stabilize a needed combination of $\{0, 1\}$ stationary states. We look for this function in the following polynomial form

$$f_{coup} = k_0 + \sum_{i=1}^{p} k_i \xi_n^i, \tag{7.13}$$

where k_0, k_i are coefficients. As demonstrated in Sec. 6.4.4, the system (7.12) has five groups of stationary states. The first group includes all stationary states that are equal to zero or one (written in row-wise)

$$\underline{\boldsymbol{\xi}}_{st}^{10} = (\{\xi_i = 0\}|\ i = 1, \dots, p), \quad \underline{\boldsymbol{\xi}}_{st}^{11} = (\{\xi_i = 1\}|\ i = 1, \dots, p). \tag{7.14}$$

The second group collects all states, where the number of the "1" stationary states is more than one

$$\underline{\boldsymbol{\xi}}_{st}^{2} = (\{\xi_i = 0\}, \{\xi_j = 1\}|\ i, j = 1, \dots, p, \ \ i \neq j, \ \ 1 < N(j) < p), \tag{7.15}$$

whereas the third group collects such states, where the number of "1" stationary states is equal to one.

$$\underline{\boldsymbol{\xi}}_{st}^{3} = (\{\xi_i = 0\}, \{\xi_j = 1\}|\ i, j = 1, \dots, p, \ \ i \neq j, \ \ N(j) = 1). \tag{7.16}$$

Finally, the rest of stationary states consists of the r-states which can be segmented into two groups. The fourth and fifth groups collect the states, where the number of r-states is equal to one or correspondingly is more than one (all other states are equal to zero or one).

$$\underline{\boldsymbol{\xi}}^4_{st} = (\{\xi_i = \{0,1\}\}, \{\xi_j = r\}|\ i,j = 1,\dots,p,\ \ i \neq j,\ \ N(j) = 1) \tag{7.17}$$

$$\underline{\boldsymbol{\xi}}^5_{st} = (\{\xi_i = \{0,1\}\}, \{\xi_j = r\}|\ i,j = 1,\dots,p,\ \ i \neq j,\ \ 1 < N(j) \leq p) \tag{7.18}$$

This separation of the stationary states simplifies the linear stability analysis. The Jacobian of the system (7.12) has the following form

$$\begin{pmatrix} k_1(\xi^1 - (\xi^1)^2) + \ (f_c)(1-2\xi^1) + 2\xi^1 & k_2(\xi^1 - (\xi^1)^2) & \dots \\ k_1(\xi^2 - (\xi^2)^2) & k_2(\xi - (\xi^2)^2) + (f_c)(1-2\xi^2) + 2\xi^2 & \dots \\ \dots & \dots & \dots \end{pmatrix} \tag{7.19}$$

where $\xi^i = \xi^i_{st}$. The eigenvalues of this matrix, according to stationary states, can be also separated into

$$\lambda_i(\underline{\boldsymbol{\xi}}^{10}_{st}) = (k_0, k_0, \dots),\quad \lambda_i(\underline{\boldsymbol{\xi}}^{11}_{st}) = (2 - k_0 - \sum_{i=1}^{p} k_i, 2 - k_0 - \sum_{i=1}^{p} k_i, \dots) \tag{7.20a}$$

$$\lambda_i(\underline{\boldsymbol{\xi}}^2_{st}) = (\{k_0 + \sum_j k_j|\ \forall(\underline{\boldsymbol{\xi}}^2_{st})_j = 1\}|(\underline{\boldsymbol{\xi}}^2_{st})_i = 0, \tag{7.20b}$$

$$\{2 - k_0 - \sum_j k_j|\ \forall(\underline{\boldsymbol{\xi}}^2_{st})_j = 1\}|(\underline{\boldsymbol{\xi}}^2_{st})_i = 1\ ), \tag{7.20c}$$

$$\lambda_i(\underline{\boldsymbol{\xi}}^3_{st}) = (\{k_0 + k_j|(\underline{\boldsymbol{\xi}}^3_{st})_j = 0\}|(\underline{\boldsymbol{\xi}}^3_{st})_i = 0,\ \{2 - k_0 - k_j|(\underline{\boldsymbol{\xi}}^3_{st})_j = 1\}|(\underline{\boldsymbol{\xi}}^2_{st})_i = 1). \tag{7.20d}$$

The notation in (7.20) $\sum_j k_j|\ \forall(\underline{\boldsymbol{\xi}}^2_{st})_j = 1$ means that only such k_j will be collected, whose corresponding ξ_j are equal to one. Similarly $\{k_0 + k_j|(\underline{\boldsymbol{\xi}}^3_{st})_j = 0\}|(\underline{\boldsymbol{\xi}}^3_{st})_i = 0$ means that the expression in bracket occurs if the corresponding ξ_i is equal to zero.

In Sec. 6.4.4 the goal was to implement the voting approach with two alternatives. Collective decision making was implemented by stabilization all "0" and "1"-states. In this case all agents receive simultaneously either 0 or 1 and so synchronize collective behavior. In the CDM for assignment plan there are essentially more alternatives, therefore the stabilization of $\underline{\boldsymbol{\xi}}^{10}_{st}$, $\underline{\boldsymbol{\xi}}^{11}_{st}$ is not suitable here. Instead we perform here a selection approach (see [12]), where all alternative patterns will be encoded by one of $\underline{\boldsymbol{\xi}}^3_{st}$. Then, after appropriate coordinate transformation, this encoded pattern (that wins the competition) will be converted in the form of the corresponding nonoptimal plan (from this viewpoint the system (7.12) represents the mode equations). Thereby the goal is to stabilize the third group of stationary states $\underline{\boldsymbol{\xi}}^3_{st}$, whereas all other stationary states should remain unstable.

The stability condition for the stationary state $\underline{\boldsymbol{\xi}}^3_{st}$ can be expressed as

$$|\lambda_i(\underline{\boldsymbol{\xi}}^{10}_{st})| > 1,\ \ |\lambda_i(\underline{\boldsymbol{\xi}}^{11}_{st})| > 1,\ \ |\lambda_i(\underline{\boldsymbol{\xi}}^2_{st})| > 1,\ \ |\lambda_i(\underline{\boldsymbol{\xi}}^3_{st})| \leq 1. \tag{7.21}$$

If the conditions (7.21) are satisfied, the condition $|\lambda_i(\underline{\boldsymbol{\xi}}^4_{st})| > 1$ is also satisfied. It follows from the Jacobian (7.19), where $(\xi^i - (\xi^i)^2)$ for $\xi^i = 0, 1$ is equal to zero. Therefore as result we get a triangular matrix, where at least one element on the main diagonal is larger than one. However, to prove the condition $|\lambda_i(\underline{\boldsymbol{\xi}}^5_{st})| > 1$ is more difficult, especially for high dimensional systems (this step is performed numerically). General remark is that in case one stationary state from the fifth group becomes stable, it can be absorbed by specifically chosen initial conditions. The unequalities (7.21), taking onto account the remarks made in Secs. 6.4.4 and 6.4.6.2, are solved by using Maple and we get

$$k_0 = 2, k_i = 1/2, k = -2, \tag{7.22}$$

where index i in k_i is equal to the number of equations that the function (7.13) is inserted in. The rest of coefficients k is equal to each other and denoted simply as k. As a result of the steps (7.14)-(7.22) we get the coupled system (7.12), where in a long time dynamics only one of the variables is equal to one, whereas all other variables become equal to zero. This "1" variable denotes a pattern that is recognized. The state variable ξ^i that will win (and correspondingly a recognized pattern) is defined by initial conditions. The system (7.12) describes a behavior of patterns, whereas we need a system that will describe a behavior of agents position. In this context this system has a role of mode equations that describe a behavior of separated modes. Let q be an agent's position and remembering that

$$q^i = \underline{V}_1\xi^1 + \underline{V}_2\xi^2 + \cdots + \underline{V}_p\xi^p = (\underline{\underline{\mathbf{V}}}\,\underline{\boldsymbol{\xi}})_i, \quad i = 1, \ldots, m, \tag{7.23}$$

where $\underline{V}$ represents a stored pattern that consists of m components and p is the number of the stored patterns, we can transform the system (7.12) from the "patterns" form into the "positions" form. The needed for transformation inverse matrix $\underline{\underline{\mathbf{V}}}^{-1}$ of the $(p \times m)$ matrix $\underline{\underline{\mathbf{V}}}$ can be obtained in the way described in [204]. Here the "pseudo-inverse" adjoint vectors are represented as superposition of the transposed vectors $\underline{\underline{\mathbf{V}}}^T$

$$\underline{\underline{\mathbf{V}}}^{-1} = \underline{\underline{\mathbf{A}}}\,\underline{\underline{\mathbf{V}}}^T, \tag{7.24}$$

where $\underline{\underline{\mathbf{A}}}$ is a matrix of coefficients. Taking into account that $\underline{\underline{\mathbf{V}}}^{-1}\underline{\underline{\mathbf{V}}} = \underline{\underline{\mathbf{I}}}$ the matrix $\underline{\underline{\mathbf{A}}}$ can be obtained as

$$\underline{\underline{\mathbf{A}}} = [\underline{\underline{\mathbf{V}}}^T\underline{\underline{\mathbf{V}}}]^{-1} \tag{7.25}$$

The matrix $[\underline{\underline{\mathbf{V}}}^T\underline{\underline{\mathbf{V}}}]$ represents a low-dimensional $p \times p$ matrix, that can be inverted without difficulties. Now with the transformation (7.23) the system (7.12) yields

$$\underline{\mathbf{q}}_{n+1} = \underline{\underline{\mathbf{V}}}^{-1}[(k_0\underline{\underline{\mathbf{I}}} + \underline{\underline{\mathbf{K}}}\,\underline{\underline{\mathbf{V}}}\,\underline{\mathbf{q}}_n)(\underline{\underline{\mathbf{V}}}\,\underline{\mathbf{q}}_n - [\underline{\underline{\mathbf{V}}}\,\underline{\mathbf{q}}_n]^2)] + \underline{\underline{\mathbf{V}}}^{-1}[\underline{\underline{\mathbf{V}}}\,\underline{\mathbf{q}}_n]^2, \tag{7.26}$$

where the main diagonal of the matrix $\underline{\underline{\mathbf{K}}}$ consists of the coefficients k_1, whereas all other elements are equal to k. In the equation (7.26) all operations in the square brackets should be performed in component form, e.g., $[\underline{\mathbf{a}}]^2 = (a_1^2, a_2^2, \ldots)$. The m-dimensional system (7.26) represents the positions of each working step, and should be incorporated (in components form) into each analytical agent.

7.3.3 Spatial eigenvectors and initial conditions

As mentioned in Sec. 7.3.1 a pattern recognition is a dynamic process $\underline{\mathbf{q}}_0 \to \underline{\mathbf{q}}_n \to \underline{\mathbf{V}}_i$, determined by (7.26), where $\underline{\mathbf{q}}_0$ an input (distorted) pattern and $\underline{\mathbf{V}}_i$ is a spatial eigenvector that represents a prestored pattern. Moreover, a form of eigenvectors and initial conditions determines the recognition process. Therefore the first point concerns representations of spatial eigenvectors $\underline{\mathbf{V}}_i$ as well as input patterns $\underline{\mathbf{q}}_0$.

Representation of spatial eigenvectors and initial conditions belongs to the step of practical implementation, i.e. there exists more than one successful solution. The performed numerical experiments have demonstrated three possible ways: each eigenvector represents agent's position, complete plan even with empty positions or only specific components of a plan like a length. Each of these variants carry out the process of recognition and therefore can be adopted for implementation. Below we discuss the first variant because it is the most suitable process to be executed by a group of distributed agents.

In this case an assignment plan can be represented numerically in form of agents start positions in manufacturing queue on a corresponding machine. The most simple way consists in separation a position and a machine by the decimal point, e.g., 3.4 means the third position on the fourth machine. All agents (and correspondingly all working steps) are represented as a sequence of a length m. These real values are collected into a vector. In the example of assignment plan there are 55 agents (5 workpieces, each with 11 working steps), therefore $m = 55$. In Figs. 7.8, 7.9 we have collected several examples of assignment plans that are used for spatial eigenvectors.

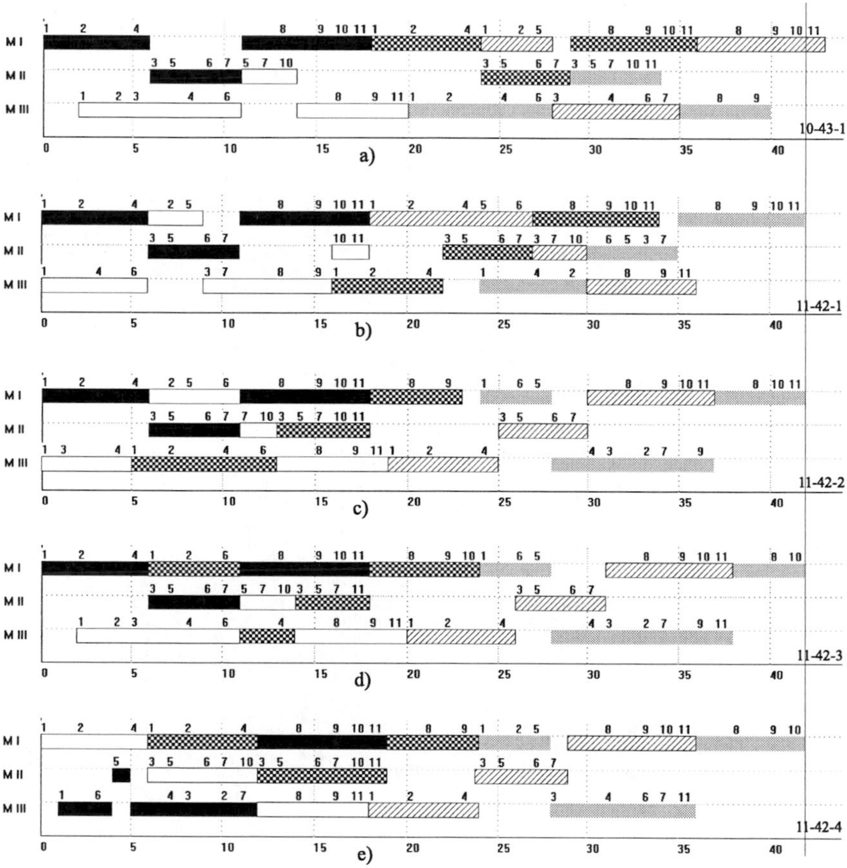

Figure 7.8: Nonoptimal assignment plans of equal length (partly from [48]). **(a)** Assignment plan of length 43 with minimal transportation cost 10; **(b)-(e)** Four examples of assignment plans, where number of transportation is increased by one and length of common plan is reduced by one.

The eigenvectors and initial conditions representing an assignment plan should be normalized as suggested e.g., in [204]. The point is that the normal form of transcritical bifurcation has a bounded attraction basin. If the initial values overpass these boundaries

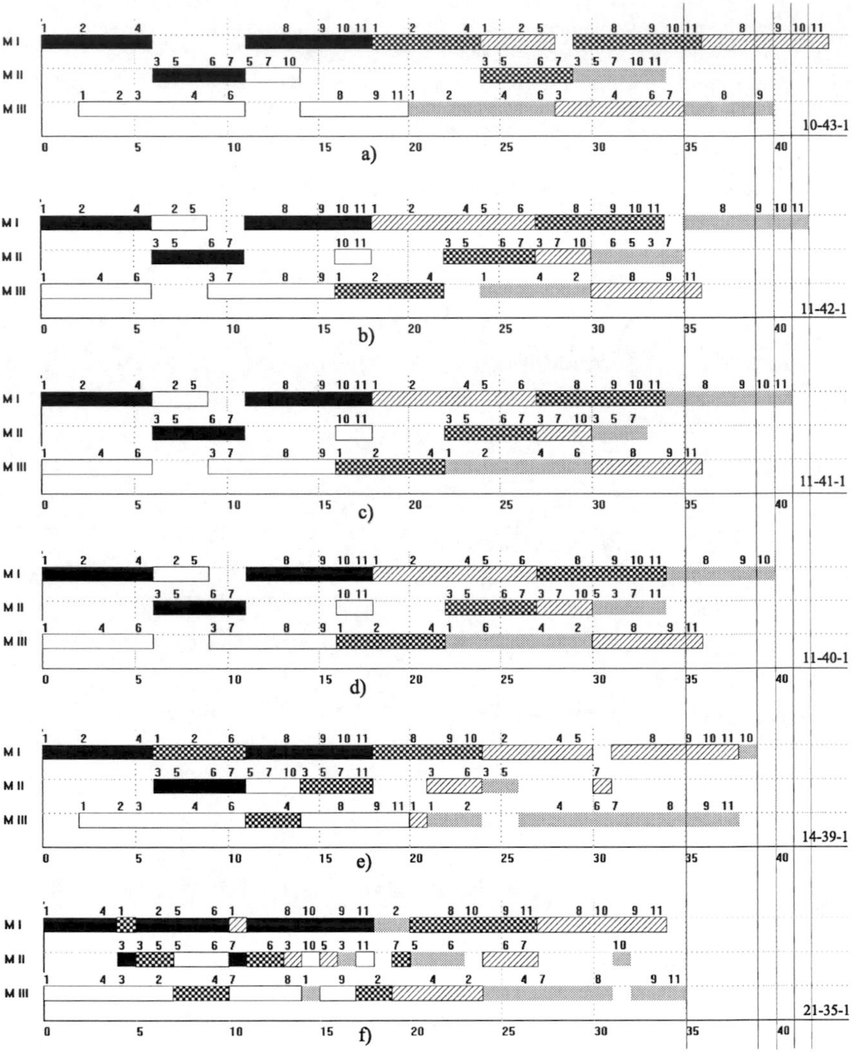

Figure 7.9: Nonoptimal assignment plans of different length (partly from [48]). **(a)** Assignment plan of length 43 with minimal transportation cost 10; **(b)-(f)** Nonoptimal assignment plans, where number of transportation is increased and length of common plan is reduced.

the system loses dissipation.

The recognition process should take into account three following points. Firstly, it has to recognize all forbidden plans (e.g., a plan that oversteps a time limit) and to control the recognition $\underline{\mathbf{q}}_0 \rightarrow \underline{\mathbf{q}}_n \rightarrow \underline{\mathbf{V}}_i$, so that $\underline{\mathbf{V}}_i$will never equal to one of these forbidden plans. Secondly, it should recognize a plan with optimal cost in the actual situation. For example, an optimal plan has a length 43 with cost 10 (see Fig. 7.8, 7.9) and a time limit is equal to 42. There exist alternative plans with length 42, 41, 40 (cost 11), 39 (cost 14), 35 (cost 21) and so on. It is clear that only the first group (plans 42,41,40) has the optimal cost. Finally, if a part of an input plan is already executed (manufactured), the recognition process should choice only between non-forbidden, cost-optimal plans that have the same executed part.

In order to satisfy all these requirements, we introduce the notion of essential components known in digital image processing (see [233]). Essential components mean one or several dominant features of a pattern. In the pattern recognition these dominant features allow to separate patterns from each other. We introduce three essential components for an assignment plan: forbidden and executed positions, optimal cost of a whole plan. The essential components will be marked by a corresponding multiplying coefficient. Every

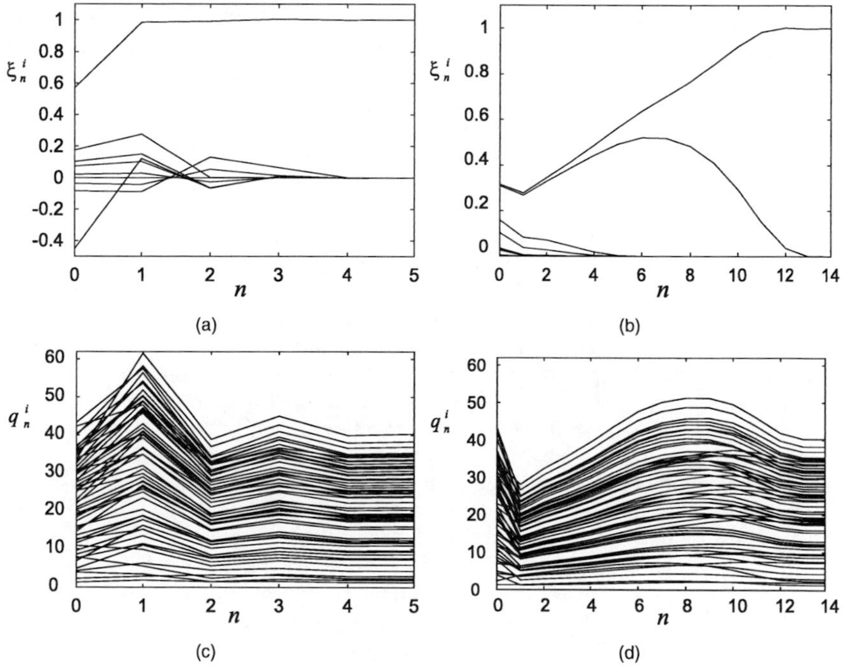

Figure 7.10: Two examples of a plan recognition, $k_{forb} = 100$, $k_{cost} = 40$. Coefficient k_{exec} depends on the number of executed steps. In the case of one-four executed steps k_{exec} can achieve 10^7-10^8; **(a), (b)** Time behavior of the mode equations (7.12); **(c), (d)** Time behavior of the output system (7.26).

forbidden position in a spatial eigenvector is multiplied by k_{forb}. This approach makes the corresponding forbidden eigenvector more different from initial vector so that it will be never recognized. Every executed position in spatial eigenvectors as well as in initial vector is multiplied by k_{exec}. This operation makes the corresponding vectors more similar so that only one of them will be recognized. Finally, we mark cost-optimal plan by multiplying all cost-nonoptimal plans with k_{cost}. In this way the cost-optimal plans will be more "attractive" for initial vectors. If all these components-based conditions are satisfied by several plans the choice between them will be performed by the principle of maximal similarity (see Fig. 7.7). Examples of the recognition is shown in Fig. 7.10.

7.4 Summary

In many respects, the procedure of collective decision-making provides a systematic way to achieve the ordered behavior of all components in a distributed system. Therefore, as pointed out in some reports (see [72]), this procedure is part of the more general mechanisms underlying distributed systems. Here we highlight distributed problem-solving [220], in the particular case where the problem being solved is to form a plan [221].

The distributed plan itself represents the sequence of its steps and all agents bargain to reach a common agreement about each step of the plan. This case is often referred to as distributed planning for distributed plans (see [5]). Evidently, the actions of all agents should be consentient, otherwise the problem cannot be solved. For the suggested approach to guarantee such consentience, it is expected that the concept of the analytical agent could be useful. The evaluation criteria used in such procedures, as well as the dynamical systems, can differ from those suggested in Sec. 6.4. The final point is concerned with the complexity of analytical and algorithmic agents. The example in Secs. 7.1 demonstrates that the degrees of complexity of the algorithmic agents used there differs. Generation of an assignment plan requires essentially more "intelligence" from an agent than does a simple movement on a virtual plane. These abilities are provided by the algorithmic components of the corresponding agents. However, the ability for collective decision- making, needed in both cases, was in fact unchanged. These procedures, based on the same type of equation, differ only in their implementation, where in the second case the pattern recognition approach has been applied. This "constant" degree of complexity of the analytical component points to an important fact, namely that analytical approaches can be applied to real problems with a high degree of complexity. For example, in robotics the problem of synchronization is no less important than in assignment-planning. In other solutions to this problem, the advantage of analytical methods lies in the predictability of group behavior, in the compact form of interactions, in reliability to noise and in several cases in security of communication. However, the primary application field is where algorithmic methods (which require a highly-developed computational architecture) cannot be used, for example in biotechnology, at the cellular or organism levels, in chemical systems, nanorobotics, and so on. The equations in these systems can be implemented by available physical or chemical processes, which points to further investigations with promising perspectives.

Chapter 8

Summary

We know nothing at all. All our knowledge
is the knowledge of school children.
The real nature of things we shall never know.
Albert Einstein

The collective activity of distributed systems is of huge practical importance; accordingly, the task of analyzing and controlling them has become especially relevant. This thesis is intended to find ways towards a solution of this task primarily considered from an analytical viewpoint. The main observation underlying this work is in the collective behavior that arises. All the participants in a collective phenomenon, in most cases, exhibit a simultaneous changing of their own behavior although there is no central coordination of these changes. The absence of any form of centralized governance in such systems makes a control particular and difficult. Moreover, due to the nonlinear interactions among system participants, the number of degrees of freedom of the overall system is essentially increased, so that analysis requires special reduction procedures.

The spontaneous changes of behavior can be viewed in the sense of the qualitative changes that every system-participant undergoes. In this case, the synergetics of the collective behavior may be involved in investigating this phenomenon. From the viewpoint of the synergetic concept, the collective activity and self-organization is governed by the reduced quantity - the order parameter (OP) that is distributed over the whole system. Therefore, the first focus of this thesis was on the derivation of this OP. Information contained in the order parameter describes the macroscopic states of a distributed system, explaining why and when the collective behavior arises. As proposed within this framework, in order to obtain the order parameter algorithmic and analytic reduction procedures need to be systematically performed over the distributed system. The second focus of the thesis is on using the OP to provide the desired collective activity. Two ways of employing the macroscopic information given by the order parameter were chosen. First, to use well-known control mechanisms, derived on the basis of the OP. These external mechanisms force the system to follow the predicted collective dynamics. Second, to design the internal structure of the distributed system so that it will voluntarily demonstrate the desired cooperative activity without enforced control. In particular, we would like to emphasize the results accomplished in the domains of:

- **Reduction procedures.** The mathematical models treated within this work are time-discrete. However the reduction techniques available have primarily been developed for time continuous systems. The "discrete versions" of these reduction procedures are either incompletely developed or their application causes considerable difficulties. In particular, unlike the continuous case, the parameter-dependent center manifold in the discrete case may have for example, two different unstable eigenvalues, something only rarely mentioned in the vast literature on the subject. Therefore, the discrete versions of the center manifold and normal form reduction have been investigated to obtain the different assumptions for the CM reduction. Improvement of the normal form reduction has been undertaken; the results of both approaches, as well as adiabatic elimination, have been compared from the viewpoint of accuracy of approximation and computational complexity. The derivation of the order parameter for maps has been explicitly shown.

The reduction techniques were also considered for the high-dimensional CML. It was shown that where the CML possesses an homogeneous structure of coupling, the local dynamics of such a CML is primarily governed by the OP contained in initial maps. Therefore, in addition to conventional reduction approaches, this "dimensional" reduction is expected to be applied to the CML.

- **Control using delayed feedback.** For control purposes, we used a time delay feedback control, first suggested by Pyragas [141]. This technique is "natural" for implementation in a distributed environment, moreover it selectively influences the dynamics of the controlled system. The proposal is to use the macroscopic information provided by the order parameter to calculate the parameters of this control scheme. Moreover, by making the original system n-iterated, one can introduce the delayed feedback function into the system, then by returning the system back to its original form, we obtain the control that can stabilize n-periodical motion. The suggested technique allows the derivation of both the local and global control feedback that is successful in controlling the bifurcation as well as the spatio-temporal chaos (see [161]).
- **Control using interaction modifications.** Perhaps the most interesting result towards achieving a desired collective activity in a distributed system lies in constructing the interactions among system-participants. The underlying idea is to equip the distributed system with the interactions (couplings) that primarily satisfy the determined restrictions (hardware requirements, nature of interactions, allowable values that the interacting systems may achieve, and so on). Performing the algorithmic simplification and thereby obtaining a mathematical model, we can then derive the order parameter pointing to all the possible collective phenomena the distributed system can exhibit. Using the suggested approach, the desired collective activity may be fixed by means of an appropriate tuning of the interactions (see [192]).
- **Collective decision-making.** Use of OP in achieving the desired collective activity has a "natural" example in the problem of collective decision-making. A bifurcation points to one of several possible collective decisions; thus, decision-making can be analytically investigated. The problem of collective decision- making has many solutions, but even where the common knowledge needed for the collective agreement is not attainable for distributed system, the procedure of common decision-making is still open to improvement. Our suggestion is to use the order parameter already contained in every system to synchronize all agents, thereby guaranteeing that all achieve a collective decision in a finite time. This approach is shown in two full-size examples originating in robotics and industrial manufacturing. The technique proposed is published in [48], [193], [209], [252], [253], .

The final remark concerns the general principles underlying the phenomenon of self-organization in the context of **swarm intelligence** [19]. This phenomenon can be considered at different levels of abstraction. At the visible macroscopic level, group intelligence lies in distributed information processing, in coordinating distributed activity of agents in compliance with the needs of the overall group, in building new distributed algorithms, and so on. "Swarm" self-organization on a macroscopic level represents a fascinating phenomenon demonstrated by "simple" and "ordinary" agents. However, as pointed out by some authors, the collective behavior is not simply the sum of each participant's behavior; it substantially emerges at the society level [251].

The existence of society points to existence of *social laws* acting in the group at the mesoscopic level. Social laws can be understood in the sense of central control programs that govern the behavior of an agent group. For example, in an ant colony optimization, the social law for a group is to follow the shortest path; the macroscopic effect is the emergence of globally optimal pathways. However the question remains how these central social laws arise in a distributed group. At the microscopic level, each agent can be represented as consisting of different "internal" components; not only separate structural or functional components, but also their different combinations. From the viewpoint of synergetics, the agents' "internal" components, interacting on the microscopic level thereby causes the abstract social law at the abstract mesoscopic level.

Social laws are a useful abstraction, allowing us to describe a group's behavior in a centralized way. Quantities similar to the social laws, or characterizing the social laws (for example the order parameter), do not exist in the group, but are a centralized abstraction. It is easy to represent how social laws, influencing the behavior of every agent, lead to the phenomenon of swarm intelligence at the macroscopic level. In any group there are several social laws; the order parameter, discussed in this thesis, is only one. Why are these social laws needed? In modifying group intelligence at the macroscopic level, the interaction among agents (or their internal components) at the microscopic level must also be modified. However, due to the nonlinear nature of interactions and complexity, any modifications that usefully change macroscopic phenomena are enormously difficult. The abstractions introduced here are used to simplify the process of modification. Between macroscopic phenomenon and social law there exists an unique relationship, for example to modify the assignment plan (macroscopic level) we must change the principle of collective decision- making (mesoscopic level). After that, a "technical question" remains, namely, which changes of interactions will (at the microscopic level) cause a change in collective decision-making.

These ideas lead to the following conclusions. First, the centralized social law acting in a distributed group is nowhere contained in an explicit form, it is always dynamically "generated", which means that this law cannot be extracted from the behavior of an individual agent. Second, in order to modify the distributed macroscopic group intelligence, the distributed interactions between agents, at the microscopic level, must be modified. This is possible by means of abstract centralized social laws. Finally, the order parameter is an example of these social laws, controlling saltatory changes of group dynamics, such as collective decisions. In this thesis, the role and use of the order parameter has been considered in detail and as a primary focus. Other social laws exist which could be the focus of further investigations.

Appendix A

A.1 The high-order terms of NF and the problem of equivalency

In this section we consider the question about an importance of high-order terms in the normal form. In this overview we follow closely the main ideas suggested by Golubitsky and Schaeffer. Describing this problem, we use the term *finite determinancy* from the singular theory, that is best characterized by the following question: "To what extent do the low-order terms in the Taylor series expansion of a bifurcation problem determine its qualitative behavior, regardless of the high-order terms that may be present?[114]". Answering this question, it is necessary to introduce the idea about the qualitative equivalency in small.

In order to exemplify the problem of equivalency, let us consider two following equations

$$g(\tilde{x},\tilde{\lambda}) = \frac{\tilde{x}^3}{6} - (2\tilde{\lambda}-1)\tilde{x} + k\tilde{x}^4 + O(5), \tag{1.1a}$$

$$h(x,\lambda) = x^3 - \lambda x, \tag{1.1b}$$

where k is some coefficient, $\tilde{\lambda} > 0$ and $\lambda > 0$. Here is the important task to investigate, whether the bifurcation problems (1.1a) and (1.1b) are equivalent *in small*? Eqs. (1.1a) and (1.1b) differ in two aspects: first, the different coefficient at x^3 and x and second, occurrence of the term $\tilde{x}^4$ in Eq. (1.1b). Performing the following linear change of coordinates

$$\tilde{\lambda} = \tilde{\lambda}\frac{6^{\frac{2}{3}}}{12} + \frac{1}{2}, \quad \tilde{x} = \tilde{x}6^{\frac{1}{3}}, \tag{1.2a}$$

we get hereby the Eq. (1.1b) up to the order four. Therefore the first difference has no effect on the qualitative picture of the solution set. Now setting coefficient k so that

$$g(\tilde{x},\tilde{\lambda}) = \tilde{x}^3 - \tilde{\lambda}\tilde{x} + \tilde{x}^4, \tag{1.3}$$

we consider the second aspect: the influence of the term $\tilde{x}^4$. This problem can be treated attracting the implicit function theorem and the further its geometrical consideration.

Theorem 5 (Implicit Function Theorem in Finite Dimension).

Let $\underline{\mathbf{f}}$ be as $\underline{\mathbf{f}}(\underline{\mathbf{x}},\{\alpha\}) = 0$, where $\underline{\mathbf{x}} = (x_1, ..., x_m) \in \mathbb{R}^m$, $\{\alpha\} = (\alpha_1, ..., \alpha_k) \in \mathbb{R}^k$ and $\underline{\mathbf{f}} = (f_1, ..., f_m) \in \mathbb{R}^m$. Suppose that $(f_1, ..., f_m)$ are differentiable functions on a neighborhood of the point $(\underline{\mathbf{x}}_0, \{\alpha_0\})$ so that $\underline{\mathbf{f}}(\underline{\mathbf{x}}_0, \{\alpha_0\}) = 0$ and $det(d\underline{\mathbf{f}})_{\underline{\mathbf{x}}_0,\{\alpha_0\}} \neq 0$. Then there exist neighborhoods U of the point $\underline{\mathbf{x}}_0$ in $\mathbb{R}^m$ and V of the point $\{\alpha_0\}$ in $\mathbb{R}^k$ and there is a unique mapping $\varphi : V \to U$ such that for every $\alpha \in V$ $\underline{\mathbf{f}}$ has the unique solution $x = \varphi(\{\alpha\})$ in U. Furthermore, φ is differentiable.

The symbolic formulation may be written as : $\underline{\mathbf{f}}(\underline{\varphi}(\{\alpha\}), \{\alpha\}) \equiv 0, \quad \underline{\varphi}(\{\alpha_0\}) = \underline{\mathbf{x}}_0$.

This theorem has different applications (e.g. [116]), here it is shown the simplest from them. We conclude that $h_x(x_0, \lambda_0) = 0$ and $g_x(\tilde{x}_0, \tilde{\lambda}_0) = 0$ is the necessary conditions for a solutions (x_0, λ_0), $(\tilde{x}_0, \tilde{\lambda}_0)$ of h, g to be a bifurcation point. Forasmuch as $x_0 = \tilde{x}_0 = 0$, $\lambda_0 = \tilde{\lambda}_0$ and moreover g and h have unique solutions in the vicinity of (x_0, λ_0) we assume that in some small area close to (x_0, λ_0) both solutions can coincide. In order to prove it, we solve (1.1a) and (1.1b), whose solutions are shown in Fig. A.1. Expanding obtained

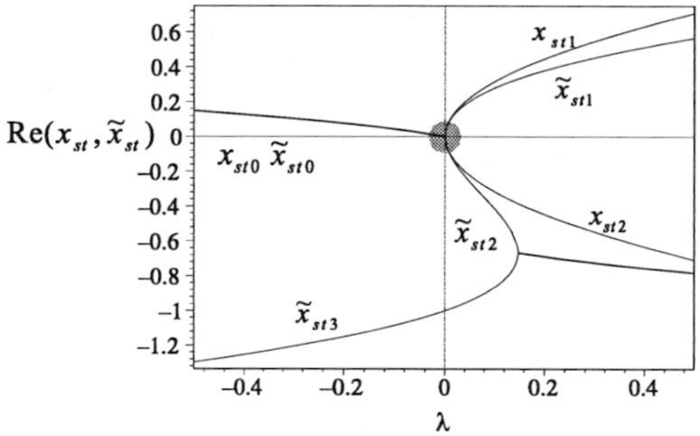

Figure A.1: The stationary states $Re(\tilde{x}_{st_{(0,1,2,3)}})$ of system (1.1a) and the stationary states $x_{st_{(0,1,2)}}$ of system (1.1b). Bold lines demonstrate the complex-conjugate areas of stationary states. Shadowed circle shows the area where $x_{st_{(0,1,2)}}$ and $\tilde{x}_{st_{(0,1,2)}}$ coincide. In this area (1.1a) and (1.1b) are equal as germs.

solutions into a Taylor series, we get the following first terms

$$\tilde{x} = \sqrt{\lambda} + O(\lambda), \quad x = \sqrt{\lambda} + O(\lambda). \tag{1.4}$$

We see, they really coincide in the terms of lowest order with respect to λ. But the further from the bifurcation point the more high-order terms will influence on the both solutions and consequently they will more differs from one another. The dependence (1.4) can be also obtained numerically by a least squares fit, taking into account the functional dependence $log(x) = \alpha * \log(\lambda) + \log(\beta)$. Choosing the logarithmically equidistant points for small interval close to λ_0, we get the values of α and β for both bifurcation problems (1.1a) and (1.1b). Therefore the second difference between (1.1a) and (1.1b) has also no effect on the qualitative behavior of the solution set *in small* and may be absorbed by a nonlinear change of coordinate. In this case (1.1a) and (1.1b) are often denoted as *equal as germs*.

In general, two bifurcation problems $g(x, \lambda)$ and $h(x, \lambda)$ are equivalent if they may be related through an equation

$$g(x,\lambda) = S(x,\lambda)h(X(x,\lambda),\lambda), \tag{1.5}$$

where S is positive and nonzero and (X,Λ) is a local diffeomorphism (see also Lie groups in e.g. [117]).

Thus the problem of equivalency of the NFs can be divided into two separate cases: the *recognition problem* for the normal form and determination of *universal unfolding.*

The first case is related with Eq. (1.5). Let $g(x,\lambda)$ be a normal form. Then the bifurcation problem $h(x,\lambda)$, that satisfies Eq. (1.5) is correspondingly equivalent to the normal form $g(x,\lambda)$, can be replaced by this NF. Therefore the NF possesses the simplest structure from all equivalent bifurcation problems and consequently can be viewed as a universal characterization of the given instability.

The second case logically follows from the first one if we consider the rest of $h(x,\lambda)$ that are not equivalent to $g(x,\lambda)$ up to the order r. These $h(x,\lambda)$ compose the unfolding of the normal form g denoted often as G. For example the following two-parameter unfolding

$$G(x,\lambda,\{\beta,\gamma\}) = x^3 - \lambda x + \beta x + \gamma x^2 \tag{1.6}$$

can not be transformed into (1.1b) by coordinate transformations (1.5). Remark that (1.6) introduces into consideration new parameters β, γ such that

$$G(x,\lambda,\{0,0\}) = h(x,\lambda). \tag{1.7}$$

Moreover $G(x,\lambda,\beta,\gamma)$ is versal because it can be scaled to $H(x,\lambda,\gamma) = x^3 - \lambda x + \gamma x^2$ by letting $\Lambda(\lambda,\beta) = \lambda - \beta$.

Definition 13.

An unfolding G of g is *versal* is every other unfolding of g factors through G. A versal unfolding of g depending on the minimum number of parameters possible is called *universal.* That minimum number is called the *codimension* of g.

A.2 The recognition problem

The recognition problem refers to the determination of all germ h that are equivalent to g. This problem can be reformulated in a following way. Let be a germ g, find small perturbations p so that $g + \varepsilon p$ is strongly equivalent to g for all small ε.

Let $\mathcal{E}_{x,\lambda}$ be the space of all functions $g : \mathbb{R}^2 \to \mathbb{R}$ and C^∞ on some neighborhood of the origin. Suppose that exist some function $S(x,\lambda,\varepsilon) \in \mathcal{E}$ and $X(x,\lambda,\varepsilon) \in \mathcal{E}$ and then for some perturbation p we can write

$$g(x,\lambda) + \varepsilon p(x,\lambda) = S(x,\lambda,\varepsilon)g(X(x,\lambda,\varepsilon),\lambda), \tag{1.8}$$

where

$$S(x,\lambda,0) \equiv 1, \;\; X(x,\lambda,0) \equiv x, \;\; X(0,0,\varepsilon) \equiv x. \tag{1.9}$$

Moreover it is expected that

$$X_x(0,0,\varepsilon) > 0, \quad S(0,0,\varepsilon) > 0. \tag{1.10}$$

Differentiating with respect to ε

$$p(x,\lambda) = S_\varepsilon(x,\lambda,t)g(X(x,\lambda,t),\lambda) + S(x,\lambda,t)g_x(X(x,\lambda,t),\lambda)X_\varepsilon(x,\lambda,t), \tag{1.11}$$

then setting $\varepsilon = 0$ and taking into account (1.9), we finally get

$$p(x,\lambda) = S_\varepsilon(x,\lambda,0)g(x,\lambda) + g_x(x,\lambda)X_\varepsilon(x,\lambda,0). \tag{1.12}$$

Here S_ε and X_ε means the partial differentiation with respect to ε.

Definition 14.

The *restricted tangent space* of a germ g, denoted as $RT(g)$ is the set of all germs p which can be written in the form

$$p(x,\lambda) = a(x,\lambda)g(x,\lambda) + b(x,\lambda)g_x(x,\lambda), \tag{1.13}$$

where $a, b \in \mathcal{E}_{x,\lambda}$ and $b(0,0) = 0$.

The following theorem imposes the condition for $g + \varepsilon p$ to be the strongly equivalent to g, when ε is small.

Theorem 6 (The recognition theorem.).

Let g and p be germs in $\mathcal{E}_{x,\lambda}$. If

$$RT(g+\varepsilon p) = RT(g) \ \ for \ \ all \ \ \varepsilon \in [0,1], \tag{1.14}$$

then $g + \varepsilon p$ is strongly equivalent to g for all $\varepsilon \in [0,1]$.

The proof of this theorem can be found in [114]. Forasmuch $b(0,0) = 0$ then there exist smooth germs $\tilde{b}$, $\tilde{c}$ such that

$$b(x,\lambda) = x\tilde{b}(x,\lambda) + \lambda\tilde{c}(x,\lambda) \tag{1.15}$$

and p has the following form

$$p(x,\lambda) = a(x,\lambda)g(x,\lambda) + (\tilde{b}(x,\lambda)x + \lambda\tilde{c}(x,\lambda))g_x(x,\lambda) \tag{1.16}$$

or in the short notation

$$p = ag + bxg_x + c\lambda g_x. \tag{1.17}$$

Here a, b, c are arbitrary functions in $\mathcal{E}_{x,\lambda}$ therefore (1.17) can be written using term *generator* as follows

$$RT(g) = \langle g, xg_x, \lambda g_x \rangle \tag{1.18}$$

More generally, the vector space $\mathcal{E}_n$, consisting of functions of n variables that may form an arbitrary linear combination with scalar coefficient, is a *ring*. If $p_1, ..., p_k$ are germs in $\mathcal{E}_n$ then the set of all combinations

$$a_1p_1 + ... + a_kp_k \tag{1.19}$$

where $a_i \in \mathcal{E}_n$ is an *ideal* $\mathcal{M}$ in $\mathcal{E}_n$. *Maximal ideal* $\mathcal{M}^k$ consists of all germs of order k. If $g \in \mathcal{E}_{x,\lambda}$ then $\mathcal{M} = \langle x, \lambda \rangle$ $\mathcal{M}^2 = \langle x^2, \lambda x, \lambda^2 \rangle$ and so on. The computational technique consists in the determination whether some power of $\mathcal{M}$ is contained in $RT(g)$. The detail of this technique can be found in [114]. Membership in $RT(g)$ for some bifurcation problems is summarized in Table A.2.

Finishing the recognition problem, we affirm that equivalence of two germs can be practically established by procedure of partial differentiation with respect to x, λ. Moreover the normal form is determined by the first nonzero derivation.

Proposition. *Let*

$$\alpha x^k + \beta\lambda, \quad k \geq 2, \tag{1.20}$$

$$\alpha x^k + \beta\lambda x, \quad k \geq 3 \tag{1.21}$$

be the normal form. Then germ $g \in \mathcal{E}_{x,\lambda}$ *is strongly equivalent to (1.20) if and only if at* $x = \lambda = 0$

$$g = \frac{\partial}{\partial x} g = \dots = \left(\frac{\partial}{\partial x}\right)^{k-1} g = 0, \;\; \alpha = sgn\left(\frac{\partial}{\partial x}\right)^k g, \;\; \beta = sgn\frac{\partial}{\partial \lambda} g, \tag{1.22}$$

and germ $g \in \mathcal{E}_{x,\lambda}$ *is strongly equivalent to (1.21) if and only if at* $x = \lambda = 0$

$$g = \frac{\partial}{\partial x} g = \dots = \left(\frac{\partial}{\partial x}\right)^{k-1} g = \frac{\partial}{\partial \lambda} g = 0, \;\; \alpha = sgn\left(\frac{\partial}{\partial x}\right)^k g, \;\; \beta = sgn\frac{\partial}{\partial \lambda}\frac{\partial}{\partial x} g \tag{1.23}$$

The flow chart of the recognition problem for such examples is shown in Table A.1

g_{xx} $\xrightarrow{\neq 0}$	g_λ $\xrightarrow{=0}$	$g_{\lambda,\lambda}$ $\xrightarrow{=0}$	$g_{\lambda,\lambda,\lambda}$ $\xrightarrow{=0}$	q $\xrightarrow{=0}$	*codim* ≥ 4
	$\downarrow \neq 0$	$\downarrow \neq 0$	$\downarrow \neq 0$	$\downarrow \neq 0$	
$\downarrow =0$	$\pm x^2 \pm \lambda$ (0)	$\pm x^2 \pm \lambda^2$ (1)	$\pm x^2 \pm \lambda^3$ (2)	$\pm x^2 \pm \lambda^4$ (3)	
g_{xxx} $\xrightarrow{\neq 0}$	g_λ $\xrightarrow{=0}$	$g_{x,\lambda}$ $\xrightarrow{=0}$	$g_{\lambda,\lambda}$ $\xrightarrow{=0}$	*codim* ≥ 4	
	$\downarrow \neq 0$	$\downarrow \neq 0$	$\downarrow \neq 0$		
$\downarrow =0$	$\pm x^3 \pm \lambda$ (1)	$\pm x^3 \pm \lambda x$ (2)	$\pm x^3 \pm \lambda^2$ (3)		
g_{xxxx} $\xrightarrow{\neq 0}$	g_λ $\xrightarrow{=0}$	$g_{x,\lambda}$ $\xrightarrow{=0}$	*codim* ≥ 4		
	$\downarrow \neq 0$	$\downarrow \neq 0$			
$\downarrow =0$	$\pm x^4 \pm \lambda$ (2)	$\pm x^4 \pm \lambda x$ (3)			
g_{xxxxx} $\xrightarrow{\neq 0}$	g_λ $\xrightarrow{=0}$	*codim* ≥ 4			
	$\downarrow \neq 0$				
$\downarrow =0$	$\pm x^5 \pm \lambda$ (3)				
codim ≥ 4					

Table A.1: The recognition problem for the singularities with codimension ≤ 3 (from [114]). Number in the parentheses after normal forms are codimensions.

A.3 The problem of universal unfolding

In the determination of the universal unfolding the $RT(g)$ and methods of linear algebra are used, therefore we show here only the central theorem and then summarize the main results for different normal forms in Table A.2.

The unfolding pertains to the perturbations of normal forms. It is implied that these perturbed NFs are not equivalent to the initial ones in sense of (1.8). Let g be in $\mathcal{E}_{x,\lambda}$; a k-parameter unfolding of g is a germ $G \in \mathcal{E}_{x,\lambda,\varepsilon}$, where $\varepsilon = (\varepsilon_1, ..., \varepsilon_k) \in \mathbb{R}^k$ such that for $\varepsilon = 0$

$$G(x, \lambda, 0) = g(x, \lambda). \tag{1.24}$$

Here G is a germ of $x, \lambda, \varepsilon_1, ..., \varepsilon_k$ defined C^∞ on a neighborhood of zero in $\mathbb{R}^{k+2}$.

Suppose that $G(x, \lambda, \varepsilon)$ is a universal unfolding of a germ $g \in \mathcal{E}_{x,\lambda}$. Then for any $q \in \mathcal{E}_{x,\lambda}$ there is the one-parameter unfolding of g

$$H(x, \lambda, \varepsilon) = g(x, \lambda) + \varepsilon q(x, \lambda). \tag{1.25}$$

Since G is universal, H factors through G

$$H(x, \lambda, \varepsilon) = S(x, \lambda, \varepsilon) G(X(x, \lambda, \varepsilon), \Lambda(\lambda, \varepsilon), A(\varepsilon)), \tag{1.26}$$

where

$$S(x, \lambda, 0) \equiv 1, \quad X(x, \lambda, 0) \equiv x, \quad \Lambda(\lambda, 0) \equiv \lambda, \quad A(0) \equiv 0. \tag{1.27}$$

Differentiating with respect to ε and setting $\varepsilon = 0$, we get

$$S_\varepsilon(x, \lambda, 0) g(x, \lambda) + g_x(x, \lambda) X_\varepsilon(x, \lambda, 0) + g_\lambda(x, \lambda) \Lambda_\varepsilon(\lambda, 0). \tag{1.28}$$

Expression (1.28) determines the *tangent space* to a germ g in $\mathcal{E}_{x,\lambda}$ denoted by $T(g)$

$$T(g) = ag + bg_x + cg_\lambda, \tag{1.29}$$

where $a, b \in \mathcal{E}_{x,\lambda}$ and $c \in \mathcal{E}_\lambda$. The following theorem gives the necessary and sufficient conditions for G to be a universal unfolding of g.

Theorem 7 (Universal unfolding theorem.).

Let g be a germ in $\mathcal{E}_{x,\lambda}$, and let G be a k-parameter unfolding of g. Then G is a universal unfolding of g if and only if

$$\mathcal{E}_{x,\lambda} = T(g) + \mathbb{R}\left\{ \frac{\partial G}{\partial \alpha_1}(x, \lambda, 0), ..., \frac{\partial G}{\partial \alpha_k}(x, \lambda, 0) \right\} \tag{1.30}$$

where $T(g)$ is the *tangent space* defined by

$$T(g) = RT(g) \oplus \mathbb{R}\{g_x, g_y, \lambda g_\lambda, ..., \lambda^l g_\lambda\} \tag{1.31}$$

The proof of this theorem can be found in [115]. Forasmuch $RT(g)$ has been computed, only straightforward linear algebra is required to compute $T(g)$. The universal unfolding theorem states that finding a universal unfolding of g is equivalent to finding a basis for a complementary subspace of $T(g)$ in $\mathcal{E}_{x,\lambda}$. Detail of computational technique can be found in [114]. $T(g)$ for some bifurcation problem is summarized in Table A.2.

Thus, the main results of problem of high-order terms and equivalency are shown in both Tables A.1, A.2 and now we go to the next point of our consideration.

Normal Form [1]	Codim	Nomenclature	$RT(h)$	$T(h)$
$\varepsilon x^2 + \delta\lambda$	0	Limit point	$\mathcal{M}^2 + \langle\lambda_u\rangle$	$\mathcal{M} + \langle\lambda_u\rangle + \mathbb{R}\{1\}$
$\varepsilon(x^2 - \lambda^2)$	1	Simple bif.	$\mathcal{M}^2 + \mathcal{M}\langle x\rangle$	$\mathcal{M} + \langle x\rangle$
$\varepsilon(x^2 + \lambda^2)$	1	Isola center	$\mathcal{M}^2 + \mathcal{M}\langle x\rangle$	$\mathcal{M} + \langle x\rangle$
$\varepsilon x^3 + \delta\lambda$	1	Hysteresis	$\mathcal{M}^3 + \langle\lambda_u\rangle$	$\mathcal{M}^2 + \langle\lambda_u\rangle + \mathbb{R}\{1\}$
$\varepsilon x^2 + \delta\lambda^3$	2	Asymmetric cusp	$\mathcal{M}^3 + \mathcal{M}\langle x\rangle$	$\mathcal{M}^2 + \langle x\rangle$
$\varepsilon x^3 + \delta\lambda x$	2	Pitchfork[2]	$\mathcal{M}^3 + \mathcal{M}\langle\lambda_u\rangle$	$\mathcal{M}^3 + \langle\lambda_u\rangle +$ $+\mathbb{R}\{x, k\varepsilon x^2 + \delta\lambda\}$
$\varepsilon x^4 + \delta\lambda$	2	Quadratic fold	$\mathcal{M}^4 + \langle\lambda_u\rangle$	$\mathcal{M}^3 + \langle\lambda_u\rangle + \mathbb{R}\{1\}$
$\varepsilon x^2 + \delta\lambda^4$	3		$\mathcal{M}^4 + \mathcal{M}\langle x\rangle$	$\mathcal{M}^3 + \langle x\rangle$
$\varepsilon x^3 + \delta\lambda^2$	3	Winged cusp	$\mathcal{M}^3 + \langle\lambda_u^2\rangle$	$\mathcal{M}^3 + \langle\lambda_u^2\rangle +$ $+\mathbb{R}\{\lambda, x^2\}$
$\varepsilon x^4 + \delta\lambda x$	3		$\mathcal{M}^4 + \mathcal{M}\langle\lambda_u\rangle$	$\mathcal{M}^4 + \langle\lambda_u\rangle +$ $+\mathbb{R}\{x, k\varepsilon x^3 + \delta\lambda\}$
$\varepsilon x^5 + \delta\lambda$	3		$\mathcal{M}^5 + \langle\lambda_u\rangle$	$\mathcal{M}^4 + \langle\lambda_u\rangle + \mathbb{R}\{1\}$
$x^2 + y^2 + \lambda,$ $2xy$	3	Hilltop bif.[3]	$\mathcal{M}^2 + \langle\lambda_u\rangle$	$(\mathcal{M}^2 + \langle\lambda_u\rangle)\overrightarrow{\mathcal{E}}_{x,\lambda} \oplus \mathbb{R}$ $\{(x,y), (-y,x), (1,0)\}$

[1] $\varepsilon, \delta = \pm 1$

[2] $x^3 + \lambda x$ subcritical case, $x^3 - \lambda x$ supercritical case, $x^2 - \lambda x$ transcritical case [114] , [103]

[3] two-dimensional bifurcation problem

Table A.2: Normal forms for singulaties of codim ≤ 3 with restricted tangent space $RT(h)$ and tangent space $T(h)$ (from [114]).

Appendix B

B.1 Resonant terms of normal forms

The calculation of normal form coefficients is shown on the example of two-dimensional map (see Sec. 3.5), where

$$\underline{x}_n = \begin{pmatrix} x_n \\ y_n \end{pmatrix}, \; \underline{\underline{\Lambda}} = \begin{pmatrix} \lambda_u & 0 \\ 0 & \lambda_s \end{pmatrix}, \; \underline{g}^{(2)}(\underline{x}_n) = \begin{pmatrix} A_2^x x_n^2 + B_2^x x_n y_n + C_2^x y_n^2 \\ A_2^y x_n^2 + B_2^y x_n y_n + C_2^y y_n^2 \end{pmatrix}. \tag{2.1}$$

Other necessary expressions for (3.63) are defined as

$$\underline{\underline{\Lambda}}\, \underline{g}^{(2)}(\underline{x}_n) = \begin{pmatrix} A_2^x \lambda_u (x_n)^2 + B_2^x \lambda_u x_n y_n + C_2^x \lambda_u (y_n)^2 \\ A_2^y \lambda_s (x_n)^2 + B_2^y \lambda_s x_n y_n + C_2^y \lambda_s (y_n)^2 \end{pmatrix}, \tag{2.2}$$

and

$$\underline{g}^{(2)}(\underline{\underline{\Lambda}}\, \underline{x}_n) = \begin{pmatrix} A_2^x (\lambda_u x_n)^2 + B_2^x \lambda_u \lambda_s x_n y_n + C_2^x (\lambda_s y_n)^2 \\ A_2^y (\lambda_u x_n)^2 + B_2^y \lambda_u \lambda_s x_n y_n + C_2^y (\lambda_s y_n)^2 \end{pmatrix}. \tag{2.3}$$

Collecting now the terms with respect to x_n, y_n, we are able to determine the resonance and the non-resonance terms in Eq. (3.63). For example for $\begin{pmatrix} \\ x_n^2 \end{pmatrix}$ in the case of $\lambda_u = 1$ the term $A_2^y(\lambda_s - \lambda_u^2) + F_{x^2} = 0$ can be eliminated by an adequate choice of $A_2^y = \dfrac{F_{x^2}}{(\lambda_u^2 - \lambda_s)}$. Considering now $\begin{pmatrix} x_n^2 \\ \end{pmatrix}$, it can be shown that the term $A_2^x(\lambda_u - \lambda_u^2) + F_{x^2} = 0$ is impossible to eliminate by choice of A_2^x because of $(\lambda_u - \lambda_u^2) = 0$ for $\lambda_u = 1$. Consequently, the second order term $\begin{pmatrix} \mu(\alpha) x_n^2 \\ \end{pmatrix}$ occurs in the structure of the normal form with $\lambda_u = 1$ and coefficient $\mu(\alpha) = F_{x^2}$. A_2^x in this case can be selected arbitrary (for example $A_2^x = 0$).

The third order terms $\underline{g}^{(3)}(\underline{x}_n)$ in (3.55) are defined as

$$\underline{g}^{(3)}(x_n) = \begin{pmatrix} A_3^x x_n^3 + B_3^x x_n^2 y_n + C_3^x x_n y_n^2 + D_3^x y_n^3 \\ A_3^y x_n^3 + B_3^y x_n^2 y_n + C_3^y x_n y_n^2 + D_3^y y_n^3 \end{pmatrix}. \tag{2.4}$$

Other necessary expressions for (3.64) are:

$$\underline{\underline{\Lambda}}\, \underline{g}^{(3)}(\underline{x}_n) = \begin{pmatrix} A_3^x x_n^3 \lambda_s + B_3^x x_n^2 y_n \lambda_s + C_3^x x_n y_n^2 \lambda_s + D_3^x y_n^3 \lambda_s \\ A_3^y x_n^3 \lambda_u + B_3^y x_n^2 y_n \lambda_u + C_3^y x_n y_n^2 \lambda_u + D_3^y y_n^3 \lambda_u \end{pmatrix} \tag{2.5}$$

and

$$\underline{g}^{(3)}(\underline{\underline{\Lambda}}\,\underline{x}_n) = \begin{pmatrix} A_3^x(x_n\lambda_s)^3 + B_3^x(x_n\lambda_s)^2 y_n\lambda_u + C_3^x x_n\lambda_s(y_n\lambda_u)^2 + D_3^x(y_n\lambda_u)^3 \\ A_3^y(x_n\lambda_s)^3 + B_3^y(x_n\lambda_s)^2 y_n\lambda_u + C_3^y x_n\lambda_s(y_n\lambda_u)^2 + D_3^y(y_n\lambda_u)^3 \end{pmatrix} \tag{2.6}$$

The resonant terms of the normal forms are shown for both $\lambda_u = 1$ and $\lambda_u = -1$ up to the fifth order in Table B.1 and B.2.

Terms	Resonant condition	$\lambda_u = 1$	$\lambda_u = -1$	$\lvert\lambda_u^{\pm}\rvert = 1$
$\begin{pmatrix} x_n^2 \\ \end{pmatrix}$	$A_2^x(\lambda_u^2 - \lambda_u) = N^{x^2} \Rightarrow A_2^x = \dfrac{N^{x^2}}{\lambda_u^2 - \lambda_u}$	(res)		
$\begin{pmatrix} \\ x_n^2 \end{pmatrix}$	$A_2^y(\lambda_u^2 - \lambda_s) = N_{x^2} \Rightarrow A_2^y = \dfrac{N_{x^2}}{\lambda_u^2 - \lambda_s}$			
$\begin{pmatrix} x_n y_n \\ \end{pmatrix}$	$B_2^x(\lambda_u\lambda_s - \lambda_u) = N^{xy} \Rightarrow B_2^x = \dfrac{N^{xy}}{\lambda_u\lambda_s - \lambda_u}$			
$\begin{pmatrix} \\ x_n y_n \end{pmatrix}$	$B_2^y(\lambda_u\lambda_s - \lambda_s) = N_{xy} \Rightarrow B_2^y = \dfrac{N_{xy}}{\lambda_u\lambda_s - \lambda_s}$	(res)		
$\begin{pmatrix} y_n^2 \\ \end{pmatrix}$	$C_2^x(\lambda_s^2 - \lambda_u) = N^{y^2} \Rightarrow C_2^x = \dfrac{N^{y^2}}{\lambda_s^2 - \lambda_u}$			
$\begin{pmatrix} \\ y_n^2 \end{pmatrix}$	$C_2^y(\lambda_s^2 - \lambda_s) = N_{y^2} \Rightarrow C_2^y = \dfrac{N_{y^2}}{\lambda_s^2 - \lambda_s}$			

Table B.1: The second order resonant terms of the normal form reduction.

Terms	Resonant condition	$\lambda_u = 1$	$\lambda_u = -1$	$\lvert\lambda_u^{\pm}\rvert = 1$
$\begin{pmatrix} x_n^3 \\ \ \end{pmatrix}$	$A_3^x(\lambda_u^3 - \lambda_u) = F^{x^3} \Rightarrow A_3^x = \dfrac{F^{x^3}}{\lambda_u^3 - \lambda_u}$	(res)		
$\begin{pmatrix} \ \\ x_n^3 \end{pmatrix}$	$A_3^y(\lambda_u^3 - \lambda_s) = F_{x^3} \Rightarrow A_3^y = \dfrac{F_{x^3}}{\lambda_u^3 - \lambda_s}$			
$\begin{pmatrix} x_n^2 y_n \\ \ \end{pmatrix}$	$B_3^x(\lambda_u^2\lambda_s - \lambda_u) = F^{x^2y} \Rightarrow B_3^x = \dfrac{F^{x^2y}}{\lambda_u^2\lambda_s - \lambda_u}$			(res)
$\begin{pmatrix} \ \\ x_n^2 y_n \end{pmatrix}$	$B_3^y(\lambda_u^2\lambda_s - \lambda_s) = F_{x^2y} \Rightarrow B_3^y = \dfrac{F_{x^2y}}{\lambda_u^2\lambda_s - \lambda_s}$	(res)		
$\begin{pmatrix} x_n y_n^2 \\ \ \end{pmatrix}$	$C_3^x(\lambda_u\lambda_s^2 - \lambda_u) = F^{xy^2} \Rightarrow C_3^x = \dfrac{F^{xy^2}}{\lambda_u\lambda_s^2 - \lambda_u}$			
$\begin{pmatrix} \ \\ x_n y_n^2 \end{pmatrix}$	$C_3^y(\lambda_u\lambda_s^2 - \lambda_s) = F_{xy^2} \Rightarrow C_3^y = \dfrac{F_{xy^2}}{\lambda_u\lambda_s^2 - \lambda_s}$			(res)
$\begin{pmatrix} y_n^3 \\ \ \end{pmatrix}$	$D_3^x(\lambda_s^3 - \lambda_u) = F^{y^3} \Rightarrow D_3^x = \dfrac{F^{y^3}}{\lambda_s^3 - \lambda_u}$			
$\begin{pmatrix} \ \\ y_n^3 \end{pmatrix}$	$D_3^y(\lambda_s^3 - \lambda_s) = F_{y^3} \Rightarrow D_3^y = \dfrac{F_{y^3}}{\lambda_s^3 - \lambda_s}$		(res)	

Table B.2: The third order resonant terms of the normal form reduction.

Appendix C

C.1 The near identity transformation with complex-conjugate unstable eigenvalues

Let us consider two-dimensional system

$$\underline{\mathbf{q}}_{n+1} = \underline{\mathbf{N}}(\underline{\mathbf{q}}_n, \{\alpha\}) \tag{3.1}$$

where $\underline{\mathbf{q}}_n = (x_n, y_n)^T$ and which has complex-conjugate eigenvalue $\lambda^{\pm}$. Then linear part of (3.1) can be rewritten as

$$\begin{pmatrix} x_{n+1} \\ y_{n+1} \end{pmatrix} = |\lambda| \begin{pmatrix} \cos(\theta) & -\sin(\theta) \\ \sin(\theta) & \cos(\theta) \end{pmatrix} \begin{pmatrix} x_n \\ y_n \end{pmatrix} + \begin{pmatrix} N_x(x_n, y_n, \{\alpha\}) \\ N_y(x_n, y_n, \{\alpha\}) \end{pmatrix}, \tag{3.2}$$

where $\theta = \arctan\left(\frac{Im(\lambda^{\pm})}{Re(\lambda^{\pm})}\right)$. Now we make the following linear transformation

$$\begin{pmatrix} x_n \\ y_n \end{pmatrix} = \frac{1}{2} \begin{pmatrix} 1 & 1 \\ -I & I \end{pmatrix} \begin{pmatrix} z_n^{\pm} \\ z_n^{\mp} \end{pmatrix}. \tag{3.3}$$

Performing required calculations, we get the following mode equation up the order four

$$\begin{aligned} z_{n+1}^{\pm} = {} & \lambda^{\pm} z_n^{\pm} + k_{20}^{\pm}(z_n^{\pm})^2 + k_{11}^{\pm} z_n^{\pm} z_n^{\mp} + k_{02}^{\pm}(z_n^{\mp})^2 + +k_{30}^{\pm}(z_n^{\pm})^3 + k_{21}^{\pm}(z_n^{\pm})^2 z_n^{\mp} + \\ & +k_{12}^{\pm} z_n^{\pm}(z_n^{\mp})^2 + k_{03}^{\pm}(z_n^{\mp})^3 + O(4). \end{aligned} \tag{3.4}$$

Next step is the nonlinear coordinate transformation (the near identity transformation) up the order four for variable $z_n^{\pm}$

$$\begin{aligned} z_n^{\pm} = {} & z_n^{\pm} + a_{20}^{\pm}(z_n^{\pm})^2 + a_{11}^{\pm} z_n^{\pm} z_n^{\mp} + a_{02}^{\pm}(z_n^{\mp})^2 + a_{30}^{\pm}(z_n^{\pm})^3 + a_{21}^{\pm}(z_n^{\pm})^2 z_n^{\mp} + \\ & +a_{12}^{\pm} z_n^{\pm}(z_n^{\mp})^2 + a_{03}^{\pm}(z_n^{\mp})^3 + O(4), \end{aligned} \tag{3.5}$$

and for $z_{n+1}^{\pm}$

$$\begin{aligned} z_{n+1}^{\pm} = {} & z_{n+1}^{\pm} + a_{20}^{\pm}(z_{n+1}^{\pm})^2 + a_{11}^{\pm} z_{n+1}^{\pm} z_{n+1}^{\mp} + a_{02}^{\pm}(z_{n+1}^{\mp})^2 + a_{30}^{\pm}(z_{n+1}^{\pm})^3 + \\ & +a_{21}^{\pm}(z_{n+1}^{\pm})^2 z_{n+1}^{\mp} + a_{12}^{\pm} z_{n+1}^{\pm}(z_{n+1}^{\mp})^2 + a_{03}^{\pm}(z_{n+1}^{\mp})^3 + O(4), \end{aligned} \tag{3.6}$$

where a are complex-conjugate coefficients. Let us assume that after performing all transformations we will get the NF without second-order terms

$$z_{n+1}^{\pm} = \lambda^{\pm} z_n^{\pm} + O(3). \tag{3.7}$$

Then we are able to replace in (3.6) different orders $z^{\pm}_{n+1}$ by $z^{\pm}_{n}$

$$(z^{\pm}_{n+1})^2 = (\lambda^{\pm})^2 (z^{\pm}_{n})^2 + O(4), \tag{3.8a}$$
$$z^{\pm}_{n+1} z^{\mp}_{n+1} = \lambda^{\pm}\lambda^{\mp} z^{\pm}_{n} z^{\mp}_{n} + O(4), \tag{3.8b}$$
$$(z^{\mp}_{n+1})^2 = (\lambda^{\mp})^2 (z^{\mp}_{n})^2 + O(4), \tag{3.8c}$$
$$(z^{\pm}_{n+1})^3 = (\lambda^{\pm})^3 (z^{\pm}_{n})^3 + O(4). \tag{3.8d}$$

Substituting (3.5) and (3.6) into (3.4) with respect to (3.8a) we remove all second order terms, setting

$$a^{\pm}_{20} = \frac{k^{\pm}_{20}}{\lambda^{\pm}(\lambda^{\pm}-1)}, \quad a^{\pm}_{11} = \frac{k^{\pm}_{11}}{\lambda^{\pm}(\lambda^{\mp}-1)}, \quad a^{\pm}_{02} = \frac{k^{\pm}_{02}}{(\lambda^{\mp})^2-\lambda^{\pm}}, \tag{3.9}$$

and affirming hereby our assumption (3.7). Considering the third order terms we arrive at the conclusion, that only the terms $(z^{\pm}_{n})^2 z^{\mp}_{n}$ can not be removed because of

$$(z^{\pm}_{n})^2 z^{\mp}_{n}: \ \ k^{\pm}_{21} + 2k^{\pm}_{20}a^{\pm}_{11} + k^{\pm}_{11}a^{\mp}_{11} + k^{\pm}_{11}a^{\pm}_{20} + 2k^{\pm}_{02}a^{\mp}_{02} + a^{\pm}_{21}\lambda^{\pm}(1-\lambda^{\pm}\lambda^{\mp}). \tag{3.10}$$

The resonant and non-resonant terms for different bifurcation problems are summarized in tables B.1 and B.2. Denoting the nonzero terms in (3.10) as $a^{\pm}_{21}$ we then get

$$a^{\pm}_{21} = k^{\pm}_{21} + \frac{k^{\pm}_{20}k^{\pm}_{11}}{\lambda^{\pm}}\left(2\frac{1}{(\lambda^{\mp}-1)} + \frac{1}{(\lambda^{\pm}-1)}\right) + \frac{k^{\pm}_{11}k^{\mp}_{11}}{\lambda^{\mp}(\lambda^{\pm}-1)} + 2\frac{k^{\pm}_{02}k^{\mp}_{02}}{(\lambda^{\pm})^2-\lambda^{\mp}} \tag{3.11}$$

or

$$a^{\pm}_{21} = k^{\pm}_{21} + \frac{k^{\pm}_{20}k^{\pm}_{11}}{\lambda^{\pm}}\left(\frac{2\lambda^{\pm}+\lambda^{\mp}-3}{(\lambda^{\mp}-1)(\lambda^{\pm}-1)}\right) + \frac{|k^{\pm}_{11}|^2}{1-\lambda^{\mp}} + 2\frac{|k^{\pm}_{02}|^2}{(\lambda^{\pm})^2-\lambda^{\mp}}. \tag{3.12}$$

Thus in consequence of transformation (3.5) (3.6) and removing all terms that possess nonzero eigenvalue expressions like in (3.9) we are able to simplify the system (3.4) and finally get the NF of the Neimark-Sacker bifurcation

$$z^{\pm}_{n+1} = \lambda^{\pm}z^{\pm}_{n} + a^{\pm}_{21}(z^{\pm}_{n})^2 z^{\mp}_{n} + O(4). \tag{3.13}$$

Transforming (3.13) into polar coordinates by letting

$$z^{\pm}_{n} = p_n e^{\pm I\phi_n}, \tag{3.14}$$

we get for the NF (3.13)

$$p_{n+1}e^{\pm\phi_{n+1}} = e^{\pm I\phi_n}\left[|\lambda^{\pm}|p_n e^{\pm I\theta} + \left(Re(a_{21}) \pm Im(a_{21})\right)(p_n)^3\right] + O(4) \tag{3.15}$$

or

$$\begin{cases} p_{n+1} = |\lambda^{\pm}|p_n + Re(a^{\pm}_{21})p^3_n + O(p^5_n) \\ \phi_{n+1} = \phi_n + \theta + Im(a^{\pm}_{21})/|\lambda^{\pm}|p^2_n + O(p^4_n) \end{cases}. \tag{3.16}$$

The first equation in (3.16) is the normal form of the pitchfork bifurcation the supercritical condition for that $Re(a^{\pm}_{21})$ has to be negative. The second equation of (3.16) is the circle map, the variable ϕ_n of which is defined modulo 2π. This equation defines rotation and number of the points that will be mapped in the bifurcation diagram of the Neimark-Sacker bifurcation.

List of Figures

List of Tables

Bibliography

Chapter 1

[1] P. Corke & S. Sukkarieh (eds.). *Field and Service Robotics.* Springer, 2006.

[2] S. Yuta, H. Asama, S. Thrun, E. Prassler & T. Tsubouchi (eds.). *Field and Service Robotics: Recent Advances in Research and Applications.* Springer, 2006.

[3] JPL Robotics. *Robotic Vehicles Group at NASA.* `http://robotics.jpl.nasa.gov/`

[4] A. Lidokhover. *Multi-Agenten-Systeme.* Vdm Verlag Dr. Müller, 2007.

[5] G. Weiss. *Multiagent systems. A modern approach to distributed artificial intelligence.* The MIT Press, 1999.

[6] E. Sahin & W. Spears (eds.). *Swarm Robotics: From sources of inspiration to domains of application.* Springer-Verlag, 2004.

[7] P. Levi & S. Kernbach (eds.). *Symbiotic Multi-Robot Organisms: Reliability, Adaptability, Evolution.* Springer-Verlag, 2010.

[8] A. Botthof & J. Pelka *Mikrosystemtechnik. Zukunftsszenarien.* Springer-Verlag, 2003.

[9] S. Fatikow & U. Rembold *Microsystem Technology and Microrobotics.* Springer-Verlag, 2007.

[10] V. Balzani, M. Venturi & A. Credi. *Molecular devices and machines - a journey into the nano world.* Wiley-VCH Verlag, Weinheim, 2003.

[11] A. Tanenbaum & M. van Steen. Distributed Systems: Principles and Paradigms. Prentice Hall, NJ, USA, 2002.

[12] W. Ebeling & R. Feistel. *Physik der Selbstorganisation und Evolution.* Akademie-Verlag, Berlin, 1986.

[13] H. Haken. *Synergetics. An Introduction.* Springer-Verlag, 1977.

[14] S. Camazine, J.-L. Deneubourg, N. Franks, J. Sneyd, G. Theraulaz & E. Bonabeau. *Self-Organization in biological systems.* Princeton University Press, 2003.

[15] C. Reynolds. *Flocks, Herds and Schools: A Distributed Behavioral Model.* Computer Graphics, 21(4), 25-34, 1987.

[16] Y. Ishida. *Distributed and autonomous sensing based on thc immune network.* Artif. Life Robotics **2**, 1-7, 1998.

[17] M. Resnick. *Turtles, termites and traffic jams. Explorations in massively parallel microworlds.* MA: MIT Press, Cambridge, 1994.

[18] E. Wilson. *The insect societies.* The Belkap Press of Harvard University Press, 1971.

[19] E. Bonabeau, M. Dorigo & G. Theraulaz. *Swarm Intelligence: From Natural to Artificial Systems.* Oxford University Press, 1999.

[20] `http://www.fotocommunity.de/pc/pc/display/17824004.`

[21] F. Heylighen. *The Science of Self-organization and Adaptivity.* The Encyclopedia of Life Support Systems, EOLSS Publishers Co. Ltd., Oxford, 2003.

[22] P. Levi & H. Haken. *Towards a Synergetic Quantum Field Theory for Evolutionary, Symbiotic Multi-Robotics.* Symbiotic Multi-Robot Organisms, P. Levi & S. Kernbach (eds.), Springer-Verlag, 25-55, 2010.

[23] H. Van Dyke Parunak. *Industrual and Practical Applications of DAI.* In: Multiagent Systems: A Modern Introduction to Distributed Artificial Intelligence, G. Weiss (ed.), MIT Press, 377-420, 1999.

[24] B. Siciliano & O. Khatib (eds.). *Springer Handbook of Robotics.* Springer-Verlag, 2008.

[25] J. Ferber. *Multi-agent Systems. Introduction to Distributed Artificial Intelligence.* Addison Wesley, 1999.

[26] T. Sandholm. *Distributed Rational Decision Making.* In: Multiagent Systems: A Modern Introduction to Distributed Artificial Intelligence, G. Weiss (ed.), MIT Press, 201-258, 1999.

[27] M. Bennewitz & W. Burgard. *Coordinating the Motions of Multiple Mobile Robots Using a Probabilistic Model.* In: Proc. of SIRS, 2000.

[28] M. Bushev. *Synergetics: chaos, order, self-organization.* World Scientific Publisher, 1994.

[29] R. Kubo, K. Matsuo & K. Kitahara. *Fluctuation and relaxation of macrovariables.* J. stat. Phys. **9**, 51, 1973.

[30] H. Haken. *Laser Theory.* Corr. printing, Springer-Verlag, 1984.

[31] F. Helbing. *Verhkersdynamics.* Springer-Verlag, 1997.

[32] B. Kerner. *Physics of Traffic: Empirical Freeway Pattern Features, Engineering Applications, and Theory (Understanding Complex Systems).* Springer-Verlag, 2004.

[33] K. Peters, A. Johansson & D. Helbing. *Swarm intelligence beyond stigmergy: Traffic optimization in ants.* Künstliche Intelligenz **4**, 11-16, 2005.

[34] F. Luna & B. Stefannson. *Economic Simulations in Swarm: Agent-Based Modelling and Object Oriented Programming.* Kluwer Academic Publishers, 2000.

[35] G. Booth. *Gecko: A continuous 2-D world for ecological modeling.* Artificial Life Journal **3**(3), 147-163, 1997.

[36] X. Tu. *Artificial animals for computer animation: biomechanics, locomotion, perception, and behaviour.* PhD thesis, University of Toronto, Canada, 1996.

[37] I-Swarm: *Intelligent Small World Autonomous Robots for Micro-manipulation.* EU Project, No FP6-2002-IST-1, 2003-2007.

[38] S. Kauffman. *The Origins of Order: Self-Organization and Selection in Evolution.* Oxford University Press, New York, 1993.

[39] S. Kauffman. *At Home in the Universe: The Search for Laws of Self-Organization and Complexity.* Oxford University Press, New York, 1995.

[40] *AnyLogic documentation.* Saint-Petersburg, 2005. `http://www.xjtek.com`

[41] Ю. Карпов. Иммитационное моделлирование систем. Санкт-Петерсбург, БХВ, 2005.

[42] S. Wiggins. *Introduction to applied nonlinear dynamical systems and chaos.* Springer-Verlag, 1990.

[43] H. Haken. *Advanced Synergetics.* Springer-Verlag, 1983.

[44] H. Haken. *Principles of Brain Functioning.* Springer-Verlag, 1996.

[45] J. Guckenheimer & P. Holmes. *Nonlinear oscillations, dynamical systems, and bifurcations of vector fields.* Springer-Verlag, 1983.

[46] T. Lüth. *Technische Multi-Agenten-Systeme.* Carl Hanser Verlag, München, Wien, 1998.

[47] J. Klein. *Breve: a 3D simulation environment for the simulation of decentralized systems and artificial life.* In: Proc. of the 8th International Conference on the Simulation and Synthesis of Living Systems (Artificial Life VIII). MIT Press, 2002. `http://www.spiderland.org/breve`

[48] S. Kornienko, O. Kornienko & J. Priese. *Application of multi-agent planning to the assignment problem.* Computers in Industry **54**(3), 273-290, 2004.

[49] S. Kernbach Structural Self-organization in Multi-Agents and Multi-Robotic Systems Logos Verlag, Berlin, 2008

Chapter 2

[50] P. Crutchfield. *The calculi of emergence: computation, dynamics and induction.* Physica **D75**, 11-54, 1994.

[51] D. Ngueyen, P. Wang & F. Hadaegh. *Self-organization of a multi-agent system in pattern formation.* In: Self-Organisation of complex Structere, F. Schweitzer (ed.), London: Gordon and Breach, 127-139, 1997.

[52] H. Nwana. *Software Agents: An Overview.* The Knowledge Engineering Review, **11**(3), 1–40, 1996.

[53] S. Bussmann & K. Schild. *Self-Organizing Manufacturing Control: An Industrial Application of Agent Technology.* In: Proc. of the ICMAS'2000, Boston, 87-94, 2000.

[54] S. Bussmannn, N. Jennings & M. Wooldridge *Multiagent Systems for Manufacturing Control. A Design Methodology.* Springer-Verlag, Springer Series on Agent Technology, 2004.

[55] H. Van Dyke Parunak. Practical and Industrial Applications of Agent-Based Industrial Technology Institute, 1998.)

[56] C. Rehtanz. *Autonomous Systems and Intelligent Agents in Power System Control and Operation.* Springer-Verlag, 2003.

[57] G. Weiss & R. Jakob. *Agentenorientierte Softwareentwicklung. Methoden und Tools.* Xpert.press Reihe, Springer-Verlag, 2004.

[58] S. Russell & P. Norvig. *Artificial Intelligence: A Modern Approach (2nd Edition).* Prentice Hall, 2003.

[59] K. Kaneko. *Relevance of dynamic clustering to biological networks.* Physica **D75**, 55-73, 1994.

[60] F. Klugl. *Multiagentensimulation. Konzepte, Werkzeuge, Anwendung.* Addison-Wesley, 2001.

[61] A. Sharkey. *Robots, insects and swarm intelligence.* Artificial Intelligence Review, DOI 10.1007/s10462-007-9057-y, 2007.

[62] W.-B. Zhang. *Synergetic economics.* Springer-Verlag, 1991.

[63] Y. Demazeau, J. Pavon, J. Corchado & J. Bajo *Advances in Intelligent and Soft Computing.* In: Proc. of the 7th International Conference on Practical Applications of Agents and Multi-Agent Systems (PAAMS'09), Vol. 55 of Advances in Intelligent and Soft Computing, Springer, 2007.

[64] V. Marik, V. Vyatkin & A. Colombo *Holonic and Multi-Agent Systems for Manufacturing.* In: Proc. of the third international conference on industrial applications of holonic and Multi-Agent Systems (HoloMAS 2007), Vol. 4659 of Lecture Notes in Computer Science, Springer, 2007.

[65] M. Wooldridge. *Intelligent agents.* In: Multiagent Systems: A Modern Introduction to Distributed Artificial Intelligence, G. Weiss (ed.), MIT Press, 27-77, 1999.

[66] M. Wooldridge & N. Jennings. *Intelligent agents: Theory and practice.* The Knowledge Egineering Review **10**(2), 115-152, 1995.

[67] M. Schillo,K. Fischer & C. Klein *The Micro-Macro Link in DAI and Sociology.* In: Proc. of the 2nd Workshop on Multi-Agent-Based Simulation, Vol. 1979 of Lecture Notes in Computer Science, 133 - 148, Springer, 2000.

[68] R. Kube & E. Bonabeau. *Cooperative transport by ants and robots.* Robotics and Autonomous Systems **30**(1/2), Elsevier, 85-101, 2000.

[69] P. Levi. *Architectures of individual and distributed autonomous agents.* In: Proc. of the IAS-2, 315-324, 1989.

[70] M. Huhns & L. Stephens. *Multiagent systems and societies of agents.* In: Multiagent Systems: A Modern Introduction to Distributed Artificial Intelligence, G. Weiss (ed.), MIT Press, 79-120, 1999.

[71] V. Trianni & M. Dorigo. *Emergent Collective Decisions in a Swarm of Robots.* In: IEEE Swarm Intelligence Symposium (SIS-05), P. Arabshahi, A. Martinoli, (eds.),IEEE Press, 241-248,2005.

[72] E. Durfee. *Distributed Problem Solving and Planing.* In: Multiagent Systems: A Modern Introduction to Distributed Artificial Intelligence, G. Weiss (ed.), MIT Press, 121-164, 1999.

[73] I. Rahwan, S. Ramchurn, N. Jennings, P. McBurney, S. Parsons & L. Sonenberg. *Argumentation-based negotiation.* The Knowledge Engineering Review, **18:4**, 343-375, Cambridge University Press, 2004.

[74] T. Sandholm. *Negotiation among self-interested computationally limited agents.* PhD thesis, University of Massachusetts, Amherst, 1996.

[75] J. Halpern & Y. Moses. *Knowledge and common knowledge in a distributed environment.* J. of the Association for Computer Machinery, **37**(3), 549-587, 1990.

[76] J. Gray. *Notes on database operating systems.* IBM Res. Rep. RJ 2188. IBM, Aug. 1987.

[77] P. Panangaden & S. Taylor. *Concurrent common knowledge: a new definition of agreement for asynchronous system.* In: Proc. of the 7th ACM symposium on principles of distributed computing. ACM, New York, 197-209, 1988.

[78] R. Fagin, & J. Halpern. *Reasoning about knowledge and probability: preliminary report.* In: Proc. of the 2nd Conf. on theoretical aspects of reasoning about knowledge. Morgan Kaufmann, San Mateo, 277-293, 1988.

[79] Y. Moses. *Resource-bounded knowledge.* In: Proc. of the 2nd conference on theoretical aspects of reasoning about knowledge. Morgan Kaufmann, San Mateo, 261-276, 1988.

[80] Scholarpedia. Swarm intelligence. `http://www.scholarpedia.org/article/Swarm_intelligence`.

[81] A. Martinoli. *Swarm Intelligence in autonomous collective robotics: from Tools to the analysis and synthesis of distributed control strategies.* PhD thesis, EPFL Lausanne, Switzerland, 1999.

[82] Open-source micro-robotic project. `http://www.swarmrobot.org/`.

[83] K. Jebens. *Development of a docking approach for autonomous recharging system for micro-robot "Jasmine".* Studienarbeit, University of Stuttgart, Germany, 2006.

[84] V. Prieto. *Development of cooperative behavioural patterns for swarm robotic scenarios.* Master Thesis, University of Stuttgart, Germany, 2006.

[85] Aristotle. *Metaphysics.* Cambridge, MA, Harvard University Press; London, William Heinemann Ltd., translated by H. Tredennick, G. Cyril Armstrong. 1933, 1989.

[86] M. Minsky. *Communication with Alien Intelligence.* Extraterrestrials: Science and Alien Intelligence, E. Regis (ed.), Cambridge University Press, 1985.

[87] Yu. Kuznetsov. *Elements of applied bifurcation theory.* Springer-Verlag, 1995.

[88] A. Lyapunov. *General problem of stability of motion.* Mathematics society of Kharkov, Kharkov, 1892.

[89] H. Poincaré. *Les Méthodes Nouvelles de la Mechanique Céleste.* Gautitier-Villars, Paris, 1892, 1893, 1899.

[90] M. Dulac. *Sur les cycles limites.* Bull. Soc. Math. France **51**, 45-188, 1923.

[91] G. Birkhoff. *Dynamical Systems.* Amer. Math. Soc. Colog. Publ., 1927.

[92] A. Kolmogorov. *La théorie générale des systéms dynamique et méchnique classique.* In: Proc. of the Int. Congr. Math., Amsterdam, North Holland, 315-333, 1954.

[93] S. Smale. *Diffeomorphisms with many periodic points.* Differential and Combinatorial Topology, S. Carins(ed.), Princeton University Press, 63-80, 1963.

[94] S. Smale. *Structurally stable systems are not dense.* Amer. J. Math. **88**, 491-496, 1966.

[95] S. Smale. *Differentiable dynamical systems.* Bull. Amer. Math. Soc. **73**, 747-817, 1967.

[96] D. Anosov. *Geodesic flows and closed Riemannian manifolds with negative curvature.* In: Proc. of Steklov Inst. **90**, 1-212, 1967. In Russian.

[97] V. Arnold. *Geometrical methods in the theory of orginary differential equations.* Springer-Verlag, 1983.

[98] P. Glansdorff & I. Prigogine. *Thermodynamics of structure, stability and fluctuations.* Wiley-Interscience Publishers, New York, 1971.

[99] I. Prigogine. *Structure, dissipation and life.* Theoretical Physics and Biology, M. Marious(ed.), North Holland Publishing Company, 1969.

[100] H. Haken. *Laser Theory.* Band XXV/2c der Reihe Encyclopedia of Physics, Springer-Verlag, 1970.

[101] H. Haken. *Synergetics. An Introduction, 3rd ed.* Springer-Verlag, 1983.

[102] H. Haken. *Synergetics. Introduction and Advanced Topics, 3rd ed.* Springer Verlag, 2004.

[103] G. Jetschke. *Mathematik der Selbstorganisation.* Friedr. Vieweg & Sohn Braunschweig, Wiesbaden, 1989.

[104] H. Haken. *Information and Self-Organisation.* Springer-Verlag, 1988.

[105] K. Kaneko & I. Tsuda. *Constructive complexity and artificial reality: an introduction.* Physica **D75**, 1-10, 1994.

[106] J. Carr. *Applications of centre manifold theory.* Springer-Verlag, 1981.

[107] J. Argyris, G. Faust & M. Haase. *Die Erforschung des Chaos: eine Einführung für Naturwissenschaftler und Ingenieure.* Vieweg-Verlag, 1985.

[108] http://monet.physik.unibas.ch/ elmer/pendulum/bif.htm.

[109] D. Arrowsmith & C. Place. *Ordinary Differential Equations. A qualitative approach with applications.* Chapman and Hall, London, New York, 1982.

[110] J. Heinbockel *Introduction to tensor calculus and continuum mechanics.* Trafford Publishing, 1996.

[111] C. Ulh, R. Friedrich & H. Haken. *Analysis of spatiotemporal signals of complex systems.* Phys. Rev. **E51**(5), 3890-3900, 1995.

[112] D. Henry. *Geometric theory of semilinear parabolic equations.* Springer Lectures Notes in Mathematics, Vol. 840, Springer-Verlag, 1981.

[113] A. Nayfeh. *Method of normal forms.* A Wiley-Interscience Publication, John Wiley & Sohn, Inc., 1993.

[114] M. Golubitsky & D. Schaeffer. *Singularities and groups in bifurcation theory, v.I.* Springer-Verlag, 1985.

[115] M. Golubitsky, I. Stewart & D. Schaeffer. *Singularities and groups in bifurcation theory, v.II.* Springer-Verlag, 1988.

[116] C. Robinson. *Dynamical systems: stability, symbolic dynamics and chaos.* CRC Press, Inc., 1995.

[117] P. Oliver. *Applications of Lie groups to differential equiations.* Springer-Verlag, 1986.

[118] K. Kaneko. *Overview of coopled map lattices.* Chaos **2**(3), 279-282, 1992.

[119] K. Kaneko. *Theory and application of coupled map lattices.* John Willey & Sons, Inc., 1993.

[120] F. Hoppensteadt & E. Izhikevich. *Synaptic organisations and dynamical properties of weakly connected neural oscillators.* Biol. Cybern. **75**, 117-127, 1996.

[121] R. Dror, C. Canavier, R. Butera, J. Clark & J. Byrne. *A mathematical criterion based on phase response curves for stability in a ring of coupled oscillators.* Biol. Cybern. **80**, 11-23, 1999.

[122] M. Georgeff & D. Kinny. *Modelling and Design of Multi-Agent System.* In: Intelligent Agents III, J. Müller, M. Wooldridge, N. Jennings (eds.), LNAI 1193, 1-20, Springer-Verlag, 1997.

[123] R. Brooks, C. Breazeal, M. Marjanovic, B. Scassellati & M. Williamson. *The Cog Project: Building a Humanoid Robot.* In: Computation for Metaphors, Analogy, and Agents, C. Nehaniv (ed.), Vol. 1562 of Lecture Notes in Artificial Intelligence, 52–87, Springer, 1999.

[124] T. Fukuda & T. Ueyama. *Cellular robotics and micro robotic systems.* In: Series in Robotics and Automated Systems, Vol. 10, World Scientific Publishing Co. Ptc. Ltd., 1994.

[125] A. Pikovsky & J. Kurths. *Do globally coupled maps really violate the law of large numbers.* Phys. Rev. Lett.**72**, 1644-1646, 1994.

[126] C. Diks, F. Takens & J. DeGoede. *Spatio-temporal chaos: A solvable model.* Physica **D104**, 269-285, 1997.

[127] S. Rasband. *Chaotic dynamics of nonlinear systems.* John Wiley & Sohn, Inc., 1990.

[128] G. Hu, J. Xiao, J. Yang, F. Xie & Z. Qu. *Synchronization of spatiotemporal chaos and its applications.* Phys. Rev. **E56**(3), 2738-2746, 1997.

[129] K. Konishi & H. Kokame. *Decentralized delayed-feedback control of a one-way coupled ring lattice.* Physica **D127**, 1-12, 1999.

[130] G. Korn & T. Korn. *Mathematical handbook.* McGraw-Hill Book Company, 1968.

[131] R. Grigoriev, M. Cross & H. Schuster. *Pinning control of spatiotemporal chaos.* Phys. Rev. Lett., **79**(15), 2795-2798, 1997.

[132] L. Kocarev & U. Parlitz. *Generalized synchronization, predictability, and equivalence of unidirectionally coupled dynamical systems.* Phys. Rev. Lett. **76**(11), 1816-1819, 1996.

[133] B. Hu & Z. Liu. *Phase synchronization of two-dimensional lattices of coupled chaotic maps.* Phys. Rev. **E62**(2), 2114-2118, 2000.

[134] J. Chen, K. Wong, Z. Chen, S. Xu & J. Shuai. *Phase synchronization in discrete chaotic systems.* Phys. Rev. **E61** (3), 2559-2561, 2000.

[135] A. Parravano & M. Cosenza. *Driven maps and emergence of ordered collective behavior in globally coupled maps.* Phys. Rev. **E58**(2), 1665-1671, 1998.

[136] R. Pinto, P Varona, A. Volkovskii, A. Szücs, H. Abarbanel & M. Rabinovich. *Synchronous behavior of two coupled electronic neurons.* Phys. Rev. **E62**(2), 2644-2656, 2000.

[137] G. Johnson, M. Löcher & E. Hunt. *Stabilized spatiotempral waves in a convectively unstable open flow system: coupled diode resonators.* Phys. Rev. **E51**(3A), R1625-R1628, 1995.

[138] S. Boccaletti, A. Farini & F. Arecchi. *Adaptive synchronization of chaos for secure communication.* Phys. Rev. **E55**(5), 4979-4981, 1997.

[139] H. Meinchard. *The algorithmic beauty of sea shells.* Springer-Verlag, 1995.

[140] Yu. Jiang, A. Antillón, P. Parmananda & J. Escalona. *Selection and stabilization of spatiotemporal patterns in two-dimensional coupled map lattices.* Phys. Rev. **E56**(3), 2568-2572, 1997.

[141] K. Pyragas. *Continuous control of chaos by self-controlling feedback.* Phys. Lett. **A170**, 421-428, 1992.

[142] Yu. Maistrenko, V. Maistrenko, A. Popovich & E. Mosekilde. *Transverse instability and riddled basin in a system of two coupled logistic maps.* Phys. Rev. **E57** (3), 2713-2724, 1998.

[143] А. Колесников, А. Гельфгат. Проектирование многокритериальных систем управления промышленными объектами. Москва, Энегроатомиздат, 1993.

[144] S. Parthasarathy & S. Sinha. *Controlling chaos in unidimensional maps using constant feedback.* Phys. Rev. **E51**(6), 6239-6242, 1995.

[145] H. Wang & E. Abed. *Bifurcation control of a chaotic system.* Automatica **31**(9), 1213-1226, 1995.

[146] E. Ott, C. Grebogi & J. Yorke. *Controlling Chaos.* Phys. Rev. Lett. **64**(11), 1196-1199, 1990.

[147] H. Friedel, R. Grauer & Ch. Marliani. *Center manifold approach to controlling chaos.* Phys. Lett. **A236**, 45-52, 1997.

[148] W. Ditto, S. Rauseo & M. Spano. *Experimental control of chaos.* Phys. Rev. Lett. **65**, 3211-3214, 1990.

Chapter 3

[149] M. Hénon. *A two-dimensional mapping with a strange attractor.* Commun. Math. Phys. **50**, 69-77, 1976.

[150] E. Lorenz. *Deterministic non-periodic flow.* J. Atmos. Sci. **20**, 130-141, 1963.

[151] G. Alessandro, P. Grassberger, S. Isola & A. Politi. *On the topology of the Hénon map.* J. Phys. A: Math. Gen. **23**, 5285-5294, 1990.

[152] M. Giovannozzi. *Analysis of the stability domain for the Hénon map.* Phys. Lett. **A182**, 255-260, 1993.

[153] R. May. *Simple mathematical models with very complicated dynamics.* Nature **261**, 459-467, 1976.

[154] J. Sandefur. *Discrete dynamical systems. Theory and application.* Calarendon Press, Oxford, 1990.

[155] A. Nayfeh & B. Balachandran. *Applied nonlinear dynamics.* A Wiley-Interscience Publication, John Willey & Sohs, Inc., 1995.

[156] J. Murray. *Lecture on Nonlinear-differential-equation models in biology.* Clarendon Press, Oxford, 1977.

[157] A. Wunderlin & H. Haken. *Generalized Ginzburg-Landau equations, slaving principle and center manifold theorem.* Z. Phys. **B44**, 135-141, 1981.

[158] E. Grigorieva, H. Haken & S. Kaschenko. *Complexity near equilibrium in model of lasers with delayed optoelectronic feedback* International Symposium on Nonlinear Theory and its Application (NOLTA'98) Crans-Montana, Switzerland, 495-498, 1998.

[159] D. Arrowsmith. *An introduction to dynamical systems.* Cambridge University Press, 1990.

[160] V. Pliss. *The reduction principle in the theory of stability of motion.* Soviet Math. **5**, 247-250, 1964.

[161] P. Levi, M. Schanz, S. Kornienko & O. Kornienko. *Application of order parameter equation for the analysis and the control of nonlinear time discrete dynamical systems.* Int. J. of Bifurcation and Chaos **9**(8), 1619-1634, 1999.

[162] S. Mane & W. Weng. *Minimal normal-form method for discrete maps.* Phys. Rev. **E48**(1), 532-542, 1993.

[163] R. Sedgewick & P. Flajolet. *An introduction to the analysis of algorithms.* Addison-Wesley, Reading MA, 1996.

Chapter 4

[164] Wikipedia. Interaction. `http://en.wikipedia.org/wiki/Interaction.`

[165] D. Pierre & A. Hübler. *A theory for adaptation and competition applied to logistic map dynamics.* Physica **D75**, 343-360, 1994.

[166] H. Chate & P. Manneville. *Emergence of effective low-dimensional dynamics in the macroscopic behavior of coupled map lattices.* Europhysics Letters, **17**(4), 291-296, 1992.

[167] Y. Gu, M. Tung, J.-M. Yuan, D. Feng & L. Narducci. *Crises and Hysteresis in coupled logistic map.* Phys. Rev. Lett., **25**(9), 701-704, 1984.

[168] S. Kernbach, E. Meister, F. Schlachter & O. Kernbach. *Adaptation and Self-adaptation of Developmental Multi-Robot Systems.* International Journal On Advances in Intelligent Systems, **3**(1&2), 121-140, 2010.

[169] N. Kataoka & K. Kaneko. *Functional dynamics I: Articulation process.* Physica **D138**, 225-250, 2000.

[170] M. Basso, A. Evangelisti, R. Genesio & A. Tesi. *On bifurcation control in time delay feedback systems.* Int. J. Bifurcation and Chaos **8**(4), 713-721, 1998.

[171] Q. Bi & P. Yu. *Computation of normal forms of differential equations associated with non-semisimple zero eigenvalues.* Int. J. of Bif.and Chaos **8**(12), 2279-2319, 1998.

[172] A. Algaba, E. Freire & E. Gamero. *Hypernormal form for the Hopf-zero bifurcation.* Int. J. of Bif. and Chaos **8**(10), 1857-1887, 1998.

[173] A.N. Kolmogorov. *Three approaches to the definition of the concept quantity of information.* Probl. Inform. Transm., Vol. 1, 1-7, 1965.

[174] W. Ebeling, J. Freund & F. Schweitzer. *Komplexe Structuren: Entropie und Information.* B.G. Teubner, Leipzig, 1998.

[175] Y. Klimontovich. *Statistical theory of open systems.* Dordrecht, Kluwer, 1995.

[176] R. Bogdanov. *Versal deformations of a singular points on the plane in the case of zero eigevalues.* Functional Anal. Appl., **9**(2), 144-145, 1975.

[177] F. Takens. *Singularities of vector fields.* Publ. Math. IHES **43**, 47-100, 1974.

Chapter 5

[178] C.-Y. Liou & S.-K. Yuan. *Error tolerant associative memory.* Biolog. Cybernetics **81**, 331-342, 1999.

[179] P. Moutarlier, B. Mirtich & J. Canny. *Shortest paths for a car-like robot to manifold in configuration space.* Int. J. of Robotics Research **15**(1), 36-60, 1996.

[180] J. Reif & H. Wang. *Social potential fields: A distributed behavioral control for autonomous robots* Robotics and Autonom. Syst. **27**, 171-194, 1999.

[181] J. Alvarenz-Ramirez. *Using nonlinear saturated feedback to control chaos: Hénon map.* Phys. Rev. **E48**(6), 3165-3167, 1993.

[182] P. Gade. *Feedback control in coupled map lattices.* Phys. Rev. **E57**(6), 7309-7312, 1998.

[183] K. Konishi, M. Ishii & H. Kokame. *Stabilizing unstable periodic points of one-dimensional nonlinear systems using delayed-feedback signals.* Phys. Rev. **E54**(4), 3455-3460, 1996.

[184] K. Pyragas. *Predictable chaos in slightly perturbed unpredictable chaotic systems.* Phys. Lett. **A181**, 203-210, 1993.

[185] J. Socolar & D. Gauthier. *Analysis and comparison of multiple-delay schemes for controlling unstable fixed points of discrete maps.* Phys. Rev. **E57**(6), 6589-6595, 1998.

[186] M. Williamson. *Rhythmic robot arm control using oscillators.* In: Proc. of the IEEE International Conference on Intelligent Robots and Systems(IROS'98), 77-83, 1998.

[187] P. Parmananda, M. Hildebrand & M. Eiswirth. *Controlling turbulence in coupled map lattice systems using feedback techniques.* Phys. Rev. **E56**(1), 239-244, 1997.

[188] H. Nakajima & Y. Ueda. *Limitation of generalized delayed feedback control.* Physica **D111**, 143-150, 1998.

[189] S. Barnett. *Matrices, methods and applications.* Clarendon Press, Oxford, 1990.

[190] H. Nakajima & Y. Ueda. *Half-period delayed feedback control for dynamical systems with symmerties.* Phys. Rev. **E58**(2), 1757-1763, 1998.

[191] А. Колесников. Синергетическая теория управления. Таганрог: ТРТУ, Москва: Энегроатомиздат, 1994.

Chapter 6

[192] O. Kornienko, S. Kornienko & P. Levi. *Collective decision making using natural self-organization in distributed systems.* In: Proc. of the International Conference on Computational Intelligence for Modelling, Control and Automation (CIMCA-2001), Las Vegas, USA, 460-471, 2001.

[193] S. Kornienko, O. Kornienko & P. Levi. *About nature of emergent behavior in micro-systems.* In: Proc. of the International Conference on Informatics in Control, Automation and Robotics (ICINCO-2004), Setubal, Portugal, 33-40, 2004.

[194] H. Nwana, L. Lee & N. Jennings. *Co-ordination in software agent systems.* BT Technology Journal, **14**(4), 1996.

[195] D. Stephens & J. Krebs *Foraging Theory.* Princeton University Press, Princeton, New Jersey, 1987.

[196] S. Kernbach, O. Kernbach & P. Levi. *New principles of coordination in large-scale micro- and molecular-robotic groups.* In: Proc. of the IARP-IEEE/RAS-EURON Joint Workshop on Micro and Nanorobotics, Paris, France, 2006.

[197] S. Kernbach, O. Kernbach & P. Levi. *Swarm embodiment - a new way for deriving emergent behavior in artificial swarms.* Autonome Mobile Systeme, P. Levi & M. Schanz (eds.), 25-31, Springer, 2005.

[198] D. Häbe. *Bio-inspired approach towards collective decision making in robotic swarms.* Diplomarbeit, Univerität Stuttgart, Germany, 2007.

[199] T. Kancheva. *Adaptive role dynamics in energy foraging behavior of a real micro-robotic swarm.* Diplomarbeit, Univerität Stuttgart, Germany, 2007.

[200] S. Kernbach, O. Kernbach & P. Levi. *Collective AI: context awareness via communication.* In: Proc. of the 19th international joint conference on artificial intelligence (IJCAI-2005), Edinburgh, Scotland, 1464-1470, 2005.

[201] A. Attarzadeh. *Development of an Advanced Power Management for Autonomous Micro-Robots.* Master Thesis, Univerität Stuttgart, Germany, 2007.

[202] R. Sedgewick. *Algorithms in C++.* Addison-Wesley, 1998

[203] *Swarm documentation.* Santa Fe Institute, 1999. http://www.swarm.org/wiki/Swarm_main_page.

[204] H. Haken. *Synergetic computers and cognition.* Springer Verlag, New York, Berlin, Heidelberg, Tokyo, 1991.

[205] U. Krause & T. Nesemann. *Differenzegleichungen und diskrete dynamische systeme.* B.G. Teubner, Stuttgart, Leipzig, 1999.

[206] S. Sinha. *Random coupling of chaotic maps leads to spatiotemporal synchronization.* Phys. Rev. **E66**(1), 016209, 2002.

[207] H. Atmanspachera & H. Scheingraber. *Stabilization of causally and non- causally coupled map lattices.* Physica **A**, 435447, 2005.

[208] Description of the micro-robot "Jasmine". Handbook, Uni-Stuttgart, 2006-20010. http://ipvszope.informatik.uni-stuttgart.de/ipvs/abteilungen/bv/forschung/projekte/Collective-robotics/Micro-robot_Jasmine/Description_Jasmine/en

[209] S. Kernbach, V. Nepomnyashchikh, T. Kancheva & O. Kernbach. *Specialization and generalization of robot behavior in swarm energy foraging.* Mathematical and Computer Modelling of Dynamical Systems, Taylor & Francis, 2011.

[210] I. Gerlovin, V. Ovsyankin, B. Stroganov & V. Zapasskii. *Coherent transients in semiconductor nanostructures as a basis for optical logical operations.* Nanotechnology, **11**, 383-386, 2000.

[211] D. DiVincenzo. *The physical implementation of quantum computation.* Preprint quant-ph/0002077 at http://xxx.lanl.gov

[212] GOLEM: *Bio-inspired Assembly Process for Mesoscale Products and Systems.* EU Project, NMP-2004-3.4.1.2-1, 2006-2009.

[213] M. Eigen. *Self-organization of matter and the evolution of biological macromolecules.* Die Naturwissenschaften **58**, 465-523, 1971.

[214] M. Eigen, J. McCasill & P. Schuster. *Molecular quasi-species.* Journal of Physical Chemistry **92**(24), 6881-6891, 1988.

[215] M. Eigen & P. Schuster. *The hypercycle - Part B: The abstract hypercycle.* Die Naturwissenschaften **65**, 7-41, 1978.

[216] M. Wooldridge. *An Introduction to Multiagent Systems.* John Wiley & Sons, 2002.

[217] N. Jennings. *Coordination Techniques for Distributed Artificial Intelligence.* In: Foundations of Distributed Artificial Intelligence, G.O'Hare & N. Jennings,(eds.), John Wiley& Sons, 187-210, 1996.

[218] S. Kornienko, O. Kornienko & P. Levi. *Minimalistic approach towards communication and perception in microrobotic swarms.* In: Proc. of the IEEE/RSJ International Conference on Intelligent Robots and System (IROS-2005), Edmonton, Alberta, Canada, 4006-4011, 2005.

[219] B. Alberts, A. Johnson, J. Lewis, M. Raff, K. Roberts & P. Walter. *Molecular Biology of the Cell.* Garland Science, 2002.

[220] R. Davis & R. Smith. *Negotiation as a metaphor for distributed problem solving.* Artificial Intelligence, 20(1), 63-109, 1983.

[221] E. Durfee. *Coordination of distributed problem solvers.* Kluwer Academic Press, Boston, 1988.

[222] H. Wiendahl. *Wandlungsfähigkeit.* wt Werkstattstechnik Jahrgang **92**, H.4, 122-127, 2002.

[223] P. Peeters, T. Heikkila, S. Bussman, J. Wyns, P. Valckenaers & H. van Brussel. *Novel manufacturing system requirements in automated, line-oriented discrete assembly.* In: Proc. of 4th IMS-WG Workshop, 10, Nancy, 1999.

[224] S. Bussmann, N. Jennings, M. Wooldridge. *Multiagent systems for manufacturing control: A design methodology.* In: Series on Agent Technology, Springer-Verlag, Germany, 2004.

[225] V. Dorrsam. *Materialflussrientierte Leistungsanalyse einstufiger Produktionssysteme.* Dissertation, Universität Karlsruhe, 1999.

[226] M. Pinedo. *Scheduling: Theory, Algorithms and Systems.* Prentice Hall, 1995.

[227] S. Graves, A. Rinnooy Kan & P. Zipkin. *Logistics of Production and Inventory.* Handbook in Operations Research and Management Science, Vol. 4,North Holland, 1993.

[228] J. Blazewicz, W. Domschke & E. Pesch. *The Job Shop Scheduling Problem: Conventional and New Solution Techniques.* European Journal of Operational Research, **93**, 1-33, 1996.

[229] K. Alicke. *Modellierung und Optimierung von mehrstufigen Umschlagsystemen.* Dissertation, Universität Karlsruhe, 1999.

[230] R. Bellman & S. Dreyfus. *Applied dynamic programming.* Princeton, New Jersey, 1962.

[231] M. Becht, J. Klarmann & M. Muscholl. *Software Demo of ROPE: Role Oriented Programming Environment for Multiagent Systems.* In: Proc. of the Agent 99, Seattle, Washington, USA, 1999.

[232] D. Corne, M. Dorigo & F. Glover (eds.). *New ideas in optimization.* McGraw-Hill, 1999.

[233] R. Gonzalez & R. Woods *Digital Image Processing.* Addison-Wesley Publishing Company, 1992.

[234] G. Ashworth. *Navigation Control Apparatus and Methods for Autonomous Vehicles.* United States Patent # 5,321,614, June 14, 1994.

[235] M. Bestehorn & H. Haken. *Traveling waves and pulses in a two-dimensional large-aspect-ratio system.* Phys. Rev. **A42**, 7195-7203, 1990.

[236] Bundesverkehrsministerium. *Verkehr in Zahlen.* D. Institut f. Wirtschaftsforschung, Berlin, 1995.

[237] J. Eggert & J. L. van Hemmen. *Unifying framework for neuronal assembly dynamics.* Phys. Rev. **E61**(2), 1855-1874, 2000.

[238] H. Haken, M. Schanz & J. Starke. *Treatment of combinatorial optimization problems using selection equations with cost terms. Part I. Two-dimensional assignment problems.* Physica **D 134**, 227-241, 1999.

[239] G. Kaplan. *Industrial electronics.* IEEE Spectrum **1**, 68-72, 1999.

[240] K. Konishi, M. Hirai & H. Kokame. *Decentralized delayed-feedback control of a coupled map model for open flow.* Phys. Rev. **E58**(3), 3055-3059, 1998.

[241] D. Larkin & G. Wilson. *Object-oriented Programming and the Objective-C Language.* NeXT software Inc., 1995

[242] P. Levi, N. Oswald, M. Becht, T. Buchheim, G. Hetzel, G. Kindermann, R. Lafrenz & M. Schule. *CoPS-Team Description.* RoboCup 2000, Melbourne, 2000.

[243] P. Maes. *Designing Autonomous Agents.* MIT Press, Cambridge, MA, 1990.

[244] J.D. Murray. *Mathematical Biology.* Clarendon Press, Oxford, 1989

[245] K. Pyragas & A. Tamaševičius. *Experimental control of chaos by delayed self-controlling feedback.* Phy. Lett. **A180**, 99-102, 1993.

[246] P. Schuster & K. Sigmund. *Self-organisation of biological macromolecules and evolutionary stable strategies.* Dyn. of synerg. systems, H. Haken, ed., Springer, 1980.

[247] S. Strogatz. *Nonlinear dynamics and chaos: with applications to physics, biology, chemistry, and engineering.* Addison-Wesley, 1994.

[248] S. Sudeshna. *Hierarchical globally coupled systems.* Phys. Rev. **E57**(5), 5217, 1998.

[249] W. Wischert, A. Wunderlin, A.Pelster, M. Oliver & J. Groslambert. *Delay-induced instabilities in nonlinear feedback systems.* Phys. Rev. **E49**(1), 203-219, 1994.

[250] P. Woafo, H.B. Fotsin & J.C. Chedjou. *Dynamics of two nonlinearly coupled oscillators.* Physica Scripta **57**, 195-200, 1998.

Chapter 8

[251] J. M. Pasteels, J. Deneubourg & S. Goss. *Self-organization mechanisms in ant societies.* In: From individual to collective behavior in social insects, J.M. Pasteels & J. Deneubourg (eds.), Birkuser Verlag, 155-175, 1987.

[252] S. Kernbach, O. Kernbach & P. Levi. *Multi-agent repairer of damaged process plans in manufacturing environment.* In: Proc. of the 8th International Conference on Intelligent Autonomous Systems (IAS04), 10-13 March, Amsterdam, The Netherlands, 285-494, 2004.

[253] S. Kernbach, R. Thenius, O. Kernbach & T. Schmickl. *Re-embodiment of Honeybee Aggregation Behavior in an Artificial Micro-Robotic System.* Adaptive Behavior, 17(3), 237-259, 2009.